T0192507

Time Series Modelling
with
Unobserved Components

Time Series Modelling *with* Unobserved Components

Matteo M. Pelagatti
University of Milano-Bicocca, Italy

CRC Press is an imprint of the
Taylor & Francis Group, an **informa** business
A CHAPMAN & HALL BOOK

CRC Press
Taylor & Francis Group
6000 Broken Sound Parkway NW, Suite 300
Boca Raton, FL 33487-2742

First issued in hardback 2019
First issued in paperback 2021

ISBN 13: 978-1-03-209843-2 (pbk)
ISBN-13: 978-1-4822-2500-6 (hbk)

Visit the Taylor & Francis Web site at
http://www.taylorandfrancis.com

and the CRC Press Web site at
http://www.crcpress.com

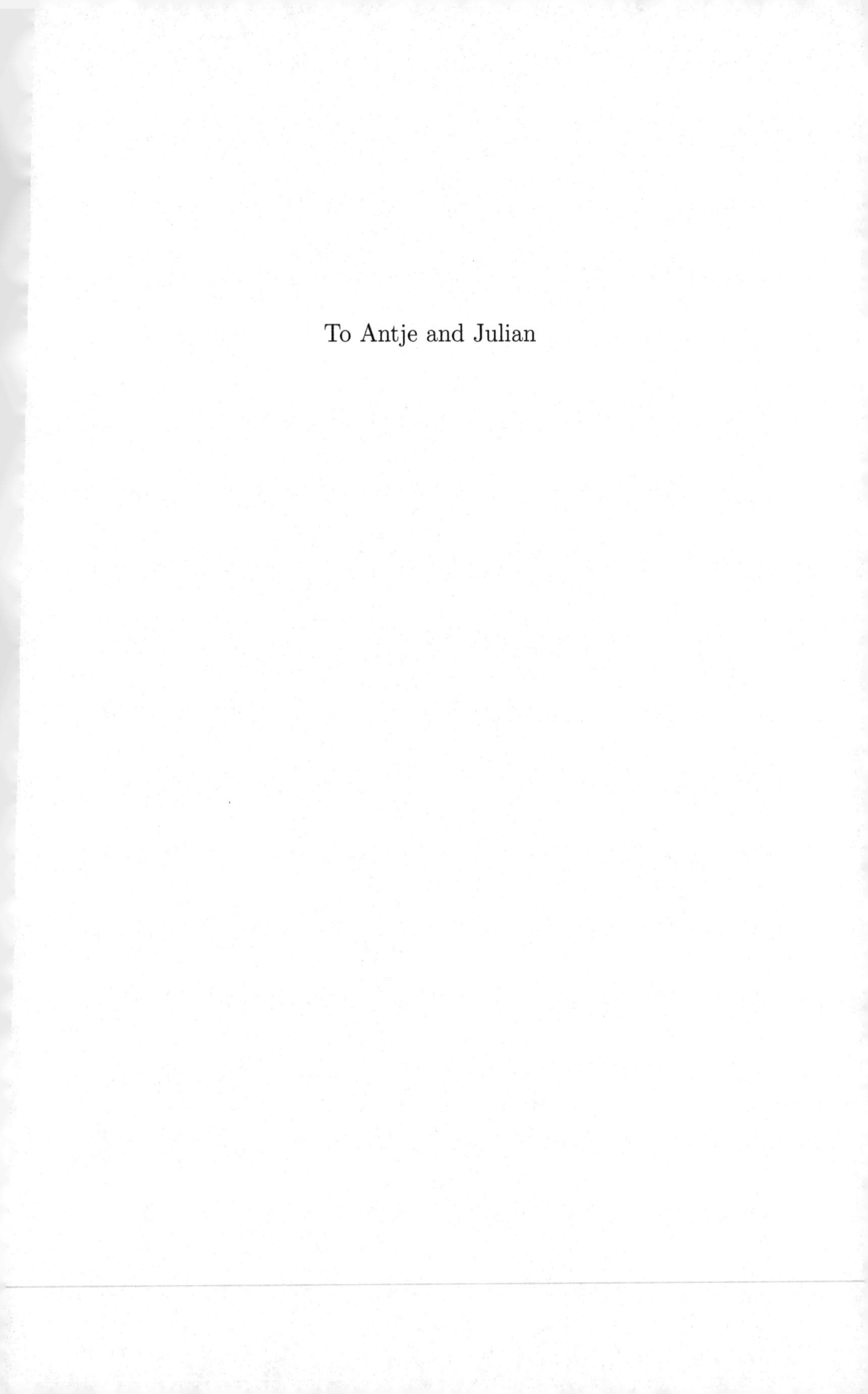

To Antje and Julian

Contents

List of figures

List of symbols

x	Scalar deterministic quantity.
$\boldsymbol{x}$	Deterministic (column) vector.
$\mathbf{0}$	Column vector of zeros.
$\mathbf{1}$	Column vector of ones.
$\mathbf{X}$	Deterministic matrix.
$\mathbf{X}^\top$	Transposition of $\mathbf{X}$.
$\mathbb{T}\mathrm{r}(\mathbf{X})$	Trace of the square matrix $\mathbf{X}$.
$\vert\mathbf{X}\vert$ or $\det(\mathbf{X})$	Determinant of the square matrix $\mathbf{X}$.
$\mathrm{diag}(\mathbf{X})$	Vector containing the elements on the main diagonal of the square matrix $\mathbf{X}$.
$\mathrm{vec}(\mathbf{X})$	Vector obtained by staking the columns of the matrix $\mathbf{X}$.
$\mathbf{I}_p$	Identity matrix of dimensions $p \times p$.
$\mathbf{O}_p$	Matrix of zeros of dimensions $p \times p$.
X	Scalar random quantity.
$\boldsymbol{X}$	Random (column) vector.
$\{X_t\}$	Random scalar sequence (also stochastic process in discrete time).
$\{X(t)\}$	Scalar stochastic process in continuous time (also random function).
$\{\boldsymbol{X}_t\}$	Random vector-valued sequence (also vector-valued stochastic process in discrete time).
$\{\boldsymbol{X}(t)\}$	Stochastic vector-valued process in continuous time (also vector-valued random function).
$(\Omega, \mathcal{F}, P)$	Probability space, where Ω is the sample space (a set whose generic element is indicated with ω), $\mathcal{F}$ a σ-algebra of events (subsets of Ω), P a probability measure.
$\mathbb{E}[X]$ or $\mathbb{E}X$	Expectation of the random variable X.

$\mathbb{P}[Y\|X_1, \ldots, X_m]$ or $\mathbb{P}[Y\|\boldsymbol{X}]$	Best linear predictor of the random variable Y based on the random variables $\{X_1, \ldots, X_m\}$ (projection of Y onto the linear span of $\{1, X_1, \ldots, X_m\}$). The second notation is to be interpreted in the same way provided the elements of $\boldsymbol{X}$ are $\{X_1, \ldots, X_m\}$.
$\mathbb{V}\text{ar}(X)$	Variance of the random variable X.
$\mathbb{V}\text{ar}(\boldsymbol{X})$	Covariance matrix of the random vector $\boldsymbol{X}$.
$\mathbb{C}\text{ov}(X, Y)$	Covariance between the random variables X and Y.
$\mathbb{C}\text{ov}(\boldsymbol{X}, \boldsymbol{Y})$	Matrix of the covariances of the elements of $\boldsymbol{X}$ with the elements of $\boldsymbol{Y}$: $\mathbb{E}[(\boldsymbol{X} - \mathbb{E}\boldsymbol{X})(\boldsymbol{Y} - \mathbb{E}\boldsymbol{Y})^\top]$.
$\mathbb{C}\text{or}(X, Y)$	Correlation between the random variables X and Y.
$\mathbb{C}\text{or}(\boldsymbol{X}, \boldsymbol{Y})$	Matrix of the correlations of the elements of $\boldsymbol{X}$ with the elements of $\boldsymbol{Y}$.
$\mathcal{N}(\boldsymbol{\mu}, \boldsymbol{\Sigma})$	Multivariate normal distribution with mean vector $\boldsymbol{\mu}$ and covariance matrix $\boldsymbol{\Sigma}$.
χ^2_m	Chi-square distribution with m degrees of freedom.
$\boldsymbol{X}_n \xrightarrow{d} \mathcal{D}$	Convergence in distribution of $\boldsymbol{X}_n$ to $\mathcal{D}$ (which is a place-holder for a distribution such as $\mathcal{N}$ or χ^2_m).
$\text{WN}(0, \sigma^2)$	White noise sequence with mean 0 and variance σ^2. (White noise sequences are zero-mean by definition; here we keep the mean parameter for conformity with the next two notational entries.)
$\text{IID}(\mu, \sigma^2)$	Independent identically distributed random sequence with mean μ and variance σ^2.
$\text{NID}(\mu, \sigma^2)$	Normally independently distributed random sequence with mean μ and variance σ^2.
$\mathbb{B}$	Backward shift (also lag) operator: $\mathbb{B}X_t = X_{t-1}$.
$\mathbb{I}(\text{condition})$	Indicator: it takes the value 1 when the condition is true and 0 otherwise.
$\mathbb{N}$	Set of natural numbers.
$\mathbb{Z}$	Set of integer numbers.
$\mathbb{R}$	Set of real numbers.
$\mathbb{C}$	Set of complex numbers.
$\lfloor x \rfloor$	Floor of x: the largest integer equal or smaller than x.
$x \wedge y$	$\min(x, y)$.
$x \vee y$	$\max(x, y)$.

Preface

On the 25th birthday of the path-breaking book *Forecasting, Structural Time Series Models and the Kalman Filter* written by Professor Andrew Harvey, I was reflecting on the relatively scarce diffusion of unobserved component models (UCM) among practitioners outside the academic community.

From (not only) my own experience, I know that UCM have many advantages over more popular forecasting techniques based on regression analysis, exponential smoothing and ARIMA. Indeed, being based on how humans naturally conceive time series, UCM are simple to specify, the results are easy to visualise and communicate to non-specialists (for example to the boss) and their forecasting performance is generally very competitive. Moreover, various types of outliers can easily be identified, missing values are effortlessly managed and working contemporaneously with time series observed at different frequencies presents no problem.

I concluded that the limited spread of UCM among practitioners could be attributed to one or more of the following causes:

1. Lack of mainstream software implementing UCM,
2. Few books on UCM and mostly academic rather than practical,
3. Limited number of university classes in which UCM are taught.

For a long time the only ready-to-use software package for UCM was STAMP, whose first version is contemporaneous with Harvey's book. STAMP is really an excellent software, but I am afraid its use outside academia is rather limited. However, in the last few years UCM procedures have started to appear in software systems such as SAS (since version 8.9) and Stata (since version 12) with a larger audience also outside the academic community. Thus, Point 1 seems to be at least partially resolved and in the future it is likely that more and more software packages will offer UCM procedures.

As for Point 2, for more than ten years the only two books on UCM were Harvey (1989) and West and Harrison (1989, from a Bayesian perpective), with the only exception being the volume by Kitagawa and Gersch (1996) which deals with similar topics but with a different approach. Again, these books are appreciated by academic scholars but are not that accessible to practitioners. The first decade of the new millennium witnessed the introduction of two new volumes on state space modelling: Durbin and Koopman (2001) and Commandeur and Koopman (2007). The first one is a great book, but quite technical, with an emphasis on state space methods rather than on modelling. The second one is introductory and, although very clear, it lacks some topics needed in economic time series modelling.

Finally, Point 3 is not easy to verify, but it is certainly linked to the first two points: if the first two causes are solved, then it is likely that the number of courses covering UCM will increase both in and outside universities.

Now, if the product (the UCM) is excellent, the supporting technology (the software) is available, then probably the product has to be popularised in a different way. So, I reflected on how a book should be designed to achieve this goal and concluded that such a book should

- Focus on the UCM approach rather than on general state space modelling
- Be oriented toward the applications
- Review the available software
- Provide enough theory to let the reader understand what's under the hood
- Keep the rigour to a level that is appropriate for academic teaching

This book has been written with those aims in mind but, of course, I am not the one who can judge if they were achieved.

Prerequisites. It is assumed that the reader has a basic knowledge of calculus, matrix algebra, probability and statistical inference at the level commonly met in the first year of undergraduate programmes in statistics, economics, mathematics, computer science and engineering.

Structure. The book is organised in three parts.

The first one covers propaedeutic time series and prediction theory, which the reader acquainted with time series analysis can skip. Unlike many other books on time series, I put the chapter on prediction at the beginning, because the problem of predicting is not limited to the field of time series analysis.

The second part introduces the UCM, presents the state space form and the related algorithms, and provides practical modelling strategies to build and select the UCM which best fits the needs of the time series analyst.

The third part presents some real-world applications with a chapter that focusses on business cycle analysis. Despite the seemingly economic-centric scope of the business cycle chapter, its content centres on the construction of band-pass filters using UCM, and this has obvious applications in many other fields. The last chapter reviews software packages that offer ready-to-use procedures for UCM and systems that are popular among statisticians and econometricians and that allow general estimation of models in state space form.

Website. Information, corrections, data and code are available at the book's website

`http://www.ucmbook.info`

I am grateful to all those readers who want to share their comments and signal errors in the book so that corrections can be placed on the site.

Acknowledgements. My gratitude goes to Professors Andrew Harvey, Siem Jan Koopman and Tommaso Proietti, whose ideas inspired this book. I thank my colleagues at Bicocca, Lisa Crosato, Gianna Monti, Alessia Paccagnini and Biancamaria Zavanella who read and corrected parts of the manuscript. Finally, many thanks go to my family who tolerated my numerous nights, weekends and holidays working on the manuscript.

Matteo M. Pelagatti
Milan, Italy

Part I

Statistical prediction and time series

Chapter 1

Statistical Prediction

A *statistical prediction* is a guess about the value of a random variable Y based on the outcome of other random variables $X_1, \dots, X_m$. Thus, a *predictor*[1] is a (measurable) function, say $p(\cdot)$, of the random variables $X_1, \dots, X_m$. In order to select an optimal predictor, we need a *loss function*, say $\ell(\cdot)$, which maps the prediction error to its cost. In principle, the loss function has to be determined case by case, but we can harmlessly assume that if the prediction error is zero also the loss is zero and that $\ell(\cdot)$ is non-decreasing in the absolute value of the prediction error. Indeed, it is reasonable to assume that an exact guess of the outcome of Y will induce no losses, while the greater the prediction error, the higher the cost.

A loss function can be symmetric about zero (i.e., $\ell(-x) = \ell(x)$) or asymmetric. In the former case positive and negative errors of equal modulus produce the same loss, while in the latter case a different weight is given to positive and negative prediction errors. While there can be many reasons for the loss function to be asymmetric (cf. Example 1.1), the most used loss functions are generally symmetric. In particular the quadratic loss function $\ell_2(x) = x^2$ is for practical reasons the most frequently used in time series analysis and statistics in general.

Example 1.1 (Asymmetric loss function).
Suppose that for a firm that produces sunglasses, the variable costs of producing one pair of glasses is 1 Euro, and its wholesale value is 20 Euro. If one pair of sunglasses is produced but not sold the firm will have to pay 1 Euro per piece for storage and recycling costs.

Every year the firm has to decide how many pairs of glasses to produce and in order to do this it needs a prediction of sunglasses sales for that year. The cost of the prediction error will be higher if the predicted sales are lower than the actual, in fact for each produced pair of glasses the cost of not selling them is just 2 Euro (production cost plus storage/recycling) while the cost of not producing them when they would

[1]Notice that the term *predictor* is also commonly used for each of the variables $X_1, \dots, X_m$ on which the prediction is based. We will avoid this second meaning of the term to prevent misunderstandings.

sell is 19 Euro of lost profits (20 Euro of lost sales revenues minus 1 Euro of production cost).

In formulas, let us call Y the unknown future value of the sunglasses demand, $\hat{Y}$ its prediction and $E = Y - \hat{Y}$ the prediction error. Then, the loss function for this problem is given by

$$\ell(E) = \begin{cases} 19E & \text{for } E \geq 0 \\ 2E & \text{for } E < 0. \end{cases}$$

Intuitively, by observing this cost function one can expect that it is convenient to build predictions that tend to be positively biased so that negative errors (less costly) are more frequent than positive ones (more costly).

A predictor is optimal if it minimises the *expected loss* (i.e., the expectation of the loss function) among the class of measurable functions.

Definition 1.1 (Optimal prediction). Let $Y, X_1, \dots, X_m$ be random variables defined on the same probability space, $\mathcal{M}$ be the class of $(X_1, \dots, X_m)$-measurable functions and $\ell(\cdot)$ be a non-negative loss function; then the predictor $\hat{Y} = \hat{p}(X_1, \dots, X_m)$, with $\hat{p} \in \mathcal{M}$, is optimal for Y with respect to the loss ℓ if

$$\mathbb{E}\,\ell\big(Y - \hat{Y}\big) = \min_{p \in \mathcal{M}} \mathbb{E}\,\ell\big(Y - p(X_1, \dots, X_m)\big).$$

In particular the quadratic loss function $\ell_2(x) = x^2$ is for practical reasons the most frequently used in time series analysis and statistics in general. By using this loss function one assumes that the loss grows quadratically with the prediction error, and positive and negative errors of the same entity correspond to equal losses. As it will become clear from the next lines, the quadratic loss function has many mathematical advantages that make it a good choice if no better reasons suggest to the use of different loss curves.

In the rest of the book, predictions will always be made with respect to the quadratic loss function unless otherwise specified.

1.1 Optimal predictor

We are now in the condition to derive the optimal predictor under the quadratic loss function.

Theorem 1.1 (Optimal predictor under quadratic loss). *Let $Y, X_1, \ldots, X_m$ be random variables with finite variance, then the optimal predictor for Y based on $X_1, \ldots, X_m$ with respect to the quadratic loss function, $\ell_2(x) = x^2$, is the conditional expectation $\mathbb{E}[Y|X_1, \ldots, X_m]$.*

Proof. We have to show that there is no other measurable function $p(X_1, \ldots, X_m)$ that has smaller expected loss than $\mathbb{E}[Y|X_1, \ldots, X_m]$. The expected (quadratic) loss of the optimal predictor (*mean squared error*) is

$$MSE_{opt} = \mathbb{E}\{Y - \mathbb{E}[Y|X_1, \ldots, X_m]\}^2.$$

If we compute the expected loss of the generic predictor and subtract and add the optimal predictor $\mathbb{E}[Y|X_1, \ldots, X_m]$ we can write the mean squared error of $p(\cdot)$ as

$$\begin{aligned} MSE_{p(\cdot)} &= \mathbb{E}\{Y - \mathbb{E}[Y|X_1, \ldots, X_m] + \mathbb{E}[Y|X_1, \ldots, X_m] - p(X_1, \ldots, X_m)\}^2 \\ &= MSE_{opt} + \mathbb{E}\{\mathbb{E}[Y|X_1, \ldots, X_m] - p(X_1, \ldots, X_m)\}^2, \end{aligned}$$

since, by Property 2. of Theorem 1.2 below, we have

$$\mathbb{E}\Big\{\Big(Y - \mathbb{E}[Y|X_1, \ldots, X_m]\Big)\Big(\mathbb{E}[Y|X_1, \ldots, X_m] - p(X_1, \ldots, X_m)\Big)\Big\} = 0.$$

Thus, $MSE_{p(\cdot)}$ is the sum of a fixed number and a non-negative term which is zero if and only if $p(X_1, \ldots, X_m) = \mathbb{E}[Y|X_1, \ldots, X_m]$ with probability one. □

The following properties of the conditional expectation will be useful throughout the entire book.

Theorem 1.2 (Properties of the conditional expectation). *Let Y, X and Z be random variables (or vectors) with finite expectation and $g(\cdot)$ a function (measurable with respect to X) such that $\mathbb{E}g(X)$ exists, then*

1. *(Linearity)* $\mathbb{E}[aY + bZ + c|X] = a\mathbb{E}[Y|X] + b\mathbb{E}[Z|X] + c$, *with* a, b, c *constants;*
2. *(Functions of conditioning variables)* $\mathbb{E}[Yg(X)|X] = \mathbb{E}[Y|X]g(X)$;
3. *(Independence with the conditioning variables)* $\mathbb{E}[Y|X] = \mathbb{E}[Y]$ *when* Y *is independent from* X;
4. *(Law of iterated expectations)* $\mathbb{E}[Y] = \mathbb{E}\{\mathbb{E}[Y|X]\}$;
5. *(Orthogonality of the prediction error)* $\mathbb{E}\{(Y - \mathbb{E}[Y|X])\, g(X)\} = 0$;
6. *(Law of total variance)* $\mathbb{V}\mathrm{ar}[Y] = \mathbb{E}[\mathbb{V}\mathrm{ar}(Y|X)] + \mathbb{V}\mathrm{ar}[\mathbb{E}[Y|X]]$.

Proof. We prove the theorem point by point.

Linearity. Being the conditional expectation, an integral, linearity of the expectation is just a consequence of the linearity of the integral.

Functions of conditioning variables. For any value x that the random variable X can take, the expectation $\mathbb{E}[Yg(X)|X = x]$ is equal to $g(x)\mathbb{E}[Y|X = x]$ as, given $X = x$, $g(X)$ becomes the constant $g(x)$. Since this holds for all the values x in the range of X, the result follows.

Independence with the conditioning variable. Under independence of Y and X, the joint distribution of (X, Y) is the product of the two marginal distributions and, thus, the (conditional) distribution of $(Y|X)$ is equal to the marginal distribution of Y. Therefore, the expectation of $(Y|X)$ and Y is equal.

Law of iterated expectations. For a general proof of Property 1, refer to any measure-theoretic book on probability (for instance Shorack, 2000, Chapter 8), we provide a proof only for the case of absolutely continuous random variables using elementary probability notions:

$$\begin{aligned}\mathbb{E}[Y] &= \int y \int f(y, x)\,\mathrm{d}x\,\mathrm{d}y = \int y \int f(y|x)f(x)\,\mathrm{d}x\,\mathrm{d}y \\ &= \int f(x) \int y f(y|x)\,\mathrm{d}y\,\mathrm{d}x = \int f(x)\,\mathbb{E}[Y|X = x]\,\mathrm{d}x \\ &= \mathbb{E}\{\mathbb{E}[Y|X]\}.\end{aligned}$$

The change in the order of integration is allowed by the assumption of finiteness of the expectations of X and Y. The reader should try to replicate this proof for X and Y discrete random variables.

Orthogonality of the prediction error. Using Properties 1, 2, and 4 we have

$$\begin{aligned}\mathbb{E}\{(Y - \mathbb{E}[Y|X])\,g(X)\} &= \mathbb{E}\left[\mathbb{E}\{(Y - \mathbb{E}[Y|X])\,g(X)|X\}\right] \\ &= \mathbb{E}\left[\mathbb{E}\{Y - \mathbb{E}[Y|X]|X\}g(X)\right] \\ &= \mathbb{E}\left[\{\mathbb{E}[Y|X] - \mathbb{E}[Y|X]\}g(X)\right] = 0\end{aligned}$$

Law of total variance. Using Property 5, we can write

$$\mathbb{V}\mathrm{ar}[Y] = \mathbb{V}\mathrm{ar}\left[Y - \mathbb{E}(Y|X) + \mathbb{E}(Y|X)\right] = \mathbb{V}\mathrm{ar}[Y - \mathbb{E}[Y|X]] + \mathbb{V}\mathrm{ar}[\mathbb{E}(Y|X)].$$

Using Property 4, the first addend after the last equal sign can be written as

$$\mathbb{V}\mathrm{ar}[Y - \mathbb{E}[Y|X]] = \mathbb{E}[(Y - \mathbb{E}[Y|X])^2] = \mathbb{E}[\mathbb{E}(Y|X - \mathbb{E}[Y|X])^2] = \mathbb{V}\mathrm{ar}[Y|X].$$

□

1.2 Optimal linear predictor

Sometimes, instead of looking for an optimal predictor among the class of all measurable functions, it can be easier to limit the search to a smaller class of functions, such as linear combinations. As will be clear from the next lines, the advantage of basing the prediction of Y on the class of linear combinations of the random variables $X_1, \ldots, X_m$ (plus a constant) is that (under quadratic loss) the covariance structure of the random variables $Y, X_1, \ldots, X_m$ is all that is needed to compute the prediction.

Let $\boldsymbol{X} = (X_1, \ldots, X_m)^\top$, $\mu_Y = \mathbb{E}[Y]$, $\boldsymbol{\mu}_X = \mathbb{E}[\boldsymbol{X}]$, $\boldsymbol{\Sigma}_{XX} = \mathbb{E}[(\boldsymbol{X} - \boldsymbol{\mu}_x)(\boldsymbol{X} - \boldsymbol{\mu}_X)^\top]$, $\boldsymbol{\Sigma}_{YX} = \mathbb{E}[(Y - \mu_Y)(\boldsymbol{X} - \boldsymbol{\mu}_X)]$ and $\boldsymbol{\Sigma}_{XY} = \boldsymbol{\Sigma}_{YX}^\top$. As the next theorem states, this information is sufficient and necessary to compute the optimal linear predictor.

Theorem 1.3 (Optimal linear predictor). *Let $Y, X_1, \ldots, X_m$ be random variables with finite variance, and let $\mathcal{L}$ be the class of linear functions $\{\beta_0 + \beta_1 X_1 + \ldots + \beta_m X_m : (\beta_0, \beta_1, \ldots, \beta_m) \in \mathbb{R}^{m+1}\}$; then the optimal predictor in the class $\mathcal{L}$ with respect to the quadratic loss function, $\ell_2(x) = x^2$, is the linear predictor*

$$\mathbb{P}[Y|X_1, \ldots, X_m] = \mu_Y + \boldsymbol{\Sigma}_{YX}\boldsymbol{\Sigma}_{XX}^{-1}(\boldsymbol{X} - \boldsymbol{\mu}_X),$$

where, if $\boldsymbol{\Sigma}_{XX}$ is singular, $\boldsymbol{\Sigma}_{XX}^{-1}$ is to be substituted with a generalised inverse. The optimal linear predictor is unique.

In time series, there is no standard symbol for the linear prediction (or linear projection), thus, we will use $\mathbb{P}[Y|X]$ that recalls the conditional expectation notation.

Notice that if $\boldsymbol{\Sigma}_{\boldsymbol{XX}}$ is singular, its generalised inverse[2] is not unique, but, as the theorem states, the projection $\mathbb{P}[Y|\boldsymbol{X}]$ will be unique. This means that there are more choices of the vector $(\beta_0, \ldots, \beta_m)$ that yield the identical prediction.

Proof. We need to minimise the following MSE with respect to $\boldsymbol{\beta}$

$$MSE_{\boldsymbol{\beta}} = \mathbb{E}[(Y - \boldsymbol{\beta}^\top\tilde{\boldsymbol{X}})(Y - \boldsymbol{\beta}^\top\tilde{\boldsymbol{X}})^\top],$$

where we have set $\tilde{\boldsymbol{X}} = (1, X_1, \ldots, X_m)^\top$. If we define $\boldsymbol{\Omega}_{XX} = \mathbb{E}[\tilde{\boldsymbol{X}}\tilde{\boldsymbol{X}}^\top]$, $\boldsymbol{\Omega}_{YX} = \mathbb{E}[Y\tilde{\boldsymbol{X}}^\top] = \boldsymbol{\Omega}_{XY}^\top$, we can write

$$MSE_{\boldsymbol{\beta}} = \mathbb{E}(Y^2) + \boldsymbol{\beta}^\top\boldsymbol{\Omega}_{XX}\boldsymbol{\beta} - 2\boldsymbol{\Omega}_{YX}\boldsymbol{\beta}.$$

By setting equal to zero the derivative with respect to $\boldsymbol{\beta}$, we obtain the system

[2]If $\boldsymbol{A}$ is a matrix, its generalised inverse is a matrix $\boldsymbol{B}$ such that $\boldsymbol{ABA} = \boldsymbol{A}$. Every matrix has at least one generalised inverse.

of *normal equations*

$$\begin{aligned}\frac{\partial}{\partial \boldsymbol{\beta}} MSE_{\boldsymbol{\beta}} &= \mathbf{0}^\top \\ 2\boldsymbol{\beta}^\top \boldsymbol{\Omega}_{XX} - 2\boldsymbol{\Omega}_{YX} &= \mathbf{0}^\top \\ \boldsymbol{\beta}^\top \boldsymbol{\Omega}_{XX} &= \boldsymbol{\Omega}_{YX},\end{aligned}$$

which is a system of $m+1$ linear equations in $n+1$ unknowns. Thus, if $\boldsymbol{\Omega}_{XX}$ is invertible there is only the solution $\boldsymbol{\beta} = \boldsymbol{\Omega}_{XX}^{-1}\boldsymbol{\Omega}_{\boldsymbol{X}Y}$; otherwise there are infinitely many choices of $\boldsymbol{\beta}$ that solve the system.

To prove the uniqueness of the optimal linear prediction also when the matrix $\boldsymbol{\Omega}_{XX}$ is non-invertible, consider two arbitrary solutions of the linear system, say $\hat{\boldsymbol{\beta}}$ and $\tilde{\boldsymbol{\beta}}$, and the distance between the predictions $\hat{\boldsymbol{\beta}}^\top \tilde{\boldsymbol{X}}$ and $\tilde{\boldsymbol{\beta}}^\top \tilde{\boldsymbol{X}}$:

$$\begin{aligned}\mathbb{E}\left[(\hat{\boldsymbol{\beta}}^\top \tilde{\boldsymbol{X}} - \tilde{\boldsymbol{\beta}}^\top \tilde{\boldsymbol{X}})(\hat{\boldsymbol{\beta}}^\top \tilde{\boldsymbol{X}} - \tilde{\boldsymbol{\beta}}^\top \tilde{\boldsymbol{X}})^\top\right] &= (\hat{\boldsymbol{\beta}}^\top - \tilde{\boldsymbol{\beta}}^\top)\boldsymbol{\Omega}_{XX}(\hat{\boldsymbol{\beta}} - \tilde{\boldsymbol{\beta}}) \\ &= (\boldsymbol{\Omega}_{YX} - \boldsymbol{\Omega}_{YX})(\hat{\boldsymbol{\beta}} - \tilde{\boldsymbol{\beta}}) = 0,\end{aligned}$$

where we used the fact that both coefficient vectors satisfy the above normal equations. This zero mean-square distance implies that $\hat{\boldsymbol{\beta}}^\top \tilde{\boldsymbol{X}}$ and $\tilde{\boldsymbol{\beta}}^\top \tilde{\boldsymbol{X}}$ are equal with probability one.

Notice that the optimal linear predictor in the theorem is expressed in a slightly different form. There, we have $\boldsymbol{\beta}$-coefficients that solve $\boldsymbol{\Sigma}_{XX}\boldsymbol{\beta}_1 = \boldsymbol{\Sigma}_X Y$ and $\beta_0 + \boldsymbol{\mu}_X^\top \boldsymbol{\beta}_1 = \mu_Y$. We can see the equivalence of the two systems of normal equations if we write $\boldsymbol{\Omega}_{XX}\boldsymbol{\beta} = \boldsymbol{\Omega}_{XY}$ in blocks as follows:

$$\begin{bmatrix} 1 & \boldsymbol{\mu}_X^\top \\ \boldsymbol{\mu}_X & \boldsymbol{\Sigma}_{XX} + \boldsymbol{\mu}_X \boldsymbol{\mu}_X^\top \end{bmatrix} \begin{bmatrix} \beta_0 \\ \boldsymbol{\beta}_1 \end{bmatrix} = \begin{bmatrix} \mu_Y \\ \boldsymbol{\Sigma}_{XY} + \boldsymbol{\mu}_X \mu_Y \end{bmatrix}.$$

From the first line we obtain $\beta_0 = \mu_Y - \boldsymbol{\mu}_X^\top \boldsymbol{\beta}_1$ and, substituting in the second block we get,

$$\boldsymbol{\mu}_X(\mu_Y - \boldsymbol{\mu}_X^\top \boldsymbol{\beta}_1) + \boldsymbol{\Sigma}_{XX}\boldsymbol{\beta}_1 + \boldsymbol{\mu}_X \boldsymbol{\mu}_X^\top \boldsymbol{\beta}_1 = \boldsymbol{\Sigma}_{XY} + \boldsymbol{\mu}_X \mu_Y,$$

which simplifies to $\boldsymbol{\Sigma}_{XX}\boldsymbol{\beta} = \boldsymbol{\Sigma}_{XY}$. □

Notice that if instead of predicting the scalar Y we need to predict the vector $\boldsymbol{Y} = (Y_1, \ldots, Y_k)^\top$, we have a set of k independent optimisations, and the prediction formula in Theorem 1.3 simply generalises to

$$\boldsymbol{Y} = \boldsymbol{\mu}_Y + \boldsymbol{\Sigma}_{YX}\boldsymbol{\Sigma}_{XX}(\boldsymbol{X} - \boldsymbol{\mu}_X).$$

We provide a list of properties that the optimal linear predictor enjoys in the more general case in which $\boldsymbol{Y}$ is a vector.

Theorem 1.4 (Properties of the optimal linear predictor). *Let all the conditions and the notation of Theorem 1.3 hold, a, b, c be constants, $\boldsymbol{Z}$ be a random vector with finite variances; then the optimal linear predictor satisfies the following properties*

1. *(Unbiasedness)* $\mathbb{E}\big[\boldsymbol{Y} - \mathbb{P}[\boldsymbol{Y}|\boldsymbol{X}]\big] = 0$;
2. *(Orthogonality of the prediction error)* $\mathbb{E}\big[(\boldsymbol{Y} - \mathbb{P}[\boldsymbol{Y}|\boldsymbol{X}])\,\boldsymbol{X}^\top\big] = \mathbf{0}$;
3. *(Mean square error of the prediction)* $MSE_{lin} = \boldsymbol{\Sigma}_{YY} - \boldsymbol{\Sigma}_{YX}\boldsymbol{\Sigma}_{XX}^{-1}\boldsymbol{\Sigma}_{XY}$;
4. *(Linearity)* $\mathbb{P}[a\boldsymbol{Y} + b\boldsymbol{Z} + c|\boldsymbol{X}] = a\mathbb{P}[\boldsymbol{Y}|\boldsymbol{X}] + b\mathbb{P}[\boldsymbol{Z}|\boldsymbol{X}] + c$;
5. *(Law of iterated projections)* $\mathbb{P}[\boldsymbol{Y}|\boldsymbol{X}] = \mathbb{P}\{\mathbb{P}[\boldsymbol{Y}|\boldsymbol{Z},\boldsymbol{X}]\big|\boldsymbol{X}\}$;
6. *(Projection on orthogonal variables) if* $\mathbb{E}(\boldsymbol{X} - \boldsymbol{\mu}_X)(\boldsymbol{Z} - \boldsymbol{\mu}_Z)^\top = \mathbf{0}$ *then*

$$\mathbb{P}[\boldsymbol{Y}|\boldsymbol{X},\boldsymbol{Z}] = \boldsymbol{\mu}_Y + \mathbb{P}[\boldsymbol{Y} - \boldsymbol{\mu}_Y|\boldsymbol{X}] + \mathbb{P}[\boldsymbol{Y} - \boldsymbol{\mu}_Y|\boldsymbol{Z}];$$

7. *(Updating)*

$$\begin{aligned}\mathbb{P}[\boldsymbol{Y}|\boldsymbol{Z},\boldsymbol{X}] &= \mathbb{P}[\boldsymbol{Y}|\boldsymbol{X}] + \mathbb{P}\big[\boldsymbol{Y} - \mathbb{P}[\boldsymbol{Y}|\boldsymbol{X}]\big|\boldsymbol{Z} - \mathbb{P}[\boldsymbol{Z}|\boldsymbol{X}]\big]\\ &= \mathbb{P}[\boldsymbol{Y}|\boldsymbol{X}] + \boldsymbol{\Sigma}_{YZ|X}\boldsymbol{\Sigma}_{ZZ|X}^{-1}(\boldsymbol{Z} - P[\boldsymbol{Z}|\boldsymbol{X}])\end{aligned}$$

with

$$\begin{aligned}\boldsymbol{\Sigma}_{YZ|X} &= \mathbb{E}\left[(\boldsymbol{Y} - \mathbb{P}[\boldsymbol{Y}|\boldsymbol{X}])(\boldsymbol{Z} - \mathbb{P}[\boldsymbol{Z}|\boldsymbol{X}])^\top\right],\\ \boldsymbol{\Sigma}_{ZZ|X} &= \mathbb{E}\left[(\boldsymbol{Z} - \mathbb{P}[\boldsymbol{Z}|\boldsymbol{X}])(\boldsymbol{Z} - \mathbb{P}[\boldsymbol{Z}|\boldsymbol{X}])^\top\right].\end{aligned}$$

If we call $MSE_{Y|X}$ the mean square error of $\mathbb{P}[\boldsymbol{Y}|\boldsymbol{X}]$, then the MSE of $\mathbb{P}[\boldsymbol{Y}|\boldsymbol{Z},\boldsymbol{X}]$ is given by

$$MSE_{Y|Z,X} = MSE_{Y|X} - \boldsymbol{\Sigma}_{YZ|X}\boldsymbol{\Sigma}_{ZZ|X}^{-1}\boldsymbol{\Sigma}_{ZY|X}.$$

Proof. For notational compactness, let $\mathbf{B}_{YX} = \boldsymbol{\Sigma}_{YX}\boldsymbol{\Sigma}_{XX}^{-1}$.

Unbiasedness.

$$\mathbb{E}\left\{\boldsymbol{Y} - \boldsymbol{\mu}_Y - \mathbf{B}_{YX}(\boldsymbol{X} - \boldsymbol{\mu}_X)\right\} = \boldsymbol{\mu}_Y - \boldsymbol{\mu}_Y - \mathbf{B}_{YX}(\boldsymbol{\mu}_X - \boldsymbol{\mu}_X) = \mathbf{0}.$$

Orthogonality.

$$\begin{aligned}&\mathbb{E}\left\{[\boldsymbol{Y} - \boldsymbol{\mu}_Y - \mathbf{B}_{YX}(\boldsymbol{X} - \boldsymbol{\mu}_X)]\boldsymbol{X}^\top\right\} =\\ &\mathbb{E}\left\{[(\boldsymbol{Y} - \boldsymbol{\mu}_Y) - \mathbf{B}_{YX}(\boldsymbol{X} - \boldsymbol{\mu}_X)][(\boldsymbol{X} - \boldsymbol{\mu}_X) + \boldsymbol{\mu}_X]^\top\right\} =\\ &\boldsymbol{\Sigma}_{YX} - \mathbf{B}_{YX}\boldsymbol{\Sigma}_{XX} + \mathbb{E}\{\boldsymbol{Y} - \boldsymbol{\mu}_Y\}\boldsymbol{\mu}_X^\top + \mathbb{E}\{\boldsymbol{Y} - \mathbb{P}[\boldsymbol{Y}|\boldsymbol{X}]\}\boldsymbol{\mu}_X^\top = 0\end{aligned}$$

Mean square error.

$$\mathbb{E}\left\{[(\boldsymbol{Y}-\boldsymbol{\mu}_Y)-\mathbf{B}_{YX}(\boldsymbol{X}-\boldsymbol{\mu}_X)][(\boldsymbol{Y}-\boldsymbol{\mu}_Y)-\mathbf{B}_{YX}(\boldsymbol{X}-\boldsymbol{\mu}_X)]^\top\right\}=$$
$$\boldsymbol{\Sigma}_{YY}+\mathbf{B}_{YX}\boldsymbol{\Sigma}_{XX}\mathbf{B}_{YX}^\top-\mathbf{B}_{YX}\boldsymbol{\Sigma}_{XY}-\boldsymbol{\Sigma}_{YX}\mathbf{B}_{YX}^\top=$$
$$\boldsymbol{\Sigma}_{YY}-\boldsymbol{\Sigma}_{YX}\boldsymbol{\Sigma}_{XX}^{-1}\boldsymbol{\Sigma}_{XY}.$$

Linearity. It is a straightforward consequence of the predictor being a linear function of $\boldsymbol{X}$.

Law of iterated projections. If we define the prediction error $\boldsymbol{E}=\boldsymbol{Y}-\mathbb{P}[\boldsymbol{Y}|\boldsymbol{Z},\boldsymbol{X}]$, we can write $\boldsymbol{Y}$ as

$$\boldsymbol{Y}=\mathbb{P}[\boldsymbol{Y}|\boldsymbol{Z},\boldsymbol{X}]+\boldsymbol{E},$$

where by Properties 1 and 2 we have $\mathbb{E}[\boldsymbol{E}\cdot\boldsymbol{X}^\top]=\mathbf{0}$. Thus, taking optimal linear predictions based on $\boldsymbol{X}$ of both sides of the above identity and using the linearity property we get

$$\mathbb{P}[\boldsymbol{Y}|\boldsymbol{X}]=\mathbb{P}\big\{\mathbb{P}[\boldsymbol{Y}|\boldsymbol{Z},\boldsymbol{X}]\big|\boldsymbol{X}\big\}+\mathbf{0}.$$

Projection on orthogonal variables.

$$\begin{aligned}P[\boldsymbol{Y}|\boldsymbol{X},\boldsymbol{Z}]&=\boldsymbol{\mu}_Y+\begin{bmatrix}\boldsymbol{\Sigma}_{YX} & \boldsymbol{\Sigma}_{YZ}\end{bmatrix}\begin{bmatrix}\boldsymbol{\Sigma}_{XX} & \mathbf{0}\\ \mathbf{0} & \boldsymbol{\Sigma}_{ZZ}\end{bmatrix}^{-1}\begin{bmatrix}\boldsymbol{X}-\boldsymbol{\mu}_X\\ \boldsymbol{Z}-\boldsymbol{\mu}_Z\end{bmatrix}\\&=\boldsymbol{\mu}_Y+\boldsymbol{\Sigma}_{YX}\boldsymbol{\Sigma}_{XX}^{-1}(\boldsymbol{X}-\boldsymbol{\mu}_X)+\boldsymbol{\Sigma}_{YZ}\boldsymbol{\Sigma}_{ZZ}^{-1}(\boldsymbol{X}-\boldsymbol{\mu}_X)\\&=\boldsymbol{\mu}_Y+\mathbb{P}[\boldsymbol{Y}-\boldsymbol{\mu}_Y|\boldsymbol{X}]+\mathbb{P}[\boldsymbol{Y}-\boldsymbol{\mu}_Y|\boldsymbol{Z}].\end{aligned}$$

Updating. First decompose $\boldsymbol{Y}$ as $\boldsymbol{Y}=\mathbb{P}[\boldsymbol{Y}|\boldsymbol{Z}]+\boldsymbol{E}$ with $\boldsymbol{E}$ orthogonal to $\boldsymbol{Z}$. Then, by taking the prediction of $\boldsymbol{Y}$ based on $\boldsymbol{X}$ and $\boldsymbol{Z}$ we get

$$\begin{aligned}\mathbb{P}[\boldsymbol{Y}|\boldsymbol{X},\boldsymbol{Z}]&=\mathbb{P}\big\{\mathbb{P}[\boldsymbol{Y}|\boldsymbol{Z}]\big|\boldsymbol{X},\boldsymbol{Z}\big\}+\mathbb{P}[\boldsymbol{E}|\boldsymbol{X},\boldsymbol{Z}]\\&=\mathbb{P}\big[\boldsymbol{\mu}_Y+\mathbf{B}_{YZ}(\boldsymbol{Z}-\boldsymbol{\mu}_Z)\big|\boldsymbol{X},\boldsymbol{Z}\big]+\mathbb{P}[\boldsymbol{E}|\boldsymbol{X}]\\&=\boldsymbol{\mu}_Y+\mathbf{B}_{YZ}(\boldsymbol{Z}-\boldsymbol{\mu}_Z)+\mathbb{P}[\boldsymbol{Y}|\boldsymbol{X}]-\mathbb{P}\big[\boldsymbol{\mu}_Y+\mathbf{B}_{YZ}(\boldsymbol{Z}-\boldsymbol{\mu}_Z)\big|\boldsymbol{X}\big]\\&=\mathbb{P}[\boldsymbol{Y}|\boldsymbol{X}]+\mathbf{B}_{YZ}(\boldsymbol{Z}-\mathbb{P}[\boldsymbol{Z}|\boldsymbol{X}])\\&=\mathbb{P}[\boldsymbol{Y}|\boldsymbol{X}]+\mathbb{P}[\boldsymbol{Y}-\boldsymbol{\mu}_Y|\boldsymbol{Z}-\mathbb{P}[\boldsymbol{Z}|\boldsymbol{X}]]\\&=\mathbb{P}[\boldsymbol{Y}|\boldsymbol{X}]+\mathbb{P}[\boldsymbol{Y}-\mathbb{P}[\boldsymbol{Y}|\boldsymbol{X}]+\mathbb{P}\big[\boldsymbol{Y}|\boldsymbol{X}]-\boldsymbol{\mu}_Y\big|\boldsymbol{Z}-\mathbb{P}[\boldsymbol{Z}|\boldsymbol{X}]\big]\\&=\mathbb{P}[\boldsymbol{Y}|\boldsymbol{X}]+\mathbb{P}\big[\boldsymbol{Y}-\mathbb{P}[\boldsymbol{Y}|\boldsymbol{X}]|\boldsymbol{Z}-\mathbb{P}[\boldsymbol{Z}|\boldsymbol{X}]\big].\end{aligned}$$

Notice that in the sequence of equalities above we used the facts: $\mathbb{P}[\boldsymbol{E}|\boldsymbol{X},\boldsymbol{Z}]=$

$\mathbb{P}[\boldsymbol{E}|\boldsymbol{Z}]$ as $\boldsymbol{E}$ is orthogonal to $\boldsymbol{Z}$; $\mathbb{P}[\boldsymbol{Z}|\boldsymbol{X},\boldsymbol{Z}] = \boldsymbol{z}$ as the best linear predictor of $\boldsymbol{Z}$ based on $\boldsymbol{Z}$ (and possibly other variables) is $\boldsymbol{Z}$, and its MSE is zero.

The updating formula for the MSE of $\mathbb{P}[\boldsymbol{Y}|\boldsymbol{X},\boldsymbol{Z}]$ is easily obtained:

$$\begin{aligned}
&\mathbb{E}\{(\boldsymbol{Y} - P[\boldsymbol{Y}|\boldsymbol{X}]) - \boldsymbol{\Sigma}_{YZ|X}\boldsymbol{\Sigma}_{ZZ|X}^{-1}(\boldsymbol{Z} - P[\boldsymbol{Z}|\boldsymbol{X}])\}\times \\
&\{(\boldsymbol{Y} - P[\boldsymbol{Y}|\boldsymbol{X}]) - \boldsymbol{\Sigma}_{YZ|X}\boldsymbol{\Sigma}_{ZZ|X}^{-1}(\boldsymbol{Z} - P[\boldsymbol{Z}|\boldsymbol{X}])\}^\top = \\
&MSE_{Y|X} - \boldsymbol{\Sigma}_{YZ|X}\boldsymbol{\Sigma}_{ZZ|X}^{-1}\boldsymbol{\Sigma}_{ZY|X}.
\end{aligned}$$

□

Property 2 of Theorem 1.4 can also be used as a definition for the optimal linear prediction; indeed it defines the same set of normal equations as the first order conditions for minimising the MSE: $\mathbb{E}[(Y - \boldsymbol{\beta}^\top \boldsymbol{X})\boldsymbol{X}^\top] = \boldsymbol{0}$, that is, $\boldsymbol{\Omega}_{YX} = \boldsymbol{\beta}^\top \boldsymbol{\Sigma}_{XX}$ (cf. proof of Theorem 1.3).

Notice that we can always decompose the random variable Y into the sum of the prediction and its error

$$Y = \mathbb{P}[Y|\boldsymbol{X}] + E,$$

where, by Properties 1 and 2 of the optimal linear predictor (Theorem 1.4), the error E has mean zero and is orthogonal to (i.e., uncorrelated with) all the predictor variables $X_1, \ldots, X_m$.

Example 1.2 (Predictions in a non-linear model).
Let

$$Y = X^3 + Z,$$

where X and Z are independent standard normal random variables. Let us compute the optimal predictor and the optimal linear predictor of Y based on X.

The optimal predictor is easily derived by applying the conditional expectation to both sides of the above equation:

$$\begin{aligned}
\mathbb{E}[Y|X] &= \mathbb{E}[X^3 + Z|X] = X^3, \\
MSE_{opt} &= \mathbb{E}[(Y - X^3)^2] = \mathbb{E}[Z^2] = 1.
\end{aligned}$$

For the optimal linear predictor we need to compute the mean of Y and covariance of Y and X:

$$\mathbb{E}[Y] = \mathbb{E}[X^3 + Z] = 0, \quad \mathbb{E}[XY] = \mathbb{E}[X^4 + XZ] = 3,$$

from which $\mathbb{P}[Y|X] = 3X$ and

$$\begin{aligned}
MSE_{lin} &= \mathbb{E}[X^3 + Z - 3X]^2 \\
&= \mathbb{E}[X^6] + \mathbb{E}[Z^2] + 9\mathbb{E}[X^2] + 6\mathbb{E}[X^4] = 28,
\end{aligned}$$

since the odd order moments of a standard normal are zero and the m-order moments with m even are given by $2^{-m/2}m!/(m/2)!$.

1.3 Linear models and joint normality

There are statistical models, like linear regression, that assume that the conditional mean of a random variable Y is a linear function of the conditioning variables, say $X_1, \ldots, X_m$, plus a constant:

$$\mathbb{E}[Y|X_1, \ldots, X_m] = \beta_0 + \beta_1 X_1 + \ldots + \beta_m X_m.$$

Of course, for these models the optimal predictor and the optimal linear predictor coincide, but there is another important class of models for which optimal linear prediction and conditional expectation are coincident: the class of *jointly normal* or *Gaussian* random variables.

Theorem 1.5 (Conditional distribution under joint normality). *Let $\boldsymbol{Y}$ and $\boldsymbol{X}$ be random vectors such that*

$$\begin{bmatrix} \boldsymbol{X} \\ \boldsymbol{Y} \end{bmatrix} \sim \mathcal{N}\left(\begin{bmatrix} \boldsymbol{\mu}_X \\ \boldsymbol{\mu}_Y \end{bmatrix}, \begin{bmatrix} \boldsymbol{\Sigma}_{XX} & \boldsymbol{\Sigma}_{XY} \\ \boldsymbol{\Sigma}_{YX} & \boldsymbol{\Sigma}_{YY} \end{bmatrix} \right),$$

then

$$\boldsymbol{Y}|\boldsymbol{X} \sim \mathcal{N}\left(\boldsymbol{\mu}_Y + \boldsymbol{\Sigma}_{YX}\boldsymbol{\Sigma}_{XX}^{-1}(\boldsymbol{X} - \boldsymbol{\mu}_X),\ \boldsymbol{\Sigma}_{YY} - \boldsymbol{\Sigma}_{YX}\boldsymbol{\Sigma}_{XX}^{-1}\boldsymbol{\Sigma}_{XY} \right),$$

where, if $\boldsymbol{\Sigma}_{XX}$ is singular, its inverse is to be replaced with a generalised inverse.

Notice that the variance of $\boldsymbol{Y}|\boldsymbol{X}$ does not depend on the particular value of $\boldsymbol{X}$ on which $\boldsymbol{Y}$ is conditioned and, thus, conditional variance and MSE coincide: this is a characteristic of the normal distribution that does not generalise to other distributions.

Proof. Let us call $\boldsymbol{\mu}$ and $\boldsymbol{\Sigma}$ the mean vector and covariance matrix of $(\boldsymbol{X}^\top, \boldsymbol{Y}^\top)^\top$. For the proof we need the easy-to-verify identity

$$\begin{bmatrix} \boldsymbol{I} & -\boldsymbol{\Sigma}_{YX}\boldsymbol{\Sigma}_{XX}^{-1} \\ \boldsymbol{0} & \boldsymbol{I} \end{bmatrix} \boldsymbol{\Sigma} \begin{bmatrix} \boldsymbol{I} & \boldsymbol{0} \\ -\boldsymbol{\Sigma}_{XX}^{-1}\boldsymbol{\Sigma}_{XY} & \boldsymbol{I} \end{bmatrix} = \begin{bmatrix} \boldsymbol{\Sigma}_{YY} - \boldsymbol{\Sigma}_{YX}\boldsymbol{\Sigma}_{XX}^{-1}\boldsymbol{\Sigma}_{XY} & \boldsymbol{0} \\ \boldsymbol{0} & \boldsymbol{\Sigma}_{XX} \end{bmatrix},$$

and taking the determinant of both sides of the identity we have

$$|\boldsymbol{\Sigma}| = |\boldsymbol{\Sigma}_{YY} - \boldsymbol{\Sigma}_{YX}\boldsymbol{\Sigma}_{XX}^{-1}\boldsymbol{\Sigma}_{XY}||\boldsymbol{\Sigma}_{XX}|.$$

From the same identity we also obtain[3]

$$\boldsymbol{\Sigma}^{-1} = \begin{bmatrix} \boldsymbol{I} & \boldsymbol{0} \\ -\boldsymbol{\Sigma}_{XX}^{-1}\boldsymbol{\Sigma}_{XY} & \boldsymbol{I} \end{bmatrix} \begin{bmatrix} (\boldsymbol{\Sigma}_{YY} - \boldsymbol{\Sigma}_{YX}\boldsymbol{\Sigma}_{XX}^{-1}\boldsymbol{\Sigma}_{XY})^{-1} & \boldsymbol{0} \\ \boldsymbol{0} & \boldsymbol{\Sigma}_{XX}^{-1} \end{bmatrix} \begin{bmatrix} \boldsymbol{I} & -\boldsymbol{\Sigma}_{YX}\boldsymbol{\Sigma}_{XX}^{-1} \\ \boldsymbol{0} & \boldsymbol{I} \end{bmatrix}.$$

[3] Recall that if $\boldsymbol{AB}$ is invertible $(\boldsymbol{AB})^{-1} = \boldsymbol{B}^{-1}\boldsymbol{A}^{-1}$; furthermore the inverse of a diagonal block matrix is a diagonal block matrix with the inverse of the respective elements on the main diagonal.

By definition, the conditional density of $\boldsymbol{Y}|\boldsymbol{X}$ is $f(\boldsymbol{y}|\boldsymbol{x}) = f(\boldsymbol{y}, \boldsymbol{x})/f(\boldsymbol{x})$, and substituting with the normal densities

$$f(\boldsymbol{y}|\boldsymbol{x}) = \frac{1}{(2\pi)^{k/2}} \frac{|\boldsymbol{\Sigma}_{XX}|}{|\boldsymbol{\Sigma}|} \frac{\exp\left\{-\frac{1}{2}\begin{bmatrix}\boldsymbol{y}_0^\top & \boldsymbol{x}_0^\top\end{bmatrix} \boldsymbol{\Sigma}^{-1} \begin{bmatrix}\boldsymbol{y}_0 \\ \boldsymbol{x}_0\end{bmatrix}\right\}}{\exp\left\{-\frac{1}{2}\boldsymbol{x}_0^\top \boldsymbol{\Sigma}_{XX}^{-1}\boldsymbol{x}_0\right\}}$$

where we have set $\boldsymbol{x}_0 = \boldsymbol{x} - \boldsymbol{\mu}_x$, $\boldsymbol{y}_0 = \boldsymbol{y} - \boldsymbol{\mu}_y$ and k equal to the number of elements of $\boldsymbol{y}$. Using the above identity for $\boldsymbol{\Sigma}^{-1}$ we obtain

$$\begin{bmatrix}\boldsymbol{y}_0^\top & \boldsymbol{x}_0^\top\end{bmatrix} \boldsymbol{\Sigma}^{-1} \begin{bmatrix}\boldsymbol{y}_0 \\ \boldsymbol{x}_0\end{bmatrix} =$$
$$(\boldsymbol{y} - \boldsymbol{\mu}_{Y|X})^\top \boldsymbol{\Sigma}_{Y|X}^{-1}(\boldsymbol{y} - \boldsymbol{\mu}_{Y|X}) + (\boldsymbol{x} - \boldsymbol{\mu}_x)^\top \boldsymbol{\Sigma}_{XX}^{-1}(\boldsymbol{x} - \boldsymbol{\mu}_x),$$

with $\boldsymbol{\mu}_{Y|X} = \boldsymbol{y} + \boldsymbol{\Sigma}_{YX}\boldsymbol{\Sigma}_{XX}^{-1}(\boldsymbol{x} - \boldsymbol{\mu}_X)$ and $\boldsymbol{\Sigma}_{Y|X} = \boldsymbol{\Sigma}_{YY} - \boldsymbol{\Sigma}_{YX}\boldsymbol{\Sigma}_{XX}^{-1}\boldsymbol{\Sigma}_{XY}$.

Substituting this result and the determinant identity in the conditional density we obtain

$$f(\boldsymbol{y}|\boldsymbol{x}) = \frac{1}{(2\pi)^{k/2}|\boldsymbol{\Sigma}_{Y|X}|} \exp\left\{(\boldsymbol{y} - \boldsymbol{\mu}_{Y|X})^\top \boldsymbol{\Sigma}_{Y|X}^{-1}(\boldsymbol{y} - \boldsymbol{\mu}_{Y|X})\right\}.$$

□

Chapter 2

Time Series Concepts

In this chapter some basic definitions and concepts of time series analysis are reviewed. The reader already acquainted with this material should skip this chapter.

2.1 Definitions

Let us define the main object of this book: the time series.

Definition 2.1 (Time series). A **time series** is a sequence of observations ordered with respect to a time index t, taking values in an index set S. If the set S contains a finite or countable number of elements we speak of *discrete-time time series* and the generic observation is indicated with the symbol y_t, while if S is a continuum we have a *continuous-time time series*, whose generic observation is represented as $y(t)$.

Even though continuous-time time series are becoming very rare in a world dominated by digital computers[1], continuous-time models are nevertheless very popular in many disciplines. Indeed, observations may be taken at approximately equispaced points in time, and in this case the discrete-time framework is the most natural, or observations may be non-equispaced and in this case continuous-time models are usually more appropriate. This book concentrates on discrete-time time series, but most of the models covered here have a continuous-time counterpart.

Since future values of real time series are generally unknown and cannot be predicted without error a quite natural mathematical model to describe their behaviour is that of a stochastic process.

[1]Digital computers are *finite-state machines* and, thus, cannot record continuous-time time series.

Definition 2.2 (Stochastic process). A **stochastic process** is a sequence of random variables defined on a probability space $(\Omega, \mathcal{F}, P)$ and ordered with respect to a time index t, taking values in an index set S.

Again, when S is numerable one speaks of *discrete-time processes* and denotes it as $\{Y_t\}_{t\in S}$, when S is a continuum we have a *continuous-time process* and represent it as $\{Y(t)\}_{t\in S}$ (sometimes also $\{Y_t\}_{t\in S}$).

By definition of random variable, for each fixed t, Y_t is a function $Y_t(\cdot)$ on the sample space Ω, while for each fixed simple event $\omega \in \Omega$, $Y_{\cdot}(\omega)$ is a function on S, or a *realization* (also *sample-path*) of a stochastic process.

As is customary in modern time series analysis, in this book we consider a time series as a finite realisation (or sample-path) of a stochastic process.

There is a fundamental difference between classical statistical inference and time series analysis. The set-up of classical inference consists of a random variable or vector X and a random selection scheme to extract simple events, say $\{\omega_1, \omega_2, \ldots, \omega_n\}$, from the sample space Ω. The observations, then, consist of the random variable values corresponding to the selected simple events: $\{x_1, x_2, \ldots, x_n\}$, where $x_i := X(\omega_i)$ for $i = 1, 2, \ldots, n$. In time series analysis, instead, we have a stochastic process $\{Y_t\}_{t\in S}$ and observe only one finite realisation of it through the extraction of a single event, say ω_1, from the sample space Ω: we have the time series $\{y_1, y_2, \ldots, y_n\}$, with $y_t := Y_t(\omega_1)$ for $t = 1, \ldots, n$. Therefore, while in classical inference we have a sample of n observations for the random variable X, in time series analysis we usually have to deal with a sample of dimension 1 with n observations corresponding to different time points of the process $\{Y_t\}$. This means that, if we cannot assume some kind of *time-homogeneity* of the process making the single sample-path "look like" a classical sample, then we cannot draw any sensible inference and prediction from a single time series. In Section 2.2 we introduce the classes of *stationary* and *integrated* processes, which are the most important time-homogeneous processes used in time series analysis.

Also in time series analysis, the normal or Gaussian distribution plays a special role. So, we conclude this section by defining the important class of *Gaussian processes*.

Definition 2.3 (Gaussian process). The process $\{Y_t\}_{t\in S}$ is Gaussian if for all the finite subsets $\{t_1, t_2, \ldots, t_m\}$ of time points in Ω the joint distribution of $(Y_{t_1}, \ldots, Y_{t_m})$ is multivariate normal.

2.2 Stationary processes

As we saw in Section 2.1, we treat a time series as a finite sample-path of a stochastic process. Unfortunately, unlike in statistical inference based on repeated random sampling, in time series analysis we have only (a part of) one observation, the time series, from the data generating process. Thus, we have to base our inference on a sample of dimension one. Generally, we do have more than one observation in the sample-path but, unless we assume some kind of *time-homogeneity* of the data generating process, every observation y_t in the time series is drawn from a different random variable Y_t.

Since most social and natural phenomena seem to evolve smoothly rather than by abrupt changes, modelling them by time-homogeneous processes is a reasonable approximation, at least for a limited period of time.

The most important form of time-homogeneity used in time series analysis is *stationarity*, which is defined as time-invariance of the whole probability distribution of the data generating process (strict stationarity), or just of its first two moments (weak stationarity).

Definition 2.4 (Strict stationarity). The process $\{Y_t\}$ is *strictly stationary* if for all $k \in \mathbb{N}$, $h \in \mathbb{Z}$, and $(t_1, t_2, \ldots, t_k) \in \mathbb{Z}^k$,

$$(Y_{t_1}, Y_{t_2}, \ldots, Y_{t_k}) \stackrel{d}{=} (Y_{t_1+h}, Y_{t_2+h}, \ldots, Y_{t_k+h})$$

where $\stackrel{d}{=}$ denotes equality in distribution.

Definition 2.5 (Weak stationarity). The process $\{Y_t\}$ is *weakly stationary* (or *covariance stationary*) if, for all $h, t \in \mathbb{Z}$,

$$\mathbb{E}(Y_t) = \mu,$$
$$\mathbb{C}\mathrm{ov}(Y_t, Y_{t-h}) = \gamma(h),$$

with $\gamma(0) < \infty$.

As customary in time series analysis, in the rest of the book the terms *stationarity* and *stationary* will be used with the meaning of *weak stationarity* and *weakly stationary*, respectively. The following obvious result on the relation between the two definitions of stationarity is given without proof.

Theorem 2.1 (Relation between strict and weak stationarity). *Let $\{Y_t\}$ be a stochastic process:*

1. *if* $\{Y_t\}$ *is strictly stationary, then it is also weakly stationary if and only if* $\mathbb{V}\text{ar}(Y_t) < \infty$;
2. *if* $\{Y_t\}$ *is a Gaussian process, then strict and weak stationarity are equivalent (i.e., one form of stationarity implies the other).*

Notice that the above definitions of stationarity assume that the process $\{Y_t\}$ is defined for $t \in \mathbb{Z}$ (i.e., the process originates in the infinite past and ends in the infinite future). This is a mathematical abstraction that is useful to derive some results (e.g., limit theorems), but the definitions can be easily adapted to the case of time series with $t \in \mathbb{N}$ or $t \in \{1, 2, \dots, n\}$ by changing the domains of t, h and k accordingly.

The most elementary (non-trivial) stationary process is the *white noise.*

Definition 2.6 (White noise)**.** A stochastic process is *white noise* if it has zero mean, finite variance, σ^2, and covariance function

$$\gamma(h) = \begin{cases} \sigma^2 & \text{for } h = 0, \\ 0 & \text{for } h \neq 0. \end{cases}$$

As the next example clarifies, white noise processes and independent identically distributed (i.i.d.) sequences are not equivalent.

Example 2.1 (White noise and i.i.d. sequences)**.**
Let $\{X_t\}$ be a sequence of *independently identically distributed* (i.i.d.) random variables and $\{Z_t\}$ be *white noise.*

The process $\{X_t\}$ is strictly stationary since the joint distribution for any k-tuple of time points is the product (by independence) of the common marginal distribution (by identical distribution), say $F(\cdot)$,

$$\Pr\{X_{t_1} \leq x_1, X_{t_2} \leq x_2, \dots, X_{t_k} \leq x_k\} = \prod_{i=1}^{k} F(x_i),$$

and this does not depend on t. $\{X_t\}$ is not necessarily weakly stationary since its first two moments may not exist (e.g., when X_t is Cauchy-distributed).

The process $\{Z_t\}$ is weakly stationary since mean and covariance are time-independent, but it is not necessarily strictly stationary since its marginal and joint distributions may depend on t even when the first two moments are time-invariant.

The function $\gamma(h)$, which characterises a weakly stationary process, is called *autocovariance function* and enjoys the following properties.

Theorem 2.2 (Properties of the autocovariance function). *Let $\gamma(\cdot)$ be the autocovariance function of a stationary process, then*

1. *(Positivity of variance)* $\gamma(0) \geq 0$,
2. *(Cauchy–Schwarz inequality)* $|\gamma(h)| \leq \gamma(0)$,
3. *(Symmetry)* $\gamma(h) = \gamma(-h)$,
4. *(Non-negative definiteness)* $\sum_{i=1}^{m} \sum_{j=1}^{m} a_i \gamma(i-j) a_j \geq 0$, $\forall m \in \mathbb{N}$ *and* $(a_1, \ldots, a_m) \in \mathbb{R}^m$.

For the proof of this theorem and in many other places in this book, we will make use of the covariance matrix of the vector of n consecutive observations of a stationary process, say $\boldsymbol{Y} := (Y_1, Y_2, \ldots, Y_n)^\top$:

$$\boldsymbol{\Gamma}_n := \begin{bmatrix} \gamma(0) & \gamma(1) & \ldots & \gamma(n-1) \\ \gamma(1) & \gamma(0) & \ldots & \gamma(n-2) \\ \vdots & \vdots & \ddots & \vdots \\ \gamma(n-1) & \gamma(n-2) & \ldots & \gamma(0) \end{bmatrix}. \tag{2.1}$$

As any covariance matrix, $\boldsymbol{\Gamma}_n$ is symmetric with respect to the main diagonal and non-negative definite but, as the reader can easily verify, $\boldsymbol{\Gamma}_n$ is also symmetric with respect to the secondary diagonal. Furthermore, the element of the matrix with indexes (i, j) equals the element with indexes $(i+1, j+1)$ (i.e., $\boldsymbol{\Gamma}_n$ is a *Toeplitz matrix*).

Proof. The first two properties are well-known properties of variance and covariance. The third property follows from stationarity and the symmetry of the arguments of the covariance:

$$\gamma(h) = \mathbb{C}\text{ov}(X_t, X_{t-h}) = \mathbb{C}\text{ov}(X_{t+h}, X_t) = \mathbb{C}\text{ov}(X_t, X_{t+h}) = \gamma(-h).$$

As for the fourth property, let $\boldsymbol{y} := (Y_1, Y_2, \ldots, Y_m)^\top$ be m consecutive observations of the stationary process with autocovariance $\gamma(\cdot)$, then for any real m-vector of constants $\boldsymbol{a}$, the random variable $\boldsymbol{a}^\top \boldsymbol{y}_m$ has variance

$$\mathbb{V}\text{ar}\left(\boldsymbol{a}^\top \boldsymbol{y}\right) = \boldsymbol{a}^\top \boldsymbol{\Gamma}_m \boldsymbol{a} = \sum_{i=1}^{m} \sum_{j=1}^{m} a_i \gamma(i-j) a_j,$$

which, being a variance, is non-negative. □

A stronger result asserts that any function on $\mathbb{Z}$ that satisfies the properties of Theorem 2.2 is the autocovariance of a stationary process, since it is always possible to build a Gaussian process with joint distributions based on such an autocovariance function.

The *autocorrelation function* (ACF) is the scale-independent version of the autocovariance function.

Definition 2.7 (Autocorrelation function (ACF)). If $\{Y_t\}$ is a stationary process with autocovariance $\gamma(\cdot)$, then its ACF is

$$\rho(h) := \mathbb{C}\text{or}(Y_t, Y_{t-h}) = \gamma(h)/\gamma(0).$$

By Theorem 2.2 the ACF satisfies the following properties:

1. $\rho(0) = 1$,
2. $|\rho(h)| \leq 1$,
3. $\rho(h) = \rho(-h)$,
4. $\sum_{i=1}^{m} \sum_{j=1}^{m} a_i \rho(i-j) a_j \geq 0$, $\forall m \in \mathbb{N}$ and $(a_1, \ldots, a_m) \in \mathbb{R}^m$.

An alternative summary of the linear dependence of a stationary process can be obtained from the *partial autocorrelation function* (PACF). The PACF measures the correlation between Y_t and Y_{t-h} after their linear dependence on the intervening random variables $Y_{t-1}, \ldots, Y_{t-h+1}$ has been removed.

Definition 2.8 (Partial autocorrelation function (PACF)). The partial autocorrelation function of the stationary process $\{Y_t\}$ is the set of correlations

$$\alpha(h) := \mathbb{C}\text{or}\big[Y_t - \mathbb{P}(Y_t|\mathbf{Y}_{t-1:t-h+1}),\, Y_{t-h} - \mathbb{P}(Y_{t-h}|\mathbf{Y}_{t-1:t-h+1})\big]$$

as function of the non-negative integer h, where $\mathbf{Y}_{t-1:t-h+1} := (Y_{t-1}, \ldots, Y_{t-h+1})^\top$.

As from the following theorem, the PACF can be derived as linear transformation of the ACF.

Theorem 2.3 (Durbin–Levinson algorithm). *Let $\{Y_t\}$ be a stationary process with mean μ and autocovariance function $\gamma(h)$; then its PACF is given by*

$$\alpha(0) = 1,$$
$$\alpha(1) = \gamma(1)/\gamma(0),$$
$$\alpha(h) = \frac{\gamma(h) - \sum_{j=1}^{h-1} \phi_{h-1,j}\gamma(h-j)}{\gamma(0) - \sum_{j=1}^{h-1} \phi_{h-1,j}\gamma(j)}, \qquad \text{for } h = 2, 3, \ldots \tag{2.2}$$

where $\phi_{h-1,j}$ denotes the j-th element of the vector $\boldsymbol{\phi}_{h-1} := \boldsymbol{\gamma}_{h-1}^\top \boldsymbol{\Gamma}_{h-1}^{-1}$ with $\boldsymbol{\gamma}_{h-1}^\top := [\gamma(1), \ldots, \gamma(h-1)]$ and $\boldsymbol{\Gamma}_{h-1}$ as in equation (2.1).

The coefficients $\boldsymbol{\phi}_h$ *can be recursively computed as*

$$\phi_{h,h} = \alpha(h), \qquad \phi_{h,j} = \phi_{h-1,j} - \alpha(h)\phi_{h-1,h-j}, \quad \textit{for } j = 1, \dots, h-1.$$

Furthermore, if we call v_{h-1} *the denominator of* $\alpha(h)$ *in equation (2.2), we can use the recursion* $v_0 = \gamma(0)$, $v_h = v_{h-1}(1 - \alpha(h)^2)$ *to compute it.*

The first part of the theorem shows how partial autocorrelations relate to autocovariances, while the second part provides recursions to efficiently compute the PACF without the need to explicitly invert the matrices $\boldsymbol{\Gamma}_{h-1}$.

Proof. The correlation of a random variable with itself is 1, and so $\alpha(0) = 1$ and as no variables intervene between Y_t and Y_{t-1}, $\alpha(1) = \rho(1) = \gamma(1)/\gamma(0)$.

In order to lighten the notation, let us assume without loss of generality that $\mathbb{E}Y_t = 0$. First, notice that by the properties of the optimal linear predictor (1 and 2 of Theorem 1.4), $\mathbb{E}(Y_t - \mathbb{P}[Y_t|\mathbf{Y}_{t-1:t-h+1}])(Y_{t-h} - \mathbb{P}[Y_{t-h}|\mathbf{Y}_{t-1:t-h+1}]) = \mathbb{E}(Y_t - \mathbb{P}[Y_t|\mathbf{Y}_{t-1:t-h+1}])Y_{t-h}$. Thus, by definition of PACF

$$\alpha(h) = \frac{\mathbb{E}(Y_t - \mathbb{P}[Y_t|\mathbf{Y}_{t-1:t-h+1}])Y_{t-h}}{\mathbb{E}(Y_t - \mathbb{P}[Y_t|\mathbf{Y}_{t-1:t-h+1}])Y_t} = \frac{\gamma(h) - \sum_{j=1}^{h-1}\phi_{h-1,j}\gamma(h-j)}{\gamma(0) - \sum_{j=1}^{h-1}\phi_{h-1,j}\gamma(j)}, \tag{2.3}$$

since $\mathbb{P}[Y_t|\mathbf{Y}_{t-1:t-h+1}] = \sum_{j=1}^{h-1}\phi_{h-1,j}Y_{t-j}$ with $\phi_{h-1,j}$ j-th element of the vector $\boldsymbol{\phi}_{h-1}^\top = [\gamma(1), \dots, \gamma(h)]\boldsymbol{\Gamma}_{h-1}^{-1}$.

Let us concentrate on the numerator of equation (2.3), but for $\alpha(h+1)$. Using the updating formula for the optimal linear predictor (Theorem 1.4, Property 7.) we can write the numerator of $\alpha(h+1)$ as

$$\begin{aligned}
&\mathbb{E}\Big(Y_t - \mathbb{P}[Y_t|\mathbf{Y}_{t-1:t-h+1}] + \\
&\quad - \mathbb{P}\big[Y_t - \mathbb{P}[Y_t|\mathbf{Y}_{t-1:t-h+1}]\big|Y_{t-h} - \mathbb{P}[Y_{t-h}|\mathbf{Y}_{t-1:t-h+1}]\big]\Big)Y_{t-h-1} \\
&= \gamma(h+1) - \sum_{j=1}^{h-1}\phi_{h-1,j}\gamma(h+1-j) + \\
&\quad - \alpha(h)\left(\gamma(1) - \sum_{j=1}^{h-1}\phi_{h-1,h-j}\gamma(h+1-j)\right),
\end{aligned}$$

since, as it can be easily checked,

$$\begin{aligned}
&\mathbb{P}\big[Y_t - \mathbb{P}[Y_t|\boldsymbol{Y}_{t-1:t-h+1}]\big|Y_{t-h} - \mathbb{P}[Y_{t-h}|\boldsymbol{Y}_{t-1:t-h+1}]\big] = \\
&\frac{\mathbb{E}\big(Y_t - \mathbb{P}[Y_t|\boldsymbol{Y}_{t-1:t-h+1}]\big)\big(Y_{t-h} - \mathbb{P}[Y_{t-h}|\boldsymbol{Y}_{t-1:t-h+1}]\big)}{\mathbb{E}\big(Y_{t-h} - \mathbb{P}[Y_{t-h}|\boldsymbol{Y}_{t-1:t-h+1}]\big)^2}\times \\
&\quad\times (Y_{t-h} - \mathbb{P}[Y_{t-h}|\boldsymbol{Y}_{t-1:t-h+1}]) = \\
&\alpha(h)(Y_{t-h} - \mathbb{P}[Y_{t-h}|\boldsymbol{Y}_{t-1:t-h+1}]) = \\
&\alpha(h)\left(Y_{t-h} - \sum_{j=1}^{h-1}\phi_{h-1,h-j}Y_{t-j}\right) = \\
&\alpha(h)\left(Y_{t-h} - \sum_{i=1}^{h-1}\phi_{h-1,i}Y_{t+i-h}\right).
\end{aligned}$$

But, by equation (2.3) we have the alternative formula for the numerator of $\alpha(h+1)$,

$$\gamma(h+1) - \sum_{j=1}^{h}\phi_{h,j}\gamma(h+1-j),$$

and equating the coefficients with the same order of autocovariance we obtain

$$\phi_{h,h} = \alpha(h), \qquad \phi_{h,j} = \phi_{h-1,j} - \alpha(h)\phi_{h-1,h-j} \quad \text{for } j = 1,\dots,h-1.$$

Let us denote with v_{h-1} the denominator of $\alpha(h)$ in equation (2.3) and repeat the reasoning for the denominator of $\alpha(h+1)$, which will be named v_h:

$$\begin{aligned}
v_h =& \mathbb{E}\Big(Y_t - \mathbb{P}[Y_t|\boldsymbol{Y}_{t-1:t-h+1}] + \\
&- \mathbb{P}\big[Y_t - \mathbb{P}[Y_t|\boldsymbol{Y}_{t-1:t-h+1}]\big|Y_{t-h} - \mathbb{P}[Y_{t-h}|\boldsymbol{Y}_{t-1:t-h+1}]\big]\Big)Y_t \\
&= v_{h-1} - \alpha(h)\left(\gamma(h) - \sum_{j=1}^{h-1}\phi_{h-1,j}\gamma(h-j)\right) \\
&= v_{h-1} - \alpha(h)^2\left(\gamma(0) - \sum_{j=1}^{h-1}\phi_{h-1,j}\gamma(j)\right) \\
&= v_{h-1} - \alpha(h)^2 v_{h-1} = v_{h-1}(1 - \alpha(h)^2),
\end{aligned}$$

where we used (2.3) for obtaining the result in the second-last line. □

Since the *population* mean μ, the autocovariances $\gamma(h)$, the autocorrelations $\rho(h)$ and the partial autocorrelations $\alpha(h)$ are generally unknown quantities, they need to be estimated from a time series. If we do not have a specific parametric model for our time series, the natural estimators for μ and $\gamma(h)$ are their sample counterparts:

$$\bar{Y}_n := \frac{1}{n}\sum_{t=1}^{n} Y_t,$$

$$\hat{\gamma}(h) := \frac{1}{n}\sum_{t=h+1}^{n}(Y_t - \bar{Y}_n)(Y_{t-h} - \bar{Y}_n).$$

Note that in the sample autocovariance the divisor is n and not $n-h$ (or $n-h-1$) as one would expect from classical statistical inference. Indeed, the latter divisor does not guarantee that the sample autocovariance function is non-negative definite. Instead, the sample autocovariance matrix, whose generic element is the above defined $\hat{\gamma}(i-j)$, can be expressed as the product of a matrix times its transpose and, therefore, is always non-negative definite. For example, define the matrix with k columns,

$$\mathbf{Y} := \begin{bmatrix} Y_1 - \bar{Y}_n & 0 & 0 & \dots & 0 \\ Y_2 - \bar{Y}_n & Y_1 - \bar{Y}_n & 0 & \dots & 0 \\ Y_3 - \bar{Y}_n & Y_2 - \bar{Y}_n & Y_1 - \bar{Y}_n & \dots & 0 \\ \vdots & \vdots & \vdots & \ddots & \vdots \\ Y_n - \bar{Y}_n & Y_{n-1} - \bar{Y}_n & Y_{n-2} - \bar{Y}_n & \dots & Y_1 - \bar{Y}_n \\ 0 & Y_n - \bar{Y}_n & Y_{n-1} - \bar{Y}_n & \dots & Y_2 - \bar{Y}_n \\ \vdots & \vdots & \vdots & \vdots & \vdots \\ 0 & 0 & 0 & 0 & Y_n - \bar{Y}_n \end{bmatrix}.$$

The autocovariance matrix containing the first $k-1$ sample autocovariances can be computed as

$$\hat{\mathbf{\Gamma}}_{k-1} = n^{-1}\mathbf{Y}^\top\mathbf{Y},$$

which is always non-negative definite.

We summarise the properties of the sample mean of a stationary process in the following theorem.

Theorem 2.4 (Properties of the sample mean). *Let $\{Y_t\}$ be a weakly stationary process with mean μ and autocovariance function $\gamma(h)$; then for the sample mean $\bar{Y}_n$ the following properties hold:*

1. *(Unbiasedness)* $\mathbb{E}\bar{Y}_n = \mu$*;*
2. *(Variance)* $\mathbb{V}\text{ar}(\bar{Y}_n) = \frac{1}{n}\sum_{h=-n+1}^{n-1}\left(1 - \frac{|h|}{n}\right)\gamma(h)$*;*
3. *(Normality) if $\{Y_t\}$ is a Gaussian process, then $\bar{Y}_n \sim \mathcal{N}(\mu, \mathbb{V}\text{ar}(\bar{Y}_n))$;*
4. *(Consistency) if $\gamma(h) \to 0$ as $h \to \infty$, then $\mathbb{E}[\bar{Y}_n - \mu]^2 \to 0$;*

5. *(Asymptotic variance) if* $\sum_{h=-\infty}^{\infty}|\gamma(h)| < \infty$, *then* $n\mathbb{E}[\bar{Y}_n - \mu]^2 \to \sum_{h=-\infty}^{\infty}\gamma(h)$;
6. *(Asymptotic normality) If* $Y_t = \mu + \sum_{j=-\infty}^{\infty}\psi_j Z_{t-j}$ *with* $Z_t \sim \text{IID}(0,\sigma^2)$, $\sum_{j=-\infty}^{\infty}|\psi_j| < \infty$ *and* $\sum_{j=-\infty}^{\infty}\psi_j \neq 0$, *then*

$$\sqrt{n}(\bar{Y}_n - \mu) \xrightarrow{d} \mathcal{N}\left(0, \sum_{h=-\infty}^{\infty}\gamma(h)\right).$$

Proof.
Unbiasedness. $\mathbb{E}\bar{Y}_n = n^{-1}\sum_{t=1}^{n}\mathbb{E}Y_t = \mu$.
Variance.

$$\begin{aligned}\mathbb{V}\text{ar}(\bar{Y}_n) &= \frac{1}{n^2}\sum_{i=1}^{n}\sum_{j=1}^{n}\text{Cov}(Y_i,Y_j) = \frac{1}{n^2}\sum_{i=1}^{n}\sum_{j=1}^{n}\gamma(i-j)\\ &= \frac{1}{n^2}\sum_{h=-n+1}^{n-1}(n-|h|)\gamma(h) = \frac{1}{n}\sum_{h=-n+1}^{n-1}\left(1-\frac{|h|}{n}\right)\gamma(h).\end{aligned}$$

Normality. The normality of the mean follows from the assumption of joint Gaussianity of $(Y_1,\dots,Y_n)$.
Consistency. In order to obtain the (mean-square) consistency of $\bar{Y}_n$, the quantity $\mathbb{E}(\bar{Y}_n - \mu) = \mathbb{V}\text{ar}(\bar{Y}_n)$ has to converge to zero. A sufficient condition for this to happen is $\gamma(h) \to 0$ as h diverges. In fact, in this case we can always fix a small positive ε and find a positive integer N such that for all $h > N$, $|\gamma(h)| < \varepsilon$. Therefore, for $n > N + 1$

$$\begin{aligned}\mathbb{V}\text{ar}(\bar{Y}_n) &= \frac{1}{n}\sum_{h=-n+1}^{n-1}\left(1-\frac{|h|}{n}\right)\gamma(h) \le \frac{1}{n}\sum_{h=-n+1}^{n-1}|\gamma(h)|\\ &= \frac{1}{n}\sum_{h=-N}^{N}|\gamma(h)| + \frac{2}{n}\sum_{h=N+1}^{n-1}|\gamma(h)| \le \frac{1}{n}\sum_{h=-N}^{N}|\gamma(h)| + 2\varepsilon.\end{aligned}$$

As n diverges the first addend converges to zero (it is a finite quantity divided by n), while the second addend can be made arbitrarily small so that, by the very definition of limit, $\mathbb{V}\text{ar}(\bar{Y}_n) \to 0$.
Asymptotic variance. After multiplying the variance of $\bar{Y}_n$ times n, we have the following inequalities:

$$n\mathbb{E}[\bar{Y}_n - \mu]^2 = \sum_{h=-n+1}^{n-1}\left(1-\frac{|h|}{n}\right)\gamma(h) \le \sum_{h=-n+1}^{n-1}|\gamma(h)|.$$

Therefore, a sufficient condition for the asymptotic variance to converge as $n \to \infty$ is that $\sum_{h=-n+1}^{n-1} |\gamma(h)|$ converges. Furthermore, by the Cesàro theorem

$$\lim_{n\to\infty} \sum_{h=-n+1}^{n-1} \left(1 - \frac{|h|}{n}\right) \gamma(h) = \lim_{n\to\infty} \sum_{h=-n+1}^{n-1} \gamma(h).$$

Asymptotic normality. This is the central limit theorem for linear processes, for the proof see Brockwell and Davis (1991, Sec. 7.3), for instance. □

When $\{Y_t\}$ is a Gaussian process, also the sample mean is normal for any sample size. If the process is not Gaussian, the distribution of the sample mean can be approximated by a normal only if a central limit theorem (CLT) for dependent processes applies; in the Theorem 2.4 we provide a CLT for linear processes, but there are alternative results under weaker conditions (in particular under *mixing* conditions).

For the sample autocorrelation $\hat{\rho}(h) := \hat{\gamma}(h)/\hat{\gamma}(0)$ we provide the following result without proof, which is rather lengthy and cumbersome and can be found in Brockwell and Davis (2002, Sec. 7.3).

Theorem 2.5 (Asymptotic distribution of the sample ACF). *Let* $\{Y_t\}$ *be the stationary process,*

$$Y_t = \mu + \sum_{j=-\infty}^{\infty} \psi_j Z_{t-j}, \qquad \text{IID}(0, \sigma^2).$$

If $\sum_{j=-\infty}^{\infty} |\psi_j| < \infty$ *and either one of the following conditions holds*

- $\mathbb{E}Z_t^4 < \infty$,
- $\sum_{j=-\infty}^{\infty} \psi_j^2 |j| < \infty$;

then for each $h \in \{1, 2, \ldots\}$

$$\sqrt{n}\left(\begin{bmatrix} \hat{\rho}(1) \\ \vdots \\ \hat{\rho}(h) \end{bmatrix} - \begin{bmatrix} \rho(1) \\ \vdots \\ \rho(h) \end{bmatrix}\right) \xrightarrow{d} \mathcal{N}(\mathbf{0}, \boldsymbol{V}),$$

with the generic (i,j)*-th element of the covariance matrix* $\boldsymbol{V}$ *being* $v_{ij} = \sum_{k=1}^{\infty} [\rho(k+i) + \rho(k-i) - 2\rho(i)\rho(k)][\rho(k+j) + \rho(k-j) - 2\rho(j)\rho(k)]$.

A corollary to this theorem is that, when a process is $\text{IID}(0, \sigma^2)$ the sample autocorrelations at different lags are asymptotically independent and $\sqrt{n}\hat{\rho}(h)$ converges in distribution to a standard normal. This result is used in the following *portmanteau* test statistics for the null hypothesis $Y_t \sim \text{IID}(0, \sigma^2)$:

Box–Pierce $Q_{BP}(h) = n\sum_{k=1}^{h}\hat{\rho}(k)^2$;

Ljung–Box $Q_{LB}(h) = n(n+2)\sum_{k=1}^{h}\hat{\rho}(k)^2/(n-k)$.

Corollary 2.6 (Portmanteau tests). *Under the hypothesis* $Y_t \sim \text{IID}(0,\sigma^2)$, *the test statistics* Q_{BP} *and* Q_{LB} *converge in distribution to a chi-square with h degrees of freedom.*

Proof. Since $Q_{BP}(h)$ is the sum of h asymptotically independent standard normal random variables, it converges in distribution to a chi-square with h degrees of freedom. The statistic $Q_{LB}(h)$ is asymptotically equivalent to $Q_{BP}(h)$, in fact, for each h:

$$\sqrt{\frac{n(n+2)}{n-h}}\hat{\rho}(h) - \sqrt{n}\hat{\rho}(h) = \sqrt{n}\hat{\rho}(h)\left(\sqrt{\frac{n+2}{n-h}}-1\right)$$

which converges in probability to zero as n diverges. □

The Ljung–Box statistic is more popular than the Box–Pierce because in small samples its actual distribution is closer to the asymptotic distribution (see Ljung and Box, 1978).

We conclude this section on stationary processes with a celebrated result used also as a justification for the use of the class of ARMA models as approximation to any stationary processes.

Before presenting the result we need the concept of *(linearly) deterministic processes*.

Definition 2.9 (Deterministic process). The stationary process $\{V_t\}$ is (linearly) deterministic if

$$\lim_{k\to\infty}\mathbb{E}\big(V_t - \mathbb{P}[V_t|V_{t-1},V_{t-2},\ldots,V_{t-k}]\big)^2 = 0.$$

In other words, a stationary process is deterministic if it can be predicted without error by a linear function of its (possibly) infinite past.

Example 2.2 (Two deterministic processes).
Let

$$W_t = W, \qquad V_t = X\cos(\lambda t) + Y\sin(\lambda t),$$

where W, X and Y are random variables with zero means and finite variances; furthermore, $\mathbb{V}\text{ar}(X) = \mathbb{V}\text{ar}(Y) = \sigma^2$ and $\mathbb{C}\text{ov}(X,Y) = 0$.

The two processes are stationary; in fact their mean is zero for all t and their autocovariance functions are

$$\begin{aligned}\mathbb{E}W_tW_{t-h} &= \mathbb{E}W^2,\\ \mathbb{E}V_tV_{t-h} &= \sigma^2\big[\cos(\lambda t)\cos\big(\lambda(t-h)\big) + \sin(\lambda t)\sin\big(\lambda(t-h)\big)\big]\\ &= \sigma^2\cos(\lambda h),\end{aligned}$$

which are invariant with respect to t.

Their linear predictions based on the past are

$$\begin{aligned}\mathbb{P}[W_t|W_{t-1}] &= W\\ \mathbb{P}[V_t|V_{t-1},V_{t-2}] &= 2\cos(\lambda)V_{t-1} - V_{t-2},\end{aligned}$$

(the reader is invited to derive the latter formula by computing the optimal linear prediction and applying trigonometric identities).

For the first process it is evident that the prediction (W) is identical to the outcome (W). In the second case, we need to show that the prediction is equal to the process outcome:

$$\begin{aligned}\mathbb{P}[V_t|V_{t-1},V_{t-2}] &= \cos(\lambda)V_{t-1} + \cos(2\lambda)V_{t-2}\\ &= 2\cos(\lambda)[X\cos(\lambda t-\lambda) + Y\sin(\lambda t-\lambda)]\\ &\quad - [X\cos(\lambda t-2\lambda) + Y\sin(\lambda t-2\lambda)]\\ &= X[2\cos(\lambda)\cos(\lambda t-\lambda) - \cos(\lambda t-2\lambda)]\\ &\quad + Y[2\cos(\lambda)\sin(\lambda t-\lambda) - \sin(\lambda t-2\lambda)]\\ &= X\cos(\lambda t) + Y\sin(\lambda t),\end{aligned}$$

where the last line is obtained by applying well-known trigonometric identities.

Theorem 2.7 (Wold decomposition). *Let Y_t be a stationary process, then*

$$Y_t = \sum_{j=0}^{\infty}\psi_j Z_{t-j} + V_t$$

where

1. $\psi_0 = 1$, $\sum_{j=0}^{\infty}\psi_j^2 < \infty$,
2. Z_t *is white noise,*
3. $\mathbb{C}\mathrm{ov}(Z_t, V_s) = 0$ *for all t and s,*
4. V_t *is (linearly) deterministic,*
5. $\lim_{k\to\infty}\mathbb{E}(Z_t - \mathbb{P}[Z_t|Y_t,Y_{t-1},\dots,Y_{t-k}])^2 = 0$;
6. $\lim_{k\to\infty}\mathbb{E}(V_t - \mathbb{P}[V_t|Y_s,Y_{s-1},\dots,Y_{s-k}])^2 = 0$ *for all t and s.*

The proof of this result is very natural using Hilbert space techniques (see Brockwell and Davis, 1991, for example). For a proof based only on calculus see Davidson (2000, Sec. 5.7).

The message of this theorem is that every stationary process can be seen as the sum of two orthogonal components: one, the deterministic, is perfectly predictable using a linear function of the past of the process (point 6); the other, the *purely non-deterministic*, is expressible as a (possibly) infinite linear combination of past and present observations of a white noise process Z_t. This process is the prediction error of Y_t based on its (possibly) infinite past,

$$Z_t = Y_t - \mathbb{P}[Y_t|Y_{t-1}, Y_{t-2}, \ldots],$$

and is generally termed *innovation* of the process Y_t. Thus, the coefficients ψ_j are the projection coefficients of Y_t on its past innovations Z_{t-j}:

$$\psi_j = \mathbb{E}[Y_t Z_{t-j}]/\mathbb{E}[Z_{t-j}^2].$$

As for the deterministic component, point 6 implies that it can be predicted without error also using the infinite past of $\{Y_t\}$ and not only using its own past $\{V_s\}_{s<t}$, as by the definition of deterministic process. In a single time series, the deterministic component cannot be discriminated from a deterministic mean and/or periodic function of time, such as a deterministic seasonal component, so the main focus of time series modelling is on the purely non-deterministic component. The class of ARMA models, that will be introduced in Section 2.4, provides a parsimonious approximation for purely non-deterministic stationary processes when the sequence of coefficients ψ_j approaches zero at least at a geometric rate.

In the rest of the chapter, we will consider only stationary processes that are purely non-deterministic. In particular, we will assume that a stationary process can always be represented as a *causal linear process*.

Definition 2.10 (Linear process)**.** The process $\{Y_t\}$ is a *linear process* if it has the representation

$$Y_t = \mu + \sum_{j=-\infty}^{\infty} \psi_j Z_{t-j}, \qquad \forall t \in \mathbb{Z},$$

where $Z_t \sim \mathrm{WN}(0, \sigma^2)$, μ is a constant mean, and $\{\psi_j\}$ is a sequence of coefficients such that $\sum_{j=-\infty}^{\infty} |\psi_j| < \infty$.

If $\psi_j = 0$ for all $j < 0$, then $\{Y_t\}$ is a *causal linear process*.

The absolute summability condition in the definition is slightly stronger than the square summability property resulting from the Wold decomposition[2], but, as we have already seen in Theorems 2.4 and 2.5, it brings some advantages.

The autocovariance of the linear process is

$$\gamma(h) = \sigma^2 \sum_{j=-\infty}^{\infty} \psi_j \psi_{j+h},$$

with the summation starting from $j = 0$ and $h \geq 0$ for the causal case. One advantage of the absolute summability condition is that it implies the absolute summability of the autocovariance function[3] that, according to Theorem 2.4, is sufficient for the consistency of the sample mean and the existence of its asymptotic variance.

2.3 Integrated processes

All the readers that have seen the plot of some economic, demographics or natural time series have probably realised that most real-world time series are not stationary.

Figure 2.1 depicts the famous Airline time series originally contained in the influential book by Box and Jenkins (1976). This time series in clearly nonstationary for three distinct reasons: its dispersion grows (i.e., its variance is not constant), its level increases (i.e., its mean is not constant) and the values of certain months are persistently higher (or lower) than those of other months (e.g., the value of July is always higher than that of November, thus the mean is also not constant for every t).

The heteroskedasticity can be fixed in this (and many other) series by taking the logarithm or one of the other variance-stabilizing transformations (see Chapter 6), while the stationarity in mean can often be obtained, at least approximately, by means of differentiations. In particular, one or more applications of the difference operator $\Delta\{Y_t\} := \{Y_t - Y_{t-1}\}$ can eliminate the trend, and the seasonal difference, $\Delta_s\{Y_t\} := \{Y_t - Y_{t-s}\}$, where s is the periodicity of the seasonal effect (i.e., the number of observations per year), can eliminate the nonstationarity due to the presence of seasonality as well as the trend.

---

[2]This is true, because the convergence of $\sum_{j=1}^{n} |\psi_j|$ for $n \to \infty$ implies that there is N such that all $|\psi_j|$ with $j > N$ are smaller than 1 and $\sum_{j=1}^{N} |\psi_j|$ is a finite quantity. Then, for all $j > N$, $|\psi_j| < \psi_j^2$ and the convergence of $\sum_{j=N+1}^{\infty} |\psi_j|$ implies the convergence of $\sum_{j=N+1}^{\infty} \psi_j^2$.

[3]To see why (all summation indexes range from $-\infty$ to ∞):

$$\sigma^2 \sum_h \left| \sum_j \psi_j \psi_{j+h} \right| \leq \sigma^2 \sum_j |\psi_j| \sum_h |\psi_{j+h}| \leq \sigma^2 \sum_j |\psi_j| \sum_i |\psi_i| = \sigma^2 \left(\sum_j |\psi_j| \right)^2.$$

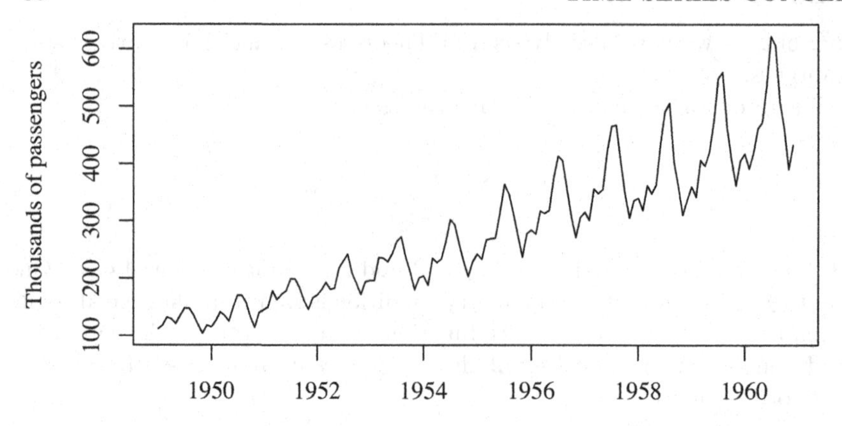

Figure 2.1 *Monthly time series of international airline passengers.*

That is the reason why processes like

$$W_t = W_{t-1} + Y_t, \qquad t \in 0, 1, \ldots, n, \tag{2.4}$$

$$V_t = V_{t-s} + Y_t, \qquad t \in 0, 1, \ldots, n, \tag{2.5}$$

with Y_t a stationary process, are so important in time series analysis. The operations (2.4) and (2.5) are called *integration* and *seasonal integration* of the process $\{Y_t\}$, and they are well defined also when the process $\{Y_t\}$ is nonstationary. As one can easily check, the differentiation of $\{W_t\}$ and the seasonal differentiation of $\{V_t\}$ return the process Y_t. Notice that a side condition such as $W_0 = W$ and $V_0 = V$, where W and V can be constants or random variables, is necessary in order for the two processes to be well defined. Unlike stationary processes, $\{W_t\}$ and $\{V_t\}$ cannot be thought of as originating in the infinite past. Of course, the processes $\{W_t\}$ and $\{V_t\}$ can be further (simply or seasonally) integrated and, in this case, more differences would be necessary to recover the driving process $\{Y_t\}$.

Let us denote with Δ^d the sequential application of the difference operator d times and Δ_s^D the repeated application of the seasonal difference operator D times, where d and D are non-negative integers.

Definition 2.11 (Integrated process)**.** Let $\{Y_t\}$ be a nonstationary process such that $\Delta^d\{Y_t\}$ is stationary, but $\Delta^{d-1}\{Y_t\}$ is nonstationary; then $\{Y_t\}$ is called *integrated of order d* and this is indicated with the notation $Y_t \sim \mathrm{I}(d)$.

In the above definition we had to specify that $d-1$ differences are not enough to make the process $\{Y_t\}$ stationary, because the difference of a stationary process is stationary (even though it introduces a "pathology" called non-invertibility that we treat in Section 2.4), and so d is the minimal number of differences that turns $\{Y_t\}$ into a stationary process.

An analogous definition can be given for *seasonally integrated* processes, but if both $\Delta\{Y_t\}$ and $\Delta_s\{Y_t\}$ make the nonstationary process $\{Y_t\}$ stationary, then the series is integrated rather than seasonally integrated. Indeed, the seasonal difference embeds the simple difference: it is straightforward to verify that

$$Y_t - Y_{t-s} = \Delta(Y_t + Y_{t-1} + \ldots + Y_{t-s+1}).$$

If W_t is an I(1) process, a further integration,

$$X_t = X_{t-1} + W_t,$$

yields an I(2) process. In fact, by applying the difference once we obtain $X_t - X_{t-1} = W_t$ which is an I(1) process.

By backward substitution it is easy to see that equation (2.4) can be written as

$$W_t = W_0 + \sum_{i=1}^{t} Y_i, \tag{2.6}$$

which makes it clear that integrating a process means taking its cumulative sum or, also, building its *partial sums process*.

Example 2.3 (Random walk with drift).
The simplest I(1) process is the *random walk* possibly with *drift*:

$$W_t = \delta + W_{t-1} + Z_t, \qquad \text{for } t = 1, 2, \ldots, n,$$

or also

$$W_t = W_0 + t\delta + \sum_{j=1}^{t} Z_j, \qquad \text{for } t = 1, 2, \ldots, n,$$

with W_0 constant or finite-variance random variable uncorrelated with Z_t for all t. The process is clearly in the forms (2.4) and (2.6) with the driving process being a white noise with non-zero mean ($Y_t = \delta + Z_t$ with $Z_t \sim \text{WN}(0, \sigma^2)$).

Using the second representation of the process, it is straightforward to obtain mean, variance and autocovariances of the random walk with drift (please try):

$$\mathbb{E}W_t = \mathbb{E}W_0 + \delta t$$
$$\mathbb{C}\text{ov}(W_s, W_t) = \mathbb{V}\text{ar}(W_0) + \min(s, t)\sigma^2.$$

Since the autocovariance function depends on both t and s and not just on their differences $t-s$, the process is clearly nonstationary. As for the mean, it is time-independent only when $\delta = 0$, but in real applications, where the first observation of the time series is arbitrary, it loses its usual meaning even when $\delta = 0$.

Moreover, the sample mean does not converge even if $\delta = 0$ and $W_0 = 0$:

$$\begin{aligned}\bar{W}_n &= n^{-1}\sum_{t=1}^{n} W_t = n^{-1}\sum_{t=1}^{n}\sum_{j=1}^{t} Z_j \\ &= n^{-1}\left[n \cdot Z_1 + (n-1)\cdot Z_2 + \ldots + 1 \cdot Z_n\right],\end{aligned}$$

whose variance is

$$\mathbb{V}\text{ar}(\bar{W}_n) = \frac{\sigma^2}{n^2}\left(n^2 + (n-1)^2 + \ldots 1^2\right) = \frac{\sigma^2(n+1)(2n+1)}{6n},$$

which diverges at rate n. In fact, it can be proved that if $\delta = 0$ and $Z_t \sim \text{IID}(0, \sigma^2)$ (this condition can be relaxed), then

$$n^{-1/2}\bar{W}_n = n^{-3/2}\sum_{t=1}^{n} W_t \xrightarrow{d} \mathcal{N}(0, \sigma^2/3).$$

From Example 2.3, it appears clear that a non-zero mean in the driving process, $\{Y_t\}$, becomes a deterministic linear trend in the I(1) process, $\{W_t\}$. We invite the reader to prove the following theorem that provides the first two moments of an integrated process.

Theorem 2.8 (Moments of the partial sums of a stationary process). *Let $\{W_t\}$ be as in equations (2.4) and (2.6), where the driving process $\{Y_t\}$ is stationary with mean δ and autocovariance function $\gamma(h)$, and W_0 is uncorrelated with $\{Y_t\}$ for all $t > 0$. Then, the first two moments of the process $\{W_t\}$ are given by*

$$\mathbb{E}W_t = \mathbb{E}W_0 + \delta t$$

$$\mathbb{C}\text{ov}(W_s, W_t) = \mathbb{V}\text{ar}(W_0) + \sum_{i=1}^{s}\sum_{j=1}^{t}\gamma(i-j).$$

Notice that the variance is given by:

$$\mathbb{V}\text{ar}(W_t) = \mathbb{V}\text{ar}(W_0) + t\sum_{j=-t+1}^{t-1}\gamma(j).$$

Thus, if $\sum_{j=-t+1}^{t-1}\gamma(j)$ converges to a finite non-zero quantity, the long-run rate of growth of the variance of an I(1) process is t, as in the random walk. We will reserve the name I(1) processes only for those processes in Theorem 2.8 for which the autocovariance function of the driving process does not sum to zero: $\sum_{j=-\infty}^{\infty} \gamma(j) \neq 0$. This choice is coherent with the definition of I(d)-ness we gave above, as the following example illustrates.

Example 2.4 (Integration of the non-invertible MA(1) process).
Let

$$Y_t = Z_t - Z_{t-1}, \qquad t \in \mathbb{Z}$$

with $Z_t \sim \mathrm{WN}(0, \sigma^2)$. As we will see in Section 2.4, this is an order-1 moving average process, or MA(1), with the characteristic of being non-invertible. It is easy to check that $\mathbb{E}Y_t = 0$ for all t and

$$\mathbb{E}[Y_t Y_{t-h}] = \gamma(h) = \begin{cases} 2\sigma^2 & \text{for } h = 0, \\ -\sigma^2 & \text{for } |h| = 1, \\ 0 & \text{for } |h| \geq 2, \end{cases}$$

showing that Y_t is stationary.

By setting $W_0 = Z_0$ and integrating Y_t for $t = 1, 2, \ldots, n$ we can write

$$W_t = \sum_{j=1}^{t} Y_j = \sum_{j=1}^{t} (Z_j - Z_{j-1}) = Z_t,$$

which is white noise and, thus, stationary. Therefore, notwithstanding that $\{W_t\}$ was built through integration, it cannot be named I(1). In fact, according to our definition of I(d)-ness, d must be the smallest number of differences that makes the process $\{W_t\}$ stationary, but since $\{W_t\}$ is already stationary it is I(0) and not I(1).[4]

Another way to check that $\{W_t\}$ is not I(1) is by noting that the autocovariance function of its driving process $\{Y_t\}$ sums to zero: $\sum_{h=\infty}^{\infty} \gamma(h) = -\sigma^2 + 2\sigma^2 - \sigma^2 = 0$.

A process like the one discussed in the example can be obtained when one takes the difference of a time series composed by a deterministic linear trend plus white noise:

$$Y_t = \alpha + \beta t + Z_t, \qquad \Delta Y_t = \beta + Z_t - Z_{t-1}.$$

For reasons discussed in Section 2.4, in this case it is statistically more successful to estimate a model for Y_t and predict Y_t directly, rather than through its differences ΔY_t.

[4]It would be reasonable to name a process like the one in the example as I(-1), since after the integration it becomes an I(0) process. However, the notation I(d) with negative d is not common and so we will avoid using it.

On the contrary if the process and its difference are given by

$$W_t = \alpha + \beta t + \sum_{j=1}^{t} Z_j, \qquad \Delta W_t = \beta + Z_t,$$

working on the difference is preferable. In this case the component $\alpha + \beta t$ is usually referred to as *linear deterministic trend*, while the component $\sum_{j=1}^{t} Z_j$ is named *stochastic trend*. Notice that the means of Y_t and W_t are both equal to $\alpha + \beta t$. Two sample paths of the processes $\{Y_t\}$ and $\{W_t\}$ are depicted in Figure 2.2. While Y_t, which is a stationary process around a linear trend, or

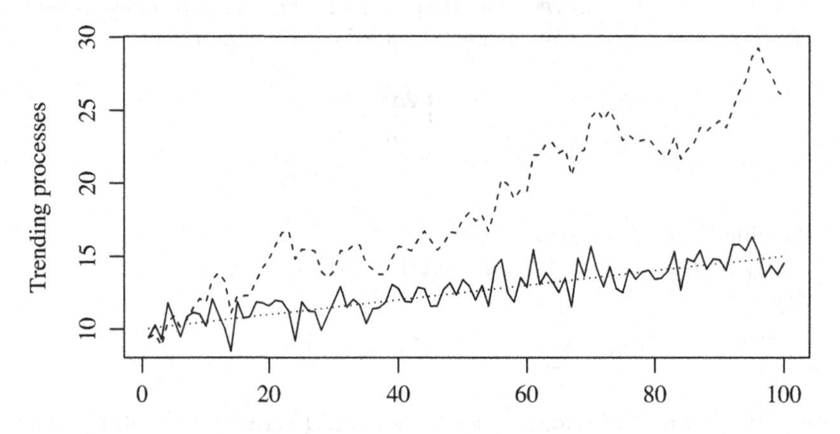

Figure 2.2 *Sample paths of the processes* $Y_t = \alpha + \beta t + Z_t$ *(solid line) and* $W_t = \alpha + \beta t + \sum_{j=1}^{t} Z_j$ *(dashed line) plotted with the common linear trend* $\alpha + \beta t$ *(dotted line), with parameters* $\alpha = 10$, $\beta = 0.05$, *and* $Z_t \sim \mathrm{NID}(0, 1)$.

simply *trend stationary*, tends to be attracted by its mean $\alpha + \beta t$, the sample paths of W_t can deviate arbitrarily far away from it.

Since integration is a linear operation, then predicting an I(d) process is straightforward. For the I(1) case, let W_t be as in equations (2.4) and (2.6), with W_0 independent of Y_t for $t > 0$, then

$$\begin{aligned}
\mathbb{E}[W_{t+k}|W_0, \ldots, W_t] &= \mathbb{E}[W_{t+k}|W_0, Y_1, \ldots, Y_t] \\
&= \mathbb{E}[W_t + Y_{t+1} + \ldots + Y_{t+k}|W_0, Y_1, \ldots, Y_t] \\
&= W_t + \mathbb{E}[Y_{t+1} + \ldots + Y_{t+k}|W_0, Y_1, \ldots, Y_t] \\
&= W_t + \mathbb{E}[Y_{t+1}|Y_1, \ldots, Y_t] + \ldots + \mathbb{E}[Y_{t+k}|Y_1, \ldots, Y_t],
\end{aligned}$$

where the same holds also if we substitute $\mathbb{P}$ to $\mathbb{E}$ (in the linear prediction case independence between W_0 and Y_t can be relaxed to zero-correlation). Notice that, since the prediction of k future realisations of an I(1) process consists in the sum of k predictions of its stationary increments, then the MSE of prediction will increase with t in the long run.

2.4 ARIMA models

In the previous section we defined I(d) processes. As long as we can reasonably assume that opportunely transformed real time series can be viewed as realisations of a stationary processes (at least approximately), we can try to model them using the class of ARMA processes, which are good approximations of any stationary process provided that its autocovariance function dies out quickly enough.

Definition 2.12 (ARMA process). The process $\{Y_t\}$ is ARMA(p,q) if it is stationary and satisfies

$$Y_t = \phi_1 Y_{t-1} + \ldots + \phi_p Y_{t-p} + Z_t + \theta_1 Z_{t-1} + \ldots + \theta_q Z_{t-q}, \qquad \forall t \in \mathbb{Z} \quad (2.7)$$

where $Z_t \sim \text{WN}(0, \sigma^2)$, $\phi_p \neq 0$, $\theta_q \neq 0$, and the equations

$$1 - \phi_1 x - \phi_2 x^2 - \phi_p x^p = 0 \quad (2.8)$$

$$1 + \theta_1 x + \theta_2 x^2 + \theta_p x^p = 0 \quad (2.9)$$

have no common roots.

The conditions in the second half of the above definition guarantee that the orders of the ARMA process cannot be harmlessly reduced: they imply that there is no other ARMA(p', q') process with either $p' < p$ or $q' < q$ that has identical (second-order) properties to the ARMA(p,q) process.

Notice that Definition 2.12 does not necessarily define a unique process, but rather asks the process $\{Y_t\}$ to satisfy the *difference equation* (2.7). Thus, we need a result for the existence and, possibly, uniqueness of such a process, where by uniqueness it is intended that there is no other such process with the same autocovariance function.

Moreover, if a solution to (2.7) exists we would like it to be meaningful not only mathematically, but also intuitively: if we see the white noise $\{Z_t\}$ as a sequence of unobservable shocks entering the (economic, natural, physical) system and determining the value of the observable process $\{Y_t\}$, then the value Y_t can only depend on present and past shocks $\{Z_t, Z_{t-1}, \ldots\}$.

Definition 2.13 (Causality). An ARMA(p,q) process $\{Y_t\}$ is *causal* if there is a sequence of constants $\{\psi_j\}$ such that $\sum_{j=0}^{\infty} |\psi_j| < \infty$ and

$$Y_t = \sum_{j=0}^{\infty} \psi_j Z_{t-j}, \qquad \forall t \in \mathbb{Z}.$$

From Wold's theorem we know that a purely non-deterministic stationary process has always such a causal representation[5], but Wold's theorem does not exclude the existence of other non-causal representations.[6]

Example 2.5 (Non-causal AR(1) process).
Let us seek a stationary solution to the difference equation

$$Y_t = 2Y_{t-1} + Z_t, \qquad \forall t \in \mathbb{Z}, \tag{2.10}$$

with $Z_t \sim \mathrm{WN}(0, \sigma^2)$. It is easy to check by direct substitution into (2.10) that

$$Y_t = -\sum_{j=1}^{\infty} \left(\frac{1}{2}\right)^j Z_{t+j},$$

solves the difference equation. Moreover, the mean of $\{Y_t\}$ is zero and its autocovariance function (the reader is invited to compute it) is

$$\gamma(h) = \left(\frac{1}{2}\right)^h \frac{\sigma^2}{3}, \qquad \forall t \in \mathbb{Z},$$

and we can conclude that $\{Y_t\}$ is stationary.

Thus, we have found a non-causal stationary solution to equation (2.10), but there is no causal stationary solution to it.

Now, Wold's theorem states that every purely non-deterministic stationary process has a representation as linear combination of present and past innovations and so there must also be one for the process $X_t := -\sum_{j=1}^{\infty} \left(\frac{1}{2}\right)^j Z_{t+j}$, which we have seen to be a stationary solution of (2.10). Indeed, the process has the representation

$$X_t = \sum_{j=0}^{\infty} \left(\frac{1}{2}\right)^j W_{t-j}, \tag{2.11}$$

where $W_t \sim \mathrm{WN}(0, \sigma^2/3)$ is the innovation sequence

$$W_t = X_t - \mathbb{P}[X_t | X_{t-1}, X_{t-2}, \ldots], \qquad \forall t \in \mathbb{Z}.$$

The reader should check that the representation (2.11) enjoys the same second-order properties (i.e., the same mean and autocovariance function) of the non-causal representation of $\{X_t\}$.

Unfortunately, the causal representation (2.11) does not satisfy the

---

[5]The absolute summability condition of causality is stronger than the square summability in Wold's theorem, and could be relaxed. We provide this definition of causality since it is the usual definition used in time series analysis.

[6]Actually, one could easily prove a Wold theorem where Y_t is a linear combination of future and present innovations built as prediction errors with respect to future values of the process: $Y_t - \mathbb{P}[Y_t | Y_{t+1}, Y_{t+2}, \ldots]$.

difference equation (2.10), which in fact does not have causal stationary solutions. However, (2.11) solves the difference equation

$$Y_t = \frac{1}{2}Y_{t-1} + W_t, \qquad \forall t \in \mathbb{Z}.$$

In fact, if the difference equation has a non-causal stationary solution one can always find another difference equation with a causal stationary solution that defines the same process (up to its second-order properties).

Theorem 2.9 (Existence and uniqueness of a causal ARMA process). *A unique causal stationary solution to equation (2.7) exists if and only if*

$$1 - \phi_1 z - \phi_2 z^2 - \ldots - \phi_p z^p \neq 0 \qquad \textit{for all } |z| \leq 1.$$

Proof. For a compact proof see Theorem 3.1.1 of Brockwell and Davis (1991); for a longer but more readable proof see Pourahmadi (2001, Sec. 4.1). □

An alternative and very common way to express the condition in the theorem is: all the roots of the p-order equation

$$1 - \phi_1 z - \phi_2 z^2 - \ldots \phi_p z^p = 0 \tag{2.12}$$

must lie outside the *unit circle*. This expression is motivated by the fact that the solutions may be complex numbers, which are generally represented in a Cartesian plane. In particular, if $z = u + vi$ is complex, with u real part and v imaginary part, the equation $|z| = \sqrt{u^2 + v^2} = 1$ defines the unit circle in the Cartesian plane and $|z| > 1$ identifies all the points outside the unit circle.

We can summarise the above results by saying that if all the roots of equation (2.12) are outside the unit circle, then the ARMA process has the MA(∞) representation,

$$Y_t = \sum_{j=0}^{\infty} Z_{t-j}.$$

Another condition that for various reasons (cf. Section 2.3) is generally imposed to ARMA processes is *invertibility*. Invertibility allows to represent an ARMA process as a linear function of its past observations plus an unpredictable component.

Definition 2.14 (Invertibility). An ARMA(p, q) process is *invertible* if there is a sequence of constants $\{\pi_j\}$ such that $\sum_{j=1}^{\infty} |\pi_j|$ and

$$Z_t = \sum_{j=0}^{\infty} \pi_j Y_{t-j}.$$

Theorem 2.10 (Invertibility). *An ARMA(p, q) process is invertible if and only if*

$$1 + \theta_1 z + \theta_2 z^2 + \ldots + \theta_q z^q \neq 0 \qquad \text{for all } |z| \leq 1.$$

Proof. See the proof of Theorem 3.1.2 in Brockwell and Davis (1991) or Section 4.2.1 of Pourahmadi (2001). □

If an ARMA process is invertible, then it can be written in the AR(∞) form,

$$Y_t = Z_t - \sum_{j=1}^{\infty} \pi_j Y_{t-j},$$

with coefficients π_j getting smaller as j increases. This allows the consistent estimation of the unobservable white noise sequence $\{Z_t\}$ from the sample path $\{Y_1, Y_2, \ldots, Y_n\}$.

Example 2.6 (Estimating the white noise sequence from a sample path of a MA(1) process).
Let $Y_1, \ldots, Y_n$ be a sample path of the MA(1) process

$$Y_t = Z_t + \theta Z_{t-1}, \qquad \forall t \in \mathbb{Z}.$$

We want to estimate the white noise sequence $\{Z_t\}$ using the observed sample path. Of course,

$$Z_t = Y_t - \theta Z_{t-1},$$

so if we could observe Z_0 we would know $Z_1, \ldots, Z_n$ as well, but Z_0 cannot be observed and must be guessed. Conditionally on my guess z_0 about the outcome of Z_0 I can compute

$$\hat{Z}_t = Y_t - \theta \hat{Z}_{t-1} \qquad \text{for } t = 1, \ldots, n,$$

obtaining

$$\hat{Z}_t = \sum_{j=0}^{t-1} (-\theta)^j Y_{t-j} + (-\theta)^t z_0.$$

Thus, the MSE of $\hat{Z}_t$ is given by

$$\begin{aligned}
&\mathbb{E}[Z_t - \hat{Z}_t]^2 \\
&= \mathbb{E}\left[\sum_{j=0}^{t-1}(-\theta)^j Y_{t-j} + (-\theta)^t Z_0 - \sum_{j=0}^{t-1}(-\theta)^j Y_{t-j} - (-\theta)^t z_0\right]^2 \\
&= \theta^{2t}\mathbb{E}[Z_0 - z_0]^2,
\end{aligned}$$

that converges to zero as t increases if and only if $|\theta| < 1$, which is the invertibility condition for a MA(1) process (the reader should verify this using Theorem 2.10).

A very useful tool when working with ARMA processes and time series in general is the *backward shift* or *lag* operator, which is named $\mathbb{B}$ in the statistical literature and $\mathbb{L}$ in the econometric literature. The effect of applying this operator to a discrete-time process is delaying the process by one unit of time: $\mathbb{B}Y_t = Y_{t-1}$. The operator operates on the whole process, so in principle it would be more correct to write $\mathbb{B}\{Y_t\} = \{Y_{t-1}\}$, but this is uncommon in the literature, and so we will maintain the mainstream notation. One nice thing about $\mathbb{B}$ is that, although the reader might have never seen it before, she or he already knows its algebra because it obeys the same rules as the algebra of real and complex variables. For example, we can define products, and so integer powers of $\mathbb{B}$:

$$\mathbb{B}^k Y_t = \mathbb{B}^{k-1}\mathbb{B}Y_t = \mathbb{B}^{k-1}Y_{t-1} = \ldots = Y_{t-k},$$

where k can be any integer number. We can define the multiplicative identity (element neutral with respect to the product) $1 := \mathbb{B}^0$, and we can define polynomials and series in $\mathbb{B}$:

$$\pi_k(B)Y_t := (\pi_0 + \pi_1\mathbb{B} + \pi_2\mathbb{B}^2 + \ldots + \pi_k\mathbb{B}^k)Y_t = \sum_{j=0}^{k}\pi_k Y_{t-k}$$

$$\pi_\infty(B)Y_t := (\pi_0 + \pi_1\mathbb{B} + \pi_2\mathbb{B}^2 + \ldots)Y_t = \sum_{j=0}^{\infty}\pi_k Y_{t-k}.$$

Using the algebra of $\mathbb{B}$ we can write an ARMA(p,q) model as

$$\phi_p(\mathbb{B})Y_t = \theta_q(\mathbb{B})Z_t,$$

with $\phi_p(\mathbb{B}) := 1 - \phi_1\mathbb{B} - \ldots - \phi_p\mathbb{B}^p$ and $\theta_q(\mathbb{B}) = 1 + \theta_1\mathbb{B} + \ldots \theta_q\mathbb{B}^q$.

The $\mathbb{B}$ operator can also be used to obtain the coefficients of the MA(∞) representation of a causal ARMA process, $Y_t = \phi_p(\mathbb{B})^{-1}\theta_q(\mathbb{B})Z_t$, and of the AR($\infty$) representation of an invertible ARMA process, $Z_t = \theta_q(\mathbb{B})^{-1}\phi_p(\mathbb{B})Y_t$. Suppose we wish to invert the operator $\phi_p(\mathbb{B})$. From the above results we can expect its inverse to be a series. Thus, we seek a series $\psi_\infty(\mathbb{B})$ such that

$$\phi_p(\mathbb{B})\psi_\infty(\mathbb{B}) = 1.$$

By multiplying and collecting the powers of $\mathbb{B}$ of the same order:

$$(1 - \phi_1\mathbb{B} - \ldots - \phi_p\mathbb{B}^p)(1 + \psi_1\mathbb{B} + \psi_2 B^2 + \ldots) = 1$$
$$1 + (\psi_1 - \phi_1)\mathbb{B} + (\psi_2 - \phi_2 - \psi_1\phi_1)\mathbb{B}^2 + (\psi_3 - \phi_3 - \psi_2\phi_1 - \psi_1\phi_2)\mathbb{B}^3 + \ldots = 1.$$

For the equality to hold all the coefficients of $\mathbb{B}^k$ for $k > 0$ must be zero, and so we obtain

$$\begin{aligned}
\psi_1 &= \phi_1 \\
\psi_2 &= \phi_2 + \psi_1\phi_1 \\
\psi_3 &= \phi_3 + \psi_2\phi_1 + \psi_1\phi_2 \\
\ldots &= \ldots.
\end{aligned}$$

The coefficients ψ_j are absolutely summable if and only if the roots of $\phi_p(\mathbb{B}) = 0$ are outside the unit circle. Therefore, under this condition the polynomial $\phi_p(\mathbb{B})$ is said to be invertible.[7]

Example 2.7 (Inverting the AR(1) and MA(1) polynomials).
Let us invert the AR(1) and MA(1) polynomials $(1 - \phi\mathbb{B})$ and $(1 + \theta\mathbb{B})$. We need a series $\psi_\infty(\mathbb{B})$ such that

$$(1 - \phi\mathbb{B})(1 + \psi_1\mathbb{B} + \psi_2\mathbb{B}^2 + \ldots) = 1.$$

By multiplying and setting to zero all the coefficients of the positive powers of $\mathbb{B}$ we obtain

$$\begin{aligned}
\psi_1 &= \phi \\
\psi_2 &= \phi\psi_1 = \phi^2 \\
\ldots &= \ldots \\
\psi_t &= \phi\psi_{t-1} = \phi^t,
\end{aligned}$$

which converges for $|\phi| < 1$. Thus, under this condition we can write

$$(1 - \phi\mathbb{B})^{-1} = \sum_{j=0}^{\infty} \phi^j \mathbb{B}^j,$$

and analogously, for $|\theta| < 1$,

$$(1 + \theta\mathbb{B})^{-1} = \sum_{j=0}^{\infty} (-\theta)^j \mathbb{B}^j.$$

[7] Do not confuse the concept of invertibility of a polynomial in $\mathbb{B}$ with the concept of invertibility of an ARMA(p, q) process. In fact the ARMA process is causal stationary if and only if the polynomial $\phi_p(\mathbb{B})$ is invertible, while it is invertible if and only if the polynomial $\theta_q(\mathbb{B})$ is invertible.

Now the MA(∞) representation of the AR(1) process can be readily obtained: $(1-\phi\mathbb{B})Y_t = Z_t$, $Y_t = (1-\phi\mathbb{B})^{-1}Z_t$, $Y_t = \sum_{j=0}^{\infty}\phi^j Z_{t-j}$. The same can be done for computing the AR(∞) representation of the MA(1): $Y_t = (1+\theta\mathbb{B})Z_t$, $(1+\theta\mathbb{B})^{-1}Y_t = Z_t$, $\sum_{j=0}^{\infty}(-\theta)^j Y_{t-j} = Z_t$.

The use of the $\mathbb{B}$ operator also reveals the relationship between an ARMA process with unit roots in the AR part and the class of I(d) processes. If the characteristic equation $\phi_p(z) = 0$ has d solutions equal to 1 (i.e., d unit roots) while the other $p-d$ solutions are outside the unit circle, then, by elementary polynomial theory, we can factorise $\phi_p(\mathbb{B})$ as

$$\phi_p(\mathbb{B}) = \varphi_{p-d}(\mathbb{B})(1-\mathbb{B})^d = \varphi_{p-d}(\mathbb{B})\Delta^d,$$

where the $p-d$ polynomial $\varphi_{p-d}(z)$ has all the roots outside the unit circle. This means, that, although there is no causal stationary process $\{Y_t\}$ that solves $\phi_p(\mathbb{B})Y_t = \theta_q(\mathbb{B})Z_t$, there is a causal stationary process $\{\Delta^d Y_t\}$ that solves $\varphi_{p-d}(\mathbb{B})\Delta^d Y_t = \theta_q(\mathbb{B})Z_t$. Thus, $\{Y_t\}$ is nonstationary, but $\Delta^d Y_t$ is, or, in symbols, $Y_t \sim \mathrm{I}(d)$.

Let us examine the second-order properties of the AR(p) and MA(q) process separately.

Theorem 2.11 (Properties of the AR(p)). *Let Y_t be the process*

$$\phi_p(\mathbb{B})Y_t = Z_t, \qquad \forall t \in \mathbb{Z},$$

with $\phi_p(x) \neq 0$ for $|x| \leq 1$ and $Z_t \sim \mathrm{WN}(0,\sigma^2)$; then $\mathbb{E}[Y_t] = 0$, its autocovariance function is given by

$$\gamma(0) = \phi_1\gamma(1) + \dots \phi_p\gamma(p) + \sigma^2,$$
$$\gamma(h) = \phi_1\gamma(h-1) + \dots + \phi_p\gamma(h-p), \qquad \text{for } h > 1,$$

which converges to zero at geometric rate (i.e., there is a $\rho \in [0,1)$ such that $|\gamma(h)| \leq \gamma(0)\rho^h$), and its partial autocorrelation function $\alpha(h)$ is zero for $h > p$.

Proof. Since the process $\{Y_t\}$ is causal stationary we know that its mean μ and autocovariance function $\gamma(h)$ are time-invariant. Thus,

$$E[Y_t] = \phi_1\mathbb{E}[Y_{t-1}] + \dots + \phi_p\mathbb{E}[Y_{t-p}] + \mathbb{E}[Z_t]$$
$$\mu = \mu(\phi_1 + \dots + \phi_p)$$

which implies that $\mu = 0$, as by assumption $1 - \phi_1 - \dots - \phi_p \neq 0$.

The sequence of autocovariances can be obtained by

$$\gamma(h) = \mathbb{E}[Y_t Y_{t-h}] = \mathbb{E}[(\phi_1 Y_{t-1} + \dots + \phi_p Y_{t-p} + Z_t)Y_{t-h}]$$
$$= \phi_1\gamma(h-1) + \dots + \phi_p + \gamma(h-p) + \mathbb{E}[Z_t Y_{t-h}],$$

where $\mathbb{E}[Z_t Y_{t-h}] = \mathbb{E}[Z_t(\phi_1 Y_{t-h-1} + \ldots + \phi_p Y_{t-h-p} + Z_{t-h})]$ is equal to σ^2 for $h = 0$ and zero for $h > 0$, since Z_t is uncorrelated with $Y_{t-1}, Y_{t-2}, \ldots$ (this can easily be assessed by exploiting the MA(∞) representation of Y_t).

For computing the PACF $\alpha(h)$ for $h > p$ we need the linear prediction of Y_t based on $Y_{t-1}, \ldots, Y_{t-h+1}$. By computing the optimal predictor for $h > p$ (i.e., the conditional expectation),

$$\begin{aligned}\mathbb{E}[Y_t|Y_{t-1}, \ldots, Y_{t-h+1}] &= \mathbb{E}[\phi_1 Y_{t-1} + \ldots + \phi_p Y_{t-p} + Z_t|Y_{t-1}, \ldots, Y_{t-h+1}] \\ &= \phi_1 Y_{t-1} + \ldots + \phi_p Y_{t-p}\end{aligned}$$

we see that this is linear in $Y_{t-1}, \ldots, Y_{t-h+1}$, and so it coincides also with the optimal linear predictor $\mathbb{P}[Y_t|Y_{t-1}, \ldots, Y_{t-h+1}]$. For $h > p$, the PACF is proportional to

$$\mathbb{E}[(Y_t - \mathbb{P}[Y_t|Y_{t-1}, \ldots, Y_{t-h+1}])Y_{t-h}] = \mathbb{E}[Z_t Y_{t-h}] = 0,$$

and, thus, $\alpha(h) = 0$ for $h > p$. □

Theorem 2.12 (Properties of the MA(q)). *Let Y_t be the process*

$$Y_t = \theta_q(\mathbb{B}) Z_t, \qquad \forall t \in \mathbb{Z},$$

with $Z_t \sim \mathrm{WN}(0, \sigma^2)$; then $\mathbb{E}[Y_t] = 0$, and its autocovariance function is given by

$$\gamma(h) = \begin{cases} \sigma^2 \sum_{j=0}^{q-h} \theta_j \theta_{j+h} & \text{for } 0 \leq h \leq q \\ 0 & \text{otherwise,} \end{cases}$$

with $\theta_0 = 1$. If the MA process is also invertible, then its partial autocorrelation function $\alpha(h)$ converges to zero at geometric rate (i.e., there is a $\rho \in [0, 1)$ such that $|\alpha(h)| \leq \rho^h$).

Proof. The proof of the first result can be obtained easily by direct application of the expectations; the second part is a consequence of the fact that any invertible MA(q) has an infinite-order AR representation with coefficients tapering off at geometric rate. However, the PACF fades out also when the MA process is non-invertible (see Triacca, 2002, for a non-invertible MA(1)), but the rate is no more geometric. □

The characterisation of AR(p) processes by a zero PACF for $h > p$ and geometrically vanishing ACF, and of MA(q) processes by a zero ACF for $h > q$ and a geometrically vanishing PACF inspired the Box–Jenkins methodology to identify an ARMA model. A flowchart of (my interpretation of) the *Box–Jenkins identification procedure* is depicted in Figure 2.3.

Often time series embedding seasonal periodicities present non-null sample

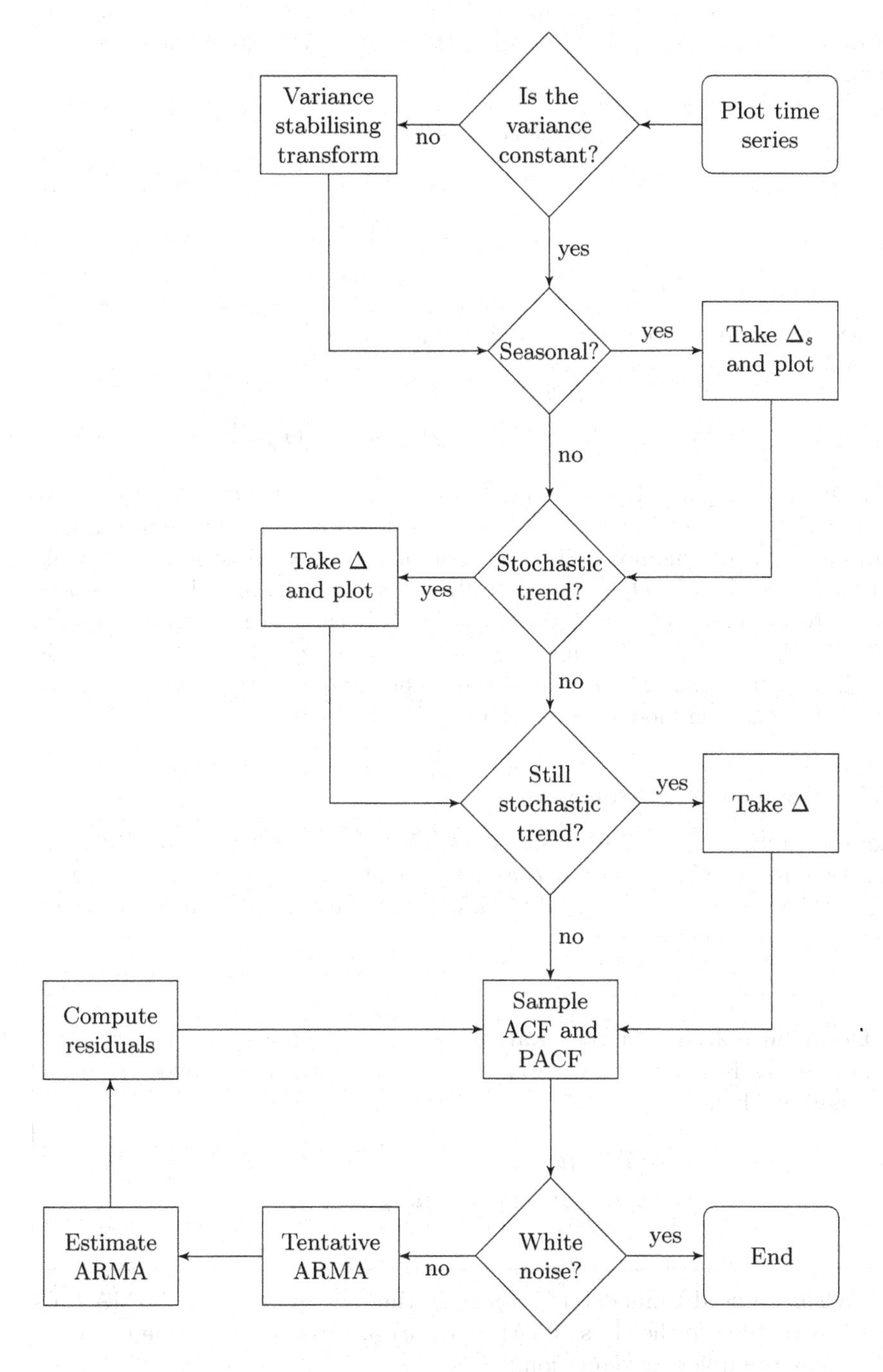

Figure 2.3 *Box–Jenkins strategy for ARIMA model identification.*

autocorrelation and partial autocorrelation at lag values h that are multiple of the seasonal period s. In order to fit this particular form of linear memory, the ARIMA(p, d, q) class,

$$\phi_p(\mathbb{B})\Delta^d Y_t = \theta_q(\mathbb{B})Z_t,$$

(i.e., the class of I(d) processes that are ARMA after d differences) is extended to embed the following seasonal ARIMA$(p, d, q)(P, D, Q)_s$ processes.

$$\phi_p(\mathbb{B})\Phi_P(\mathbb{B})\Delta^d \Delta_s^D Y_t = \theta_q(\mathbb{B})\Theta_Q(\mathbb{B})Z_t, \tag{2.13}$$

where the seasonal polynomials in $\mathbb{B}$ are defined as

$$\Phi_P(\mathbb{B}) = 1 - \Phi_1\mathbb{B}^s - \Phi_2\mathbb{B}^{2s} - \ldots - \Phi_P\mathbb{B}^{Ps}$$
$$\Theta_P(\mathbb{B}) = 1 + \Theta_1\mathbb{B}^s + \Theta_2\mathbb{B}^{2s} + \ldots + \Theta_Q\mathbb{B}^{Ps}.$$

The stationary part of process (2.13) is just an ARMA$(P + p, Q + q)$ process, but the constraints on its coefficients deriving from this multiplicative form allow a parsimonious representation of many real time series. In applications the orders P, D, Q are generally either 0 or 1. A model that has been found to fit many real-world time series well is the so-called *Airline model*: ARIMA$(0, 1, 1)(0, 1, 1)_s$. In many cases the Airline can be kept as a benchmark mode to which other models should be compared (refer to Section 6.5 for a discussion on model comparison).

2.5 Multivariate extensions

Let us consider a vector of N time series, $\{\boldsymbol{Y}_t\}$, for $t \in \mathbb{Z}$, and define the mean function $\boldsymbol{\mu}_t = \mathbb{E}\boldsymbol{Y}_t$ and *cross-covariance* function $\boldsymbol{\Gamma}_{t,s} = \mathbb{E}(\boldsymbol{Y}_t - \boldsymbol{\mu}_t)(\boldsymbol{Y} - \boldsymbol{\mu}_s)^\top$. The vector process is said to be weakly (jointly) stationary if these two functions are time-invariant.

Definition 2.15 (Jointly covariance-stationary processes). The vector process $\boldsymbol{Y}_t$ is weakly or covariance stationary, when the following two conditions hold:

$$\mathbb{E}\boldsymbol{Y}_t = \boldsymbol{\mu}$$
$$\mathbb{E}(\boldsymbol{Y}_t - \boldsymbol{\mu}_t)(\boldsymbol{Y}_{t-h} - \boldsymbol{\mu}_{t-h}) = \boldsymbol{\Gamma}(h).$$

When we need to model a vector of N time series $\{\boldsymbol{Y}_t\}$, the ARMA class can be extended to the class of VARMA(p, q) processes, that is, the processes that obey the difference equation:

$$\boldsymbol{Y}_t = \boldsymbol{\Phi}_1\boldsymbol{Y}_{t-1} + \ldots + \boldsymbol{\Phi}_p\boldsymbol{Y}_{t-p} + \boldsymbol{Z}_t + \boldsymbol{\Theta}_1\boldsymbol{Z}_{t-1} + \ldots + \boldsymbol{\Theta}_q\boldsymbol{Z}_{t-q}, \tag{2.14}$$

where $\mathbf{\Phi}_j$ and $\mathbf{\Theta}_j$ are $N \times N$ matrices of coefficients and $\boldsymbol{Z}_t$ is a vector of white noise sequences with covariance matrix $\mathbf{\Sigma}$, or, using a compact notation, $\boldsymbol{Z}_t \sim \text{WN}(\mathbf{0}, \mathbf{\Sigma})$, which is equivalent to

$$\mathbb{E}\boldsymbol{Z}_t = \mathbf{0}, \qquad \mathbb{E}[\boldsymbol{Z}_t \boldsymbol{Z}_{t-h}^\top] = \begin{cases} \mathbf{\Sigma} & \text{if } h = 0, \\ \mathbf{0} & \text{otherwise} \end{cases}.$$

As in the univariate case, we say that a VARMA process is causal if equation (2.14) has a solution as the linear combination of past and present shocks $\boldsymbol{Z}_t$:

$$\boldsymbol{Y}_t = \sum_{j=0}^{\infty} \mathbf{\Psi}_j \boldsymbol{Z}_{t-j}.$$

Theorem 2.13 (Existence and uniqueness of a causal stationary VARMA process). *Assume there is no value* $z \in \mathbb{C}$ *that solves* $\det(\mathbf{I}_N - \mathbf{\Phi}_1 z - \ldots - \mathbf{\Phi}_p z^p) = 0$ *and* $\det(\mathbf{I}_N + \mathbf{\Theta}_1 z + \ldots + \mathbf{\Theta}_p z^q) = 0$ *simultaneously. Then, a unique causal stationary solution to equation (2.14) exists if*

$$\det(\mathbf{I}_N - \mathbf{\Phi}_1 z - \ldots - \mathbf{\Phi}_p z^p) \neq 0 \qquad \textit{for all } z \in \mathbb{C} \textit{ such that } |z| \leq 1.$$

Proof. See the proof to Theorem 11.3.1 in Brockwell and Davis (1991). □

Using the $\mathbb{B}$ operator, the VARMA model is often represented as

$$\mathbf{\Phi}_p(\mathbb{B})\boldsymbol{Y}_t = \mathbf{\Theta}_q(\mathbb{B})\boldsymbol{Z}_t$$

with $\mathbf{\Phi}_p(\mathbb{B}) = \mathbf{I}_N - \mathbf{\Phi}_1\mathbb{B} - \ldots - \mathbf{\Phi}_p\mathbb{B}^p$, and $\mathbf{\Theta}_q(\mathbb{B}) = \mathbf{I}_N + \mathbf{\Theta}_1\mathbb{B} + \ldots + \mathbf{\Theta}_q\mathbb{B}^q$.

However, because of the *course of dimensionality*[8] and of identification problems[9], VARMA models are rarely used in applications, while the subclass of VAR(p) processes

$$\boldsymbol{Y}_t = \mathbf{\Phi}_1\boldsymbol{Y}_{t-1} + \ldots + \mathbf{\Phi}_p\boldsymbol{Y}_{t-p} + \boldsymbol{Z}_t, \tag{2.15}$$

is very popular, especially in economic applications.

[8]In statistics, this expression refers to the fact that in multivariate models the number of parameters to be estimated tends to increase much more quickly than the number of observations, when the number of variables to be modelled grows. For example, in the above mean-zero VARMA model there are $(p+q)N^2$ coefficients and $N(N+1)/2$ covariances that must be estimated from the data.

[9]A statistical model is not identified when there are many alternative sets of parameter values that yield an observationally equivalent model. For example the model $X \sim \mathcal{N}(a + b, \sigma^2)$ is not identified because there are infinite values of a and b that yield the same normal random variable with mean $\mu = a + b$.

Also in VAR and VARMA processes there is a relationship between the integration of the time series in the vector $\boldsymbol{Y}_t$ and the pN solutions of the characteristic equation $\boldsymbol{\Phi}_p(z) = 0$. Suppose that N roots are equal to 1, while all the other are outside the unit circle. Then, as in the univariate case there is no causal stationary process $\{\boldsymbol{Y}_t\}$ that solves equation (2.14) or (2.15), but since the VAR operator can be factorised as $\boldsymbol{\Phi}_p = \tilde{\boldsymbol{\Phi}}_{p-1}(\mathbb{B})(1-\mathbb{B})$ with $\tilde{\boldsymbol{\Phi}}_{p-1}(z)$ having all the roots outside the unit circle, then there is a causal stationary solution $\{\Delta\boldsymbol{Y}_t\}$ and so the time series in $\boldsymbol{Y}_t$ are I(1). Something very interesting happens when $\boldsymbol{\Phi}_p(z) = 0$ has m unit roots with $1 \leq m < N$. In this case it can be shown that all the N time series in $\boldsymbol{Y}_t$ are integrated of order one, but there are $r = N - m$ independent linear combinations of them that are stationary. When this happens, one says that the I(1) time series in $\boldsymbol{Y}_t$ are *cointegrated* with *cointegration rank* equal to r. In their pathbreaking article, Engle and Granger (1987) prove that a VAR(p) with m unit roots and all the other solutions outside the unit circle can be written in the vector error correction (or vector equilibrium correction) form

$$\Delta\boldsymbol{Y}_t = \boldsymbol{\Pi}\boldsymbol{Y}_{t-1} + \boldsymbol{\Lambda}_1\Delta\boldsymbol{Y}_{t-1} + \ldots + \boldsymbol{\Lambda}_{p-1}\Delta\boldsymbol{Y}_{t-p+1} + \boldsymbol{Z}_t,$$

where the matrix $\boldsymbol{\Pi}$ has rank $r = N - m$. Since $\Delta\boldsymbol{Y}_t$ is stationary, then so must be also the random quantity on the rhs of the equation. This implies that $\boldsymbol{\Pi}\boldsymbol{Y}_{t-1}$ is stationary and, therefore, the $N \times N$ matrix $\boldsymbol{\Pi}$ contains r linearly independent stationary linear combinations of the I(1) processes in $\boldsymbol{Y}_t$. From matrix algebra, we know that any rank r $N \times N$ matrix can be factorised as

$$\boldsymbol{\Pi} = \mathbf{A}\mathbf{B}^\top,$$

where $\mathbf{A}$ and $\mathbf{B}$ are $N \times r$ full-rank matrices. Thus, the columns of $\mathbf{B}$ contain r independent vectors of coefficients that make $\boldsymbol{Y}_t$ stationary: in symbols $\mathbf{B}^\top\boldsymbol{Y}_t \sim \mathrm{I}(0)$. These vectors are called *cointegration vectors*. The cointegration vectors are not unique, in fact if $\mathbf{Q}$ is a $r \times r$ full-rank matrix, then

$$\boldsymbol{\Pi} = \mathbf{A}\mathbf{B}^\top = \mathbf{A}\mathbf{Q}^{-1}\mathbf{Q}\mathbf{B}^\top,$$

and so the $N \times r$ matrix $\mathbf{B}\mathbf{Q}^\top$ also contains cointegration vectors in its columns. Indeed, we can say that the columns of the matrix $\mathbf{B}$ *span the cointegration space*.

Figure 2.4 depicts two cointegrated time series generated by the error correction process

$$\boldsymbol{Y}_t = \boldsymbol{Y}_{t-1} + \begin{bmatrix} 0 \\ -0.5 \end{bmatrix} \begin{bmatrix} 1 & -0.5 \end{bmatrix} \boldsymbol{Y}_{t-1} + \boldsymbol{Z}_t.$$

The two time series are clearly nonstationary, while the error correction term $[1 \;\; -0.5]\boldsymbol{Y}_t$ is stationary (in this case it is an AR(1) with $\phi = 0.5$).

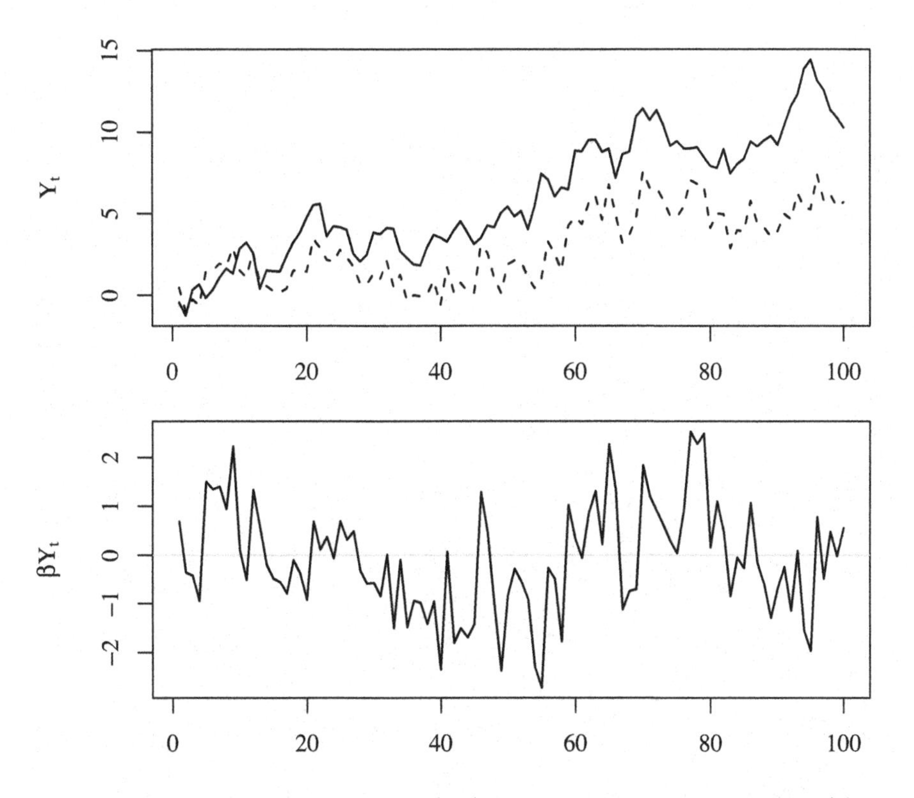

Figure 2.4 *Two simulated time series (top) generated by a cointegrated VAR(1) with cointegration vector* $\boldsymbol{\beta}^\top = [1 \quad -0.5]$, *and error correction term* $\boldsymbol{\beta}^\top \boldsymbol{Y}_t$ *(bottom).*

Part II

Unobserved components

Chapter 3

Unobserved Components Model

In this chapter we introduce the unobserved components model and describe the genesis of its main components.

3.1 The unobserved components model

A natural way humans tend to think about time series is as the sum of non directly observable components such as trend, seasonality, cycle. The seminal works of Holt (1957) and Winters (1960) on exponentially weighted moving average (EWMA) methods are early successful examples of this approach. Also official statisticians are used to conceive time series as sum of unobserved components: since the appearance of the first intra-annual time series, national statistical institutes had to provide seasonally adjusted versions of them.

The unobserved components models (UCM) treated in this book can be seen as the natural evolution of the EWMA approach into stochastic modelling. In fact, the EWMA approach provides forecasting rules without a model, while in modern time series analysis one first identifies and estimates a stochastic model that well approximate the "data generating mechanism" and then uses the optimal prediction formula for that particular model.

After the publication of the famous Box and Jenkins book in the seventies, many forecasters started thinking that ARIMA models were the best available tools for predicting time series, but real-world applications and comparisons soon demonstrated that very often EWMA techniques outperform ARIMA-based forecasting (see for example the synthesis of the results in the M3-Competition in Makridakis and Hibon, 2000).

Unobserved components models select the best features of both worlds: they share the same stochastic framework as ARIMA models (UCM can be seen as sums of ARIMA models), which allows statistical testing, the use of (possibly dynamic) regressors and the derivation of the whole distribution of future outcomes (not just point forecasts), but at the same time they are based on components as EWMA and tend to perform very well in forecasting (for example Zavanella et al., 2008). Moreover, seasonal adjustment and the prediction of future values of the unobserved components are natural "side-products" of UCM. This is also a very desirable feature since seasonality is often perceived as a nuisance by decision makers, who prefer the prediction of trends and cycles to the prediction of the raw series.

In the first UCM we present in this book the observable time series is the sum of trend, cycle, seasonality and (white) noise:

$$Y_t = \underset{\text{trend}}{\mu_t} + \underset{\text{cycle}}{\psi_t} + \underset{\text{seasonal}}{\gamma_t} + \underset{\text{noise}}{\varepsilon_t} . \tag{3.1}$$

Of course, some of these components could be skipped and some others could be added, but this is a model that fits many real-world time series and a good starting point to introduce UCM.

Except for ε_t, which is just a white noise sequence, we will see in the next sections that all the components can be seen as stochastic versions of well-known deterministic functions of time. Indeed, for extremely regular or very short time series deterministic components such as a linear trend and seasonal dummy variables can represent a decent approximation to the time series, but for most real-world time series the components evolve stochastically.

Figure 3.1 depicts the logarithm of the famous Airline time series with its deterministic linear trend and seasonal component estimated by ordinary least squares. The series is very regular for the period 1949-1960 and so the sum of the two deterministic components seems to represent a good approximation of the time series. However, it is hard to believe that the rate of growth of the number of airline passengers will always be the same (some 12% per year) and that the seasonality of travelling will never change. Furthermore,

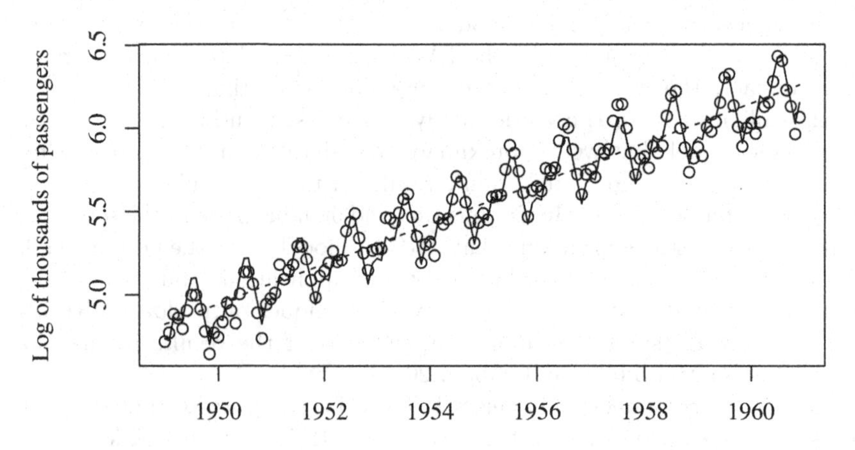

Figure 3.1 *Log of the Airline time series (points), linear trend (dashed line) and linear trend plus a deterministic seasonal component (solid line).*

assuming that these components are not evolving in time makes the forecaster underestimate the uncertainty of his/her predictions as he/she does not take into account the possible random evolution of the components in the future.

In the next sections we see how to build stochastically evolving components starting from simple deterministic functions such as those depicted in Figure 3.1. We will use the same procedure for all the components:

1. We take the "natural" deterministic component that does the job in very regular time series,
2. We write the function of the deterministic component in *incremental form* (the value at time t as function of the value at time $t-1$),
3. We add some random noise so that the component can evolve stochastically.

3.2 Trend

The trend component usually adopted in UCM is the *local linear trend* (LLT):

$$\begin{aligned} \mu_t &= \mu_{t-1} + \beta_{t-1} + \eta_t \\ \beta_t &= \beta_{t-1} + \zeta_t, \end{aligned} \tag{3.2}$$

where μ_t is the level of the series at time t, β_t is (part of) the increment of the level from time t to time $t+1$ and η_t and ζ_t are independent white noise sequences.

3.2.1 Genesis

Let us take a linear function defined on the integers,

$$\mu_t = \mu_0 + \beta_0 t, \tag{3.3}$$

where μ_0 is the intercept and β_0 the slope, and write it in incremental form (i.e., as difference equation),

$$\mu_t = \mu_{t-1} + \beta_0. \tag{3.4}$$

It is straightforward to check that equations (3.3) and (3.4) define the same linear function, in fact iterating (3.4) for $t' = 1, 2, \ldots, t$ we obtain:

$$\begin{aligned} &\mu_1 = \mu_0 + \beta_0 \\ &\mu_2 = \mu_1 + \beta_0 = \mu_0 + 2\beta_0 \\ &\ldots\ldots\ldots\ldots\ldots \\ &\mu_t = \mu_{t-1} + \beta_0 = \mu_0 + t\beta_0. \end{aligned}$$

By adding the white noise, η_t, to the rhs of equation (3.4) we obtain,

$$\mu_t = \mu_{t-1} + \beta_0 + \eta_t,$$

where the level μ_t evolves as a random walk with drift, and which can also be written as

$$\mu_t = \mu_0 + \beta_0 t + \sum_{s=1}^{t} \eta_s.$$

From this last expression we can interpret μ_t as a linear trend with a random walk intercept, but the slope remains unchanged. Thus, the prediction of μ_{t+k} given μ_t is

$$\mathbb{E}[\mu_{t+k}|\mu_t] = \mathbb{E}\left[\mu_t + \beta_0 k + \sum_{s=t+1}^{k} \eta_t \middle| \mu_t\right] = \mu_t + \beta_0 k$$

for any time t, which means that the slope of the prediction is constant whatever the last available observation μ_t. It would be nice to have also a time-varying slope in our trend, and this can be easily achieved by letting the slope evolve as a random walk:

$$\begin{aligned}\mu_t &= \mu_{t-1} + \beta_{t-1} + \eta_t \\ \beta_t &= \beta_{t-1} + \zeta_t,\end{aligned}$$

with ζ_t white noise sequence. The last equation pair defines the local linear trend, which can be interpreted as a linear trend where both intercept and slope evolve in time as random walks.

3.2.2 *Properties*

The local linear trend embeds different special cases of interest that can be obtained by fixing the values of the variances of the white noise sequences, σ^2_η, σ^2_ζ, or of the initial slope β_0 to zero.

Linear trend If we set $\sigma^2_\eta = \sigma^2_\zeta = 0$ we obtain a deterministic linear trend.

Random walk with drift If we set $\sigma^2_\zeta = 0$ the slope becomes constant and we have a random walk with drift β_0.

Random walk If we set $\sigma^2_\zeta = \beta_0 = 0$ the slope is constant and equals 0, and we obtain a random walk.

Integrated random walk If we set $\sigma^2_\eta = 0$ we obtain a very smooth trend that we can call integrated random walk. This type of trend is particularly useful when a cycle component is present in the UCM.

Figure 3.2 depicts the various trends embedded in a local linear trend generated using the same initial values and the same sequences of $\{\eta_t\}_{t=1,\dots,100}$ and $\{\zeta_t\}_{t=1,\dots,100}$.

We can see the local linear trend also as an ARIMA process. If the variance of ζ_t is greater than zero, then μ_t is an I(2) process as μ_t is nonstationary, its first difference $\Delta\mu_t = \beta_{t-1} + \eta_t$ is still nonstationary (it is the sum of a random walk and a white noise), while its second difference

$$\Delta^2\mu_t = \zeta_{t-1} + \eta_t - \eta_{t-1}$$

is stationary and MA(1)[1].

[1]To verify this statement compute the ACF of $\{\zeta_{t-1} + \eta_t - \eta_{t-1}\}$ and compare it with the ACF of a MA(1)

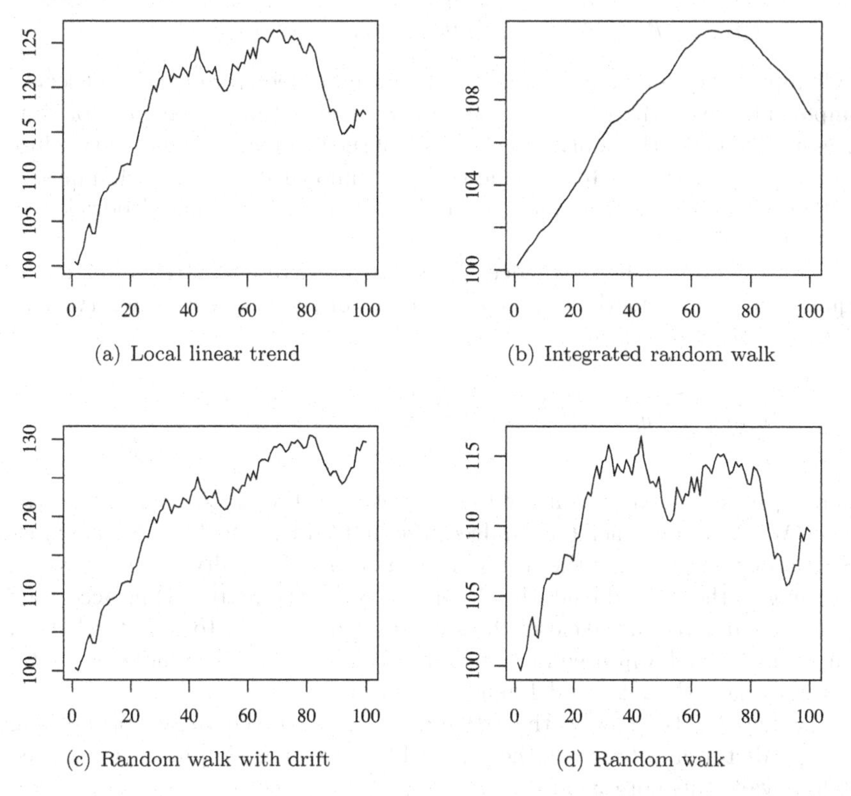

(a) Local linear trend

(b) Integrated random walk

(c) Random walk with drift

(d) Random walk

Figure 3.2 *Sample path of a local linear trend and its embedded trends.*

If we can observe μ_t and β_t, then the optimal prediction of future values of the local linear trend is given

$$\mathbb{E}[\mu_{t+k}|\mu_t, \beta_t] = \mathbb{E}\left[\mu_t + k\beta_t + \sum_{j=1}^{k}\left(\eta_{t+j} + \sum_{i=1}^{j-1}\zeta_{t+i}\right)\middle|\mu_t, \beta_t\right] = \mu_t + k\beta_t,$$

and its MSE is

$$\mathbb{E}[\mu_{t+k} - \mu_t - k\beta_t]^2 = \mathbb{E}\left[\sum_{j=1}^{k}\left(\eta_{t+j} + \sum_{i=1}^{j-1}\zeta_{t+i}\right)\right]^2 = k\sigma_\eta^2 + \frac{k(k-1)}{2}\sigma_\zeta^2.$$

Thus, the optimal predictor of the LLT at time $t + k$ based on the actual values of the level and slope at time t is linear in the prediction horizon k, and the intercept and slope of the line are just the level and slope available at time t. The MSE of the forecasts grows linearly in k if the variance of ζ_t is zero and quadratically in k otherwise.

3.2.3 Relation with Hodrick–Prescott filtering and cubic splines

In Chapter 5 we will see how to build the optimal predictor of the unobserved components given the observed time series. The optimal predictors of time series generated by the sum of special cases of the local linear trend and a white noise sequence turn out to coincide with well-known statistical techniques.

The first of these techniques is Hodrick–Prescott filtering (Hodrick and Prescott, 1997), which is very popular in those economic applications where a time series has to be detrended. The Hodrick–Prescott (HP) filter is the sequence of linear functions $\tau_t = \sum_{s=1}^{n} \alpha_{t,s} y_s$ of the time series observations $\{y_t\}_{t=1,\ldots,n}$ that solve

$$\min\left\{\sum_{t=1}^{n}(y_t - \tau_t)^2 + \lambda_{\text{HP}}\sum_{t=2}^{n-1}[(\tau_{t+1} - \tau_t) - (\tau_t - \tau_{t-1})]^2\right\},$$

where λ_{HP} is a fixed parameter that determines the smoothness of τ_t: the higher λ_{HP}, the smoother $\{\tau_t\}$. Indeed, the first addend in the objective function is a fit penalty while the second is a smoothness penalty. In other words, according to the first addend, the closer τ_t is to y_t, the better, while according to the second addend, the closer the second difference of τ_t to zero, the better. Common values of λ_{HP} used in economic applications are 6.25 for yearly data, 1,600 for quarterly data, and 129,600 for monthly data.

The trend extracted by the HP filter can be proved to be the optimal linear predictor of μ_t based on the observable time series $\{Y_t\}$ in the integrated random walk plus noise model

$$\begin{aligned}
Y_t &= \mu_t + \epsilon_t, & \epsilon_t &\sim \text{WN}\big(0, \sigma_\zeta^2 \cdot \lambda_{\text{HP}}\big)\\
\mu_t &= \mu_{t-1} + \beta_{t-1}, & &\\
\beta_t &= \beta_{t-1} + \zeta_t, & \zeta_t &\sim \text{WN}(0, \sigma_\zeta^2).
\end{aligned}$$

Example 3.1.
Figure 3.3 compares the quarterly time series of the European Union real GDP (at 2010 prices) with its trend as extracted by the HP filter with parameter $\lambda_{\text{HP}} = 1600$.

An identical signal can be obtained by extracting the μ_t component of the integrated random walk plus noise model using the smoother described in Section 5.3.2.

Now, suppose that the (observable) variable Y is related to the (observable) variable X through the equation

$$Y = f(X) + \epsilon_t,$$

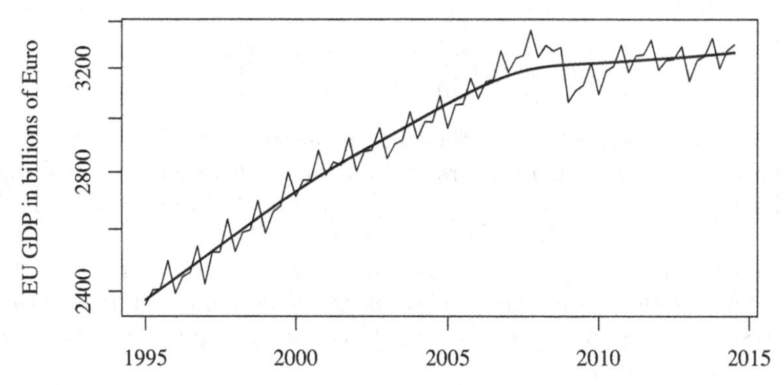

Figure 3.3 *European Union real GDP and trend extracted with HP filter with* $\lambda_{\text{HP}} = 1600$.

where $f(\cdot)$ is an unknown continuous function and $\{\epsilon_t\}$ a sequence of independent random variables with zero mean and variance $\sigma_\epsilon^2 < \infty$. If a sample of the pairs $\{x_t, y_t\}_{t=1,\ldots,n}$, such that $x_1 \leq x_2 \leq \ldots \leq x_n$, is available, then it is possible to estimate $f(\cdot)$ nonparametrically using *cubic splines* (for a classic text refer to Green and Silverman, 1993). As for the HP filter, the function $f(\cdot)$ is estimated at the sample points by minimising an objective function that compromises between the goodness of fit and smoothness of the estimated f:

$$\sum_{t=1}^{n} \left[y_t - f(x_t)\right]^2 + \lambda_S \int \left[\frac{\partial^2 f(x)}{\partial x^2}\right]^2 \mathrm{d}x.$$

It can be proved (Kohn and Ansley, 1987) that this objective function is equivalent to minus the log-likelihood of the continuous time Gaussian smooth-trend plus noise model

$$\begin{aligned} Y_t &= \mu(x_t) + \epsilon_t, & \epsilon_t &\sim \text{NID}(0, \sigma^2), \\ \mu(x_{t+1}) &= \mu(x_t) + \sqrt{\lambda_S}\sigma \int_{x_t}^{x_{t+1}} W(s)\,\mathrm{d}s, \end{aligned}$$

where $W(\cdot)$ is a standard Brownian motion (also Wiener process)[2]. If we define the time increments $\delta_t = x_{t+1} - x_t$, we can write the discrete-time version of the model as

$$\begin{aligned} Y_t &= \mu_t + \epsilon_t, & \epsilon_t &\sim \text{NID}(0, \sigma^2) \\ \mu_{t+1} &= \mu_t + \delta_t \beta_t + \eta_t, \\ \beta_{t+1} &= \beta_t + \zeta_t, \end{aligned}$$

[2]If you are not acquainted with the notion of Brownian motion you can think of it as a continuous-time Gaussian random walk with the first two moments equal to $\mathbb{E}[W(s)] = 0$, $\mathbb{E}[W(s)^2] = s$ and $\mathbb{E}[W(r)W(s)] = \min(r, s)$ and $r, s \in \mathbb{R}$.

with

$$\begin{bmatrix} \eta_t \\ \zeta_t \end{bmatrix} \sim \mathcal{N}\left(\begin{bmatrix} 0 \\ 0 \end{bmatrix}, \lambda_S \sigma^2 \begin{bmatrix} \frac{1}{3}\delta_t^3 & \frac{1}{2}\delta_t^2 \\ \frac{1}{2}\delta_t^2 & \delta_t \end{bmatrix}\right).$$

This is a slight modification of the local linear trend defined above, and inference on the unknown parameters and on the unobserved component μ_t can be done using the methods described in Chapter 5. In particular, the linear prediction of μ_t based on all the observations $\{Y_1, \ldots, Y_n\}$ is equivalent to the values produced by applying a cubic spline to the data. This UCM-based approach has the advantage of allowing the maximum likelihood estimation of the smoothness parameter, λ_S, which in standard splines methodology has to be fixed by the user.

Example 3.2 (Cubic spline estimation of a logit function).
We simulated 201 observations from

$$Y_t = f(x_t) + \epsilon_t, \qquad \epsilon_t \sim \text{NID}(0, 0.1^2),$$

with $x_1 = -100, x_2 = -99, \ldots x_{201} = 100$ and

$$f(x) = \frac{1}{1 - \exp(-x/20)}.$$

Then, we applied the UCM approach for cubic splines to the simulated observations estimating the model unknowns by maximum likelihood obtaining $\hat{\sigma}_\epsilon = 0.0922, \hat{\lambda}_S = 0.0001$.

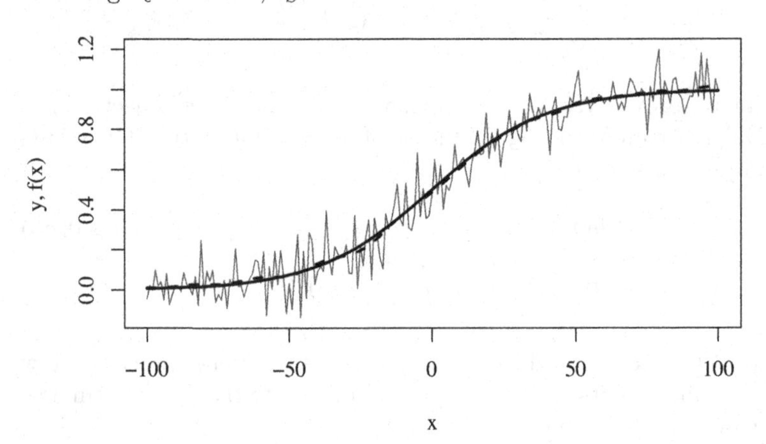

Figure 3.4 *Observation generated by $Y_t = f(x_t) + \epsilon_t$ with $\epsilon_t \sim$ NID(0, 0.01) plotted with the real $f(x_t)$ and its estimate $\hat{f}(x_t)$ based on UCM-based cubic splines.*

Figure 3.4 depicts the simulated observations together with the real function $f(x_t)$ and its estimate $\hat{f}(x_t)$, obtained by linear predicting μ_t using $\{y_1, \ldots, y_n\}$.

3.3 Cycle

In unobserved component models the stochastic cycle is the first component (ψ_t) of the bivariate VAR(1) process

$$\begin{bmatrix} \psi_t \\ \psi_t^* \end{bmatrix} = \rho \begin{bmatrix} \cos(\lambda) & \sin(\lambda) \\ -\sin(\lambda) & \cos(\lambda) \end{bmatrix} \begin{bmatrix} \psi_{t-1} \\ \psi_{t-1}^* \end{bmatrix} + \begin{bmatrix} \kappa_t \\ \kappa_t^* \end{bmatrix}, \qquad (3.5)$$

where the parameter $\rho \in [0, 1]$ is called *damping factor*, $\lambda \in [0, \pi]$ is the (central) *frequency* of the cycle and κ_t and κ_t^* are independent white noise sequences with common variance σ_κ^2. The (central) period of the cycle is given by $2\pi/\lambda$.

3.3.1 *Genesis*

The natural deterministic function for generating a cycle with frequency λ (number of oscillations per unit of time expressed in radians) is the sinusoid $R\cos(\lambda t + \phi)$, where R is the *amplitude* and ϕ is the *phase*.

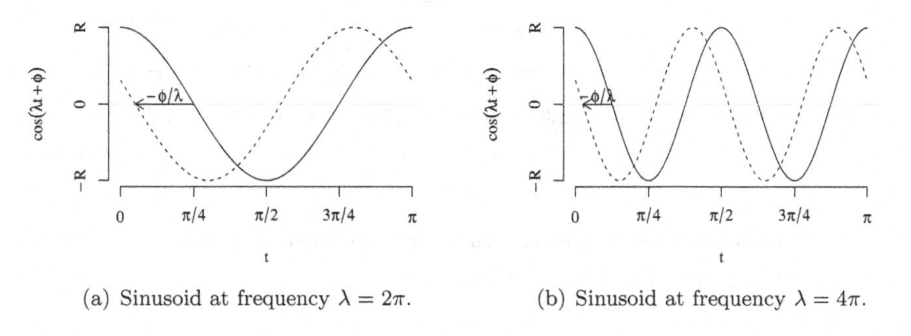

(a) Sinusoid at frequency $\lambda = 2\pi$.

(b) Sinusoid at frequency $\lambda = 4\pi$.

Figure 3.5 *Sinusoids at different frequencies and phases.*

Using the trigonometric identity of the cosine of a sum, we can obtain a useful alternative formula for the sinusoid

$$R\cos(\lambda t + \phi) = R\cos(\phi)\cos(\lambda t) - R\sin(\phi)\sin(\lambda t) = A\cos(\lambda t) + B\sin(\lambda t),$$

where we set $A = R\cos(\phi)$ and $B = -R\sin(\phi)$. We can recover phase and amplitude from A and B as (prove as exercise)[3]

$$R = \sqrt{A^2 + B^2}, \qquad \phi = \begin{cases} \arccos(A/R) & B \geq 0, \\ -\arccos(A/R) & B < 0. \end{cases}$$

[3]Many computer languages and software packages implement the function $\text{atan2}(x, y)$, which can be used to compute ϕ from A and B as $\phi = \text{atan2}(A, -B)$.

A simple geometrical way to generate a sinusoid with a given frequency (angular speed), amplitude and phase is taking a circumference of radius equal to the amplitude and letting a point move on it (for example clockwise). The sinusoid is generated by taking either the projection of the x-axis or on the y-axis.

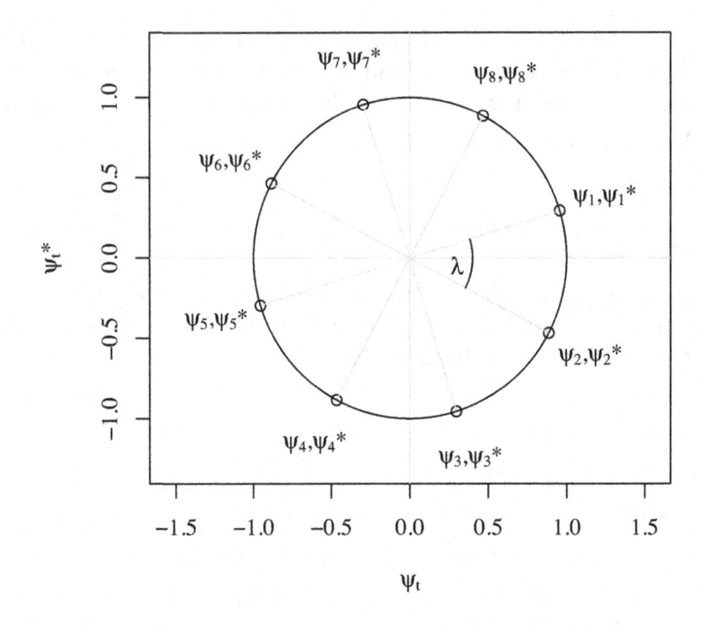

Figure 3.6 *Geometric visualisation of a sinusoid at frequency* λ.

For writing the sinusoid in incremental form we can use a *rotation matrix*

$$\mathbf{R}(\lambda) = \begin{bmatrix} \cos(\lambda) & \sin(\lambda) \\ -\sin(\lambda) & \cos(\lambda) \end{bmatrix}, \tag{3.6}$$

which applied to a point in two dimensions rotates it clockwise by an angle λ. In Figure 3.6 the point (ψ_2, ψ_2^*) is obtained by applying $\mathbf{R}(\lambda)$ to the point (ψ_1, ψ_1^*) and in general

$$\begin{bmatrix} \psi_t \\ \psi_t^* \end{bmatrix} = \mathbf{R}(\lambda) \begin{bmatrix} \psi_{t-1} \\ \psi_{t-1}^* \end{bmatrix}. \tag{3.7}$$

Notice that $\mathbf{R}(\lambda)^k = \mathbf{R}(k\lambda)$ (prove it as an exercise using the trigonometric identities for the sum) and therefore

$$\begin{bmatrix} \psi_{t+1} \\ \psi_{t+1}^* \end{bmatrix} = \mathbf{R}(\lambda) \begin{bmatrix} \psi_t \\ \psi_t^* \end{bmatrix} = \mathbf{R}(\lambda)\mathbf{R}(\lambda) \begin{bmatrix} \psi_{t-1} \\ \psi_{t-1}^* \end{bmatrix} = \ldots = \mathbf{R}(t\lambda) \begin{bmatrix} \psi_1 \\ \psi_1^* \end{bmatrix}.$$

We can now turn the sinusoid stochastic by adding some random noise to the incremental form (3.7):

$$\begin{bmatrix} \psi_t \\ \psi_t^* \end{bmatrix} = \mathbf{R}(\lambda) \begin{bmatrix} \psi_{t-1} \\ \psi_{t-1}^* \end{bmatrix} + \begin{bmatrix} \kappa_t \\ \kappa_t^*, \end{bmatrix} \tag{3.8}$$

where the two random shocks are uncorrelated white noise sequences with common variance σ_κ^2. The effects of these shocks are twofold: i) the circle on which the point (ψ_t, ψ_t^*) lands is not the same circle on which the point $(\psi_{t-1}, \psi_{t-1}^*)$ lies; ii) the angle at the origin between the two points is different than λ, and λ becomes just the expected angle.

There is only one problem left in order to be able to use the above process as a model for business cycles: ψ_t is nonstationary. In fact, the two eigenvalues of the matrix $\mathbf{R}(\lambda)$ are the complex conjugate $\cos(\lambda) \mp i \sin(\lambda)$, whose modulus is 1. Economists tend to think of the business cycle as an oscillatory response to some shock which eventually dies out over time. In the cycle defined by (3.8), instead, a single shock has a permanent effect on the cycle, and so we have to make a last change to guarantee stationarity. By multiplying the rotation matrix by a constant in the unit interval as in equation (3.5), we obtain eigenvalues in the unit circle and, therefore, a stationary stochastic cycle. The nonstationary cycle (3.8) will be used in the next section to build the stochastic seasonal component.

The sequential application of the rotation matrix $\mathbf{R}(\lambda)$, the damping factor ρ and sum of the random noise $\boldsymbol{\kappa}$ is geometrically represented in Figure 3.7. The circles around the point $\rho\mathbf{R}(\lambda)\boldsymbol{\psi}_t$ are contours of a bivariate normal density with uncorrelated components and represent the possible action of summing the random pair $\boldsymbol{\kappa}_t$ to $\rho\mathbf{R}(\lambda)\boldsymbol{\psi}_t$.

3.3.2 Properties

We summarise the main properties of the stochastic cycle (3.5) in the following theorem.

Theorem 3.1. *Let $\boldsymbol{\psi}_t$ be defined by equation (3.5) and assume $\rho \in [0, 1)$. Then, there is a unique causal and stationary solution with*

- $\mathbb{E}[\boldsymbol{\psi}_t] = \mathbf{0}$,
- *for $h \in \mathbb{Z}$,*

$$\mathbb{E}[\boldsymbol{\psi}_{t+h}\boldsymbol{\psi}_t^\top] = \frac{\rho^h \cos(h\lambda)\sigma_\kappa^2}{1-\rho^2}\mathbf{I}_2,$$

- *ψ_t is and ARMA(2,1) process with complex roots in the AR polynomial.*

The optimal predictor of the stochastic cycle at time $t + h$ based on its past up to time t is

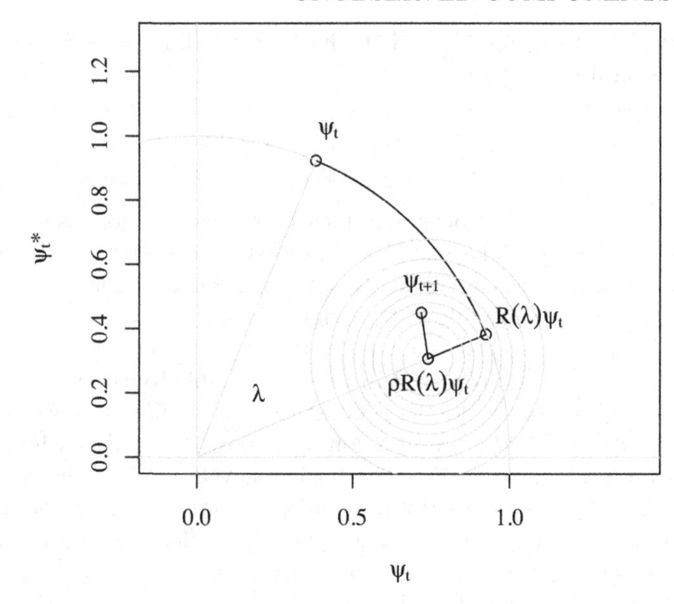

Figure 3.7 *Geometric representation of the actions of* $\mathbf{R}(\lambda)$, ρ *and the additive random error* $\boldsymbol{\kappa}_t$ *that, generate* $\boldsymbol{\psi}_{t+1}$ *starting from* $\boldsymbol{\psi}_t$.

- $\hat{\boldsymbol{\psi}}_{t+h|t} = \mathbb{E}[\boldsymbol{\psi}_{t+h}|\boldsymbol{\psi}_t, \boldsymbol{\psi}_{t-1}, \ldots] = \mathbb{E}[\boldsymbol{\psi}_{t+h}|\boldsymbol{\psi}_t] = \rho^h \mathbf{R}(h\lambda)\boldsymbol{\psi}_t$,
- *with MSE given by*

$$\mathbb{E}[(\boldsymbol{\psi}_{t+h} - \hat{\boldsymbol{\psi}}_{t+h|t})(\boldsymbol{\psi}_{t+h} - \hat{\boldsymbol{\psi}}_{t+h|t})^\top] = \sigma_\kappa^2 \frac{1-\rho^{2h}}{1-\rho^2}.$$

Proof. Since equation (3.5) defines a VAR(1), then there is a causal stationary process that solves if and only if all the eigenvalues of the matrix $\rho\mathbf{R}(\lambda)$ are in modulus smaller than 1. As can be easily verified, the two eigenvalues of $\rho\mathbf{R}(\lambda)$ are $\rho(\cos(\lambda) \mp i \sin(\lambda))$ and their modulus is ρ. Thus, the condition $|\rho| < 1$ is necessary and sufficient for such a solution to exist. The further assumption that ρ is nonnegative is just for statistical identifiability reasons.

Now that we know that the first two (marginal) moments exist we can compute them. By naming $\boldsymbol{\mu}_\psi$ the expectation of $\boldsymbol{\psi}_t$ and taking the expectation of both sides of (3.5) we obtain $\boldsymbol{\mu}_\psi = \rho\mathbf{R}(\lambda)\boldsymbol{\mu}_\psi$ which is solved only by $\boldsymbol{\mu}_\psi = \mathbf{0}$.

The covariance matrix of $\boldsymbol{\psi}_t$, $\boldsymbol{\Sigma}_\psi$, can be obtained by taking the expectation of the lhs and rhs of (3.5) multiplied by their respective transpose,

$$\mathbb{E}[\boldsymbol{\psi}_t \boldsymbol{\psi}_t^\top] = \rho^2 \mathbf{R}(\lambda)\mathbb{E}[\boldsymbol{\psi}_{t-1}\boldsymbol{\psi}_{t-1}^\top]\mathbf{R}(\lambda)^\top + \sigma_\kappa^2 \mathbf{I},$$

and, thus, solving the equation $\boldsymbol{\Sigma}_\psi = \rho^2 \mathbf{R}(\lambda)\boldsymbol{\Sigma}_\psi \mathbf{R}(\lambda)^\top + \sigma_\kappa^2 \mathbf{I}$ (notice that for a rotation matrix $\mathbf{R}(\lambda)\mathbf{R}(\lambda)^\top = \mathbf{I}$ and $\mathbf{R}(\lambda)^h = \mathbf{R}(h\lambda)$). We can try to solve this matrix equation by iteratively substituting the lhs into the rhs:

$$\begin{aligned}
\boldsymbol{\Sigma}_\psi &= \rho^2 \mathbf{R}(\lambda)\boldsymbol{\Sigma}_\psi \mathbf{R}(\lambda)^\top + \sigma_\kappa^2 \mathbf{I} \\
&= \rho^2 \mathbf{R}(\lambda)\left[\rho^2 \mathbf{R}(\lambda)\boldsymbol{\Sigma}_\psi \mathbf{R}(\lambda)^\top + \sigma_\kappa^2 \mathbf{I}\right]\mathbf{R}(\lambda)^\top + \sigma_\kappa^2 \mathbf{I} \\
&= \rho^4 \mathbf{R}(2\lambda)\boldsymbol{\Sigma}_\psi \mathbf{R}(2\lambda)^\top + (1+\rho^2)\sigma_\kappa^2 \mathbf{I} \\
&= \dots \\
&= \rho^{2h} \mathbf{R}(h\lambda)\boldsymbol{\Sigma}_\psi \mathbf{R}(h\lambda)^\top + \left(\sum_{j=0}^{h-1} \rho^{2j}\right)\sigma_\kappa^2 \mathbf{I}.
\end{aligned}$$

By letting h diverge we see that the first addend converges to zero and the summation in the second addend, which is a geometric series in ρ^2, converges to $(1-\rho^2)^{-1}$ and, thus, we obtain $\boldsymbol{\Sigma}_\psi = (1-\rho^2)^{-1}\sigma_\kappa^2 \mathbf{I}$.

By substituting equation (3.5) into itself h times we can express the observation at time $t+h$ as

$$\boldsymbol{\psi}_{t+h} = \rho^h \mathbf{R}(h\lambda)\boldsymbol{\psi}_t + \sum_{j=0}^{h-1} \rho^j \mathbf{R}(j\lambda)\boldsymbol{\kappa}_{t+h-j}, \tag{3.9}$$

and this can be used to obtain the autocovariance function, the conditional expectation and its MSE. For the autocovariance function

$$\begin{aligned}
\mathbb{E}[\boldsymbol{\psi}_{t+k}\boldsymbol{\psi}_t^\top] &= \mathbb{E}\left[\rho^h \mathbf{R}(h\lambda)\boldsymbol{\psi}_t\boldsymbol{\psi}_t^\top + \sum_{j=0}^{h-1} \rho^j \mathbf{R}(j\lambda)\boldsymbol{\kappa}_{t+h-j}\boldsymbol{\psi}_t^\top\right] \\
&= \mathbf{R}(h\lambda)\rho^h \boldsymbol{\Sigma}_\psi \cos(h\lambda) \\
&= \frac{\rho^h \cos(h\lambda)\sigma_\kappa^2}{1-\rho^2}\mathbf{I}.
\end{aligned}$$

We let the reader use equation (3.9) to derive the conditional expectation and its MSE as an exercise.

It is left to prove that the stochastic cycle is an ARMA(2,1) process. Using the backward operator and exploiting stationarity we can express $\boldsymbol{\psi}_t$ in its MA(∞) form

$$\begin{aligned}
\boldsymbol{\psi}_t &= \left[\mathbf{I} - \rho\mathbf{R}(\lambda)\mathbb{B}\right]^{-1}\boldsymbol{\kappa}_t \\
&= \frac{1}{1 - 2\rho\cos\lambda\mathbb{B} + \rho^2\mathbb{B}^2}\left[\mathbf{I} - \rho\mathbf{R}(-\lambda)\mathbb{B}\right]\boldsymbol{\kappa}_t.
\end{aligned}$$

The first row of this matrix equation reads

$$\psi_t = \frac{1}{1 - 2\rho\cos\lambda\mathbb{B} + \rho^2\mathbb{B}^2}[\kappa_t - \rho\cos(\lambda)\kappa_{t-1} + \rho\sin(\lambda)\kappa^*_{t-1}] \tag{3.10}$$

which is an ARMA(2,1) process since the sum of a MA(1) process and an uncorrelated white noise is a MA(1) process. Solving the characteristic equation of the AR part, $1 - 2\rho\cos\lambda x + \rho^2 x^2 = 0$, we obtain the complex conjugate roots $\rho^{-1}[\cos(\lambda) \mp \sin(\lambda)]$ which, of course, lie outside the unit circle. □

When $\rho = 1$ and $\mathbb{E}[\boldsymbol{\kappa}\boldsymbol{\kappa}^\top] = \mathbf{0}$, we have a deterministic cycle with frequency λ while, when $\rho = 1$ and $\mathbb{E}[\boldsymbol{\kappa}\boldsymbol{\kappa}^\top] = \sigma^2_\kappa \boldsymbol{I}$ with $\sigma^2_\kappa > 0$, we obtain a nonstationary cycle. Sometimes, in order to let the stochastic cycle embed the deterministic cycle but not the nonstationary cycle, some software packages (e.g., STAMP and SAS/ETS's proc UCM) parametrise the model using the variance of $\boldsymbol{\psi}_t$ instead of that of $\boldsymbol{\kappa}_t$ so that σ^2_κ is computed as $\sigma^2_\psi(1-\rho^2)$ and if ρ is estimated equal to 1 the disturbances' variance goes to zero and the cycle becomes deterministic.

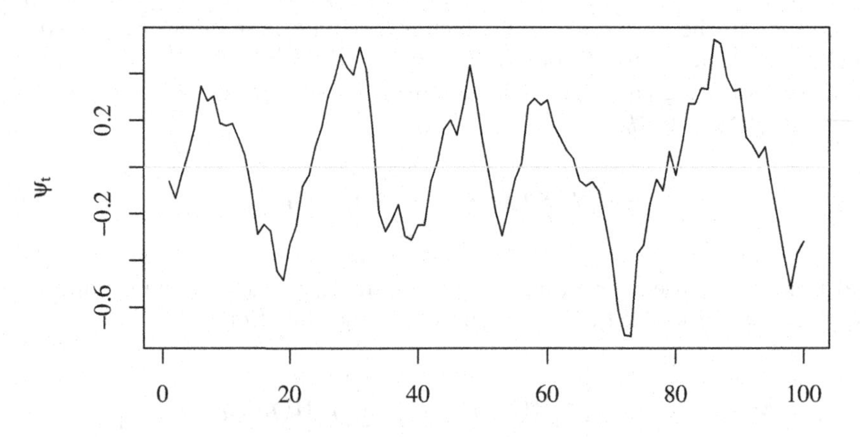

Figure 3.8 *Sample path of a stochastic cycle with* $\rho = 0.95$, $\lambda = 2\pi/20$, $\sigma^2_\kappa = 0.01$.

3.3.3 A smooth extension

By observing Figure 3.8 one may object that the sample path of the stochastic cycle appears somewhat rougher than one would expect from a business cycle. An alternative model for the cycle that increases the smoothness of the sample paths but embeds the stochastic cycle (3.5) was proposed by Harvey and Trimbur (2003). Their idea is substituting the two white noises driving the stochastic cycle with a stochastic cycle with the same characteristics. The *stochastic cycle of order m* is the first element of the 2-vector $\boldsymbol{\psi}_{m,t}$ in the recursion

$$\begin{aligned} \boldsymbol{\psi}_{1,t} &= \rho \mathbf{R}(\lambda) \boldsymbol{\psi}_{1,t-1} + \boldsymbol{\kappa}_t, & \boldsymbol{\kappa}_t &\sim \mathrm{WN}(0, \sigma_\kappa^2 \mathbf{I}) \\ \boldsymbol{\psi}_{j,t} &= \rho \mathbf{R}(\lambda) \boldsymbol{\psi}_{j,t-1} + \boldsymbol{\psi}_{j-1,t-1}, & &\text{for } j = 2, 3, \ldots, m. \end{aligned} \tag{3.11}$$

So, for $m = 1$ we obtain the stochastic cycle (3.5), while as m increases the sample paths of the higher-order cycle become smoother. Generally, m is taken from the set $\{1, 2, 3, 4\}$.

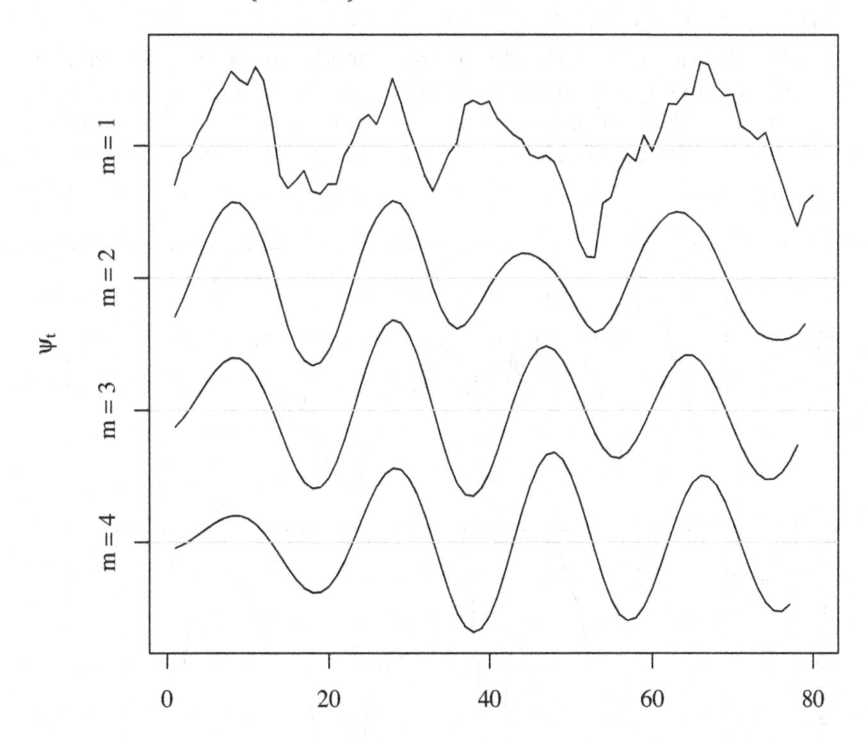

Figure 3.9 *Sample paths of standardised higher-order stochastic cycles for* $m = 1, 2, 3, 4$ *with* $\rho = 0.95$, $\lambda = 2\pi/20$. *The variance of the disturbances has been adjusted to assure that the cycles have similar scales.*

Figure 3.9 depicts m order cycles for $m = 1, 2, 3, 4$. The increment in smoothness is particularly evident when m moves from 1 to 2, while it is less significant for increments to higher orders. This suggests that $m = 2$ should be large enough for most business cycle applications.

The m order cycle can be shown to be causal stationary if and only if $|\rho| < 1$, but deriving the second-order properties of it is rather lengthy and tedious. For $m = 2$ its variance is given

$$\mathbb{V}\mathrm{ar}(\psi_{2,t}) = \frac{1 + \rho^2}{(1 - \rho^2)^3} \sigma_\kappa^2,$$

and the autocorrelation function is

$$\rho(h) = \rho^h \cos(h\lambda) \left(1 + \frac{1-\rho^2}{1+\rho^2} h\right),$$

which decays more slowly than the one for $m = 1$, which is $\rho^h \cos(h\lambda)$. The derivation of these formulae and expressions for any order m can be found in Trimbur (2005).

Example 3.3 (Higher-order cycles of U.S. investments).
We applied a model with a linear trend and cycles of different orders to the logarithm of the quarterly U.S. investments time series from 1947Q1 to 2007Q2. The estimates of the cycle component for a sub-sample of the time series are depicted in Figure 3.10.

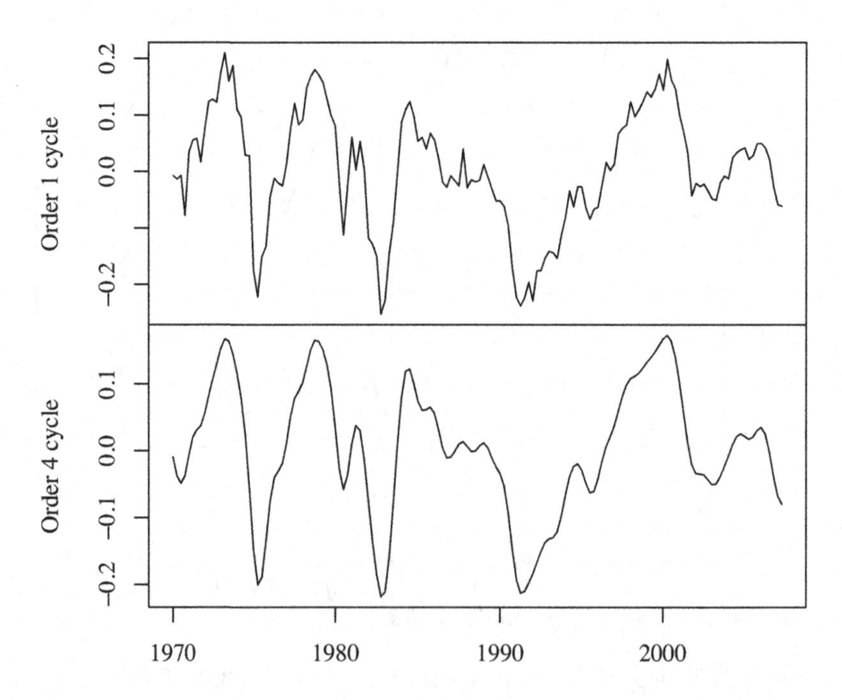

Figure 3.10 *Order 1 and order 4 cycle estimated on log of the U.S. investments time series.*

The correlation between the two estimates of the cycle component is almost 0.98, and the two signals are almost coincident; however, the 4-order cycle is significantly smoother.

Chapter 8 is completely devoted to the business cycle and its extraction.

So, the interested reader is referred to that chapter for more examples and applications.

3.4 Seasonality

The seasonal component is generally modelled using two alternative forms: the *stochastic dummy form* and the *stochastic trigonometric form*.

Let s be the periodicity of the seasonal component (e.g., $s = 12$ for monthly data and $s = 4$ for quarterly time series), the stochastic dummy form is defined by the recursion

$$\gamma_t = -\sum_{i=1}^{s-1} \gamma_{t-i} + \omega_t, \tag{3.12}$$

with $\omega_t \sim \mathrm{WN}(0, \sigma_\omega^2)$.

The trigonometric form is given by[4] $\gamma_t = \sum_{j=1}^{\lfloor s/2 \rfloor} \gamma_{j,t}$, where $\gamma_{j,t}$ is the nonstationary stochastic cycle

$$\begin{bmatrix} \gamma_{j,t} \\ \gamma^*_{j,t} \end{bmatrix} = \begin{bmatrix} \cos(2j\pi/s) & \sin(2j\pi/s) \\ -\sin(2j\pi/s) & \cos(2j\pi/s) \end{bmatrix} \begin{bmatrix} \gamma_{j,t-1} \\ \gamma^*_{j,t-1} \end{bmatrix} + \begin{bmatrix} \omega_{j,t} \\ \omega^*_{j,t} \end{bmatrix}, \tag{3.13}$$

with $j = 1, 2, \ldots, \lfloor s/2 \rfloor$ and $\boldsymbol{\omega}_{j,t} \sim \mathrm{WN}(\mathbf{0}, \sigma_\omega^2 \mathbf{I})$ for all j, with $\mathbb{E}[\boldsymbol{\omega}_{j,t}\boldsymbol{\omega}_{k,t}^\top] = \mathbf{0}$ for $j \neq k$. Notice that the coefficient matrix is the rotation matrix $\mathbf{R}(\cdot)$ defined in equation (3.6). The arguments of the sinusoids, $\{2\pi/s, 4\pi/s, \ldots, \lfloor s/2 \rfloor \pi/s\}$, are called *seasonal frequencies*. Notice that when s is even, for $j = s/2$ the second equation of the system (3.13) can be omitted as the sine of π equals zero and so the first equation of the system reduces to $\gamma_{s/2,t} = -\gamma_{s/2,t-1} + \omega_{s/2,t}$ which does not depend on the value of the second element $\gamma^*_{s/2,t}$.

An alternative way of modelling time-varying seasonal periodicities is by defining s variables that evolve as random walk whose sum is constrained to equal a zero expectation random variable. This model was introduced by Harrison and Stevens (1976). Let $\gamma_{j,t}$ represent the effect of season j at time t, then

$$\gamma_{j,t} = \gamma_{j,t-1} + \omega_{j,t}, \quad \text{for } j = 1, 2, \ldots, s,$$

where $\omega_{j,t} \sim \mathrm{WN}(0, \sigma_\omega^2)$. The restriction on the sum of the s components to obtain a mean zero variable is obtained by imposing the following covariance structure to the s white noise processes:

$$\mathbb{E}[\boldsymbol{\omega}_t \boldsymbol{\omega}_t^\top] = \sigma_\omega^2 (\mathbf{I}_s - s^{-1}\mathbf{1}\mathbf{1}^\top),$$

where $\mathbf{1}$ is a vector of ones of dimensions $s \times 1$ and the vector $\boldsymbol{\omega}_t$ contains the s white noise sequences $\omega_{j,t}$. This model is sometimes referred to as *balanced dummy variable seasonal model*, and its relations with the other two models were analysed by Proietti (2000). In particular, Proietti (2000) shows that by

[4] The symbol $\lfloor x \rfloor$ denotes the *floor* of x, that is, the greatest integer not larger than x.

opportunely adjusting the variances, this model is equivalent to the stochastic trigonometric model. For this reason and because the balanced dummy seasonal component is rarely encountered in software packages, in the next sections we limit our attention to the other two seasonal forms.

3.4.1 Genesis

By definition, a deterministic seasonal component of period s is a sequence of numbers that repeat themselves every s consecutive observations: $\gamma_t = \gamma_{t-s}$. Moreover, its values sum to zero over s consecutive observations as yearly data are not affected by any seasonality: $\sum_{j=0}^{s-1} \gamma_{t-j} = 0$ for all t. Thus, a deterministic seasonal component is a zero-mean periodic function defined on the integers, and given the first s values (that must sum to zero), all the other values can be obtained by the recursion

$$\gamma_t = -\sum_{j=1}^{s-1} \gamma_{t-j}.$$

The most straightforward way to make this component evolve stochastically over time is by adding mean-zero random shocks. This results in the stochastic dummy form of equation (3.12).

In order to derive the stochastic trigonometric form, let us start with the following result for periodic functions.

Lemma 3.2. *Let $f_t : \mathbb{Z} \mapsto \mathbb{R}$ a function of period $s \in \mathbb{Z}$. Then,*

$$f_t = \bar{f} + \sum_{j=1}^{\lfloor s/2 \rfloor} a_j \cos\left(\frac{2j\pi}{s}t\right) + b_j \sin\left(\frac{2j\pi}{s}t\right),$$

with

$$\bar{f} = \frac{1}{s}\sum_{t=1}^{s} f_t, \quad a_j = \frac{2}{s}\sum_{t=1}^{s} f_t \cos\left(\frac{2j\pi}{s}t\right), \quad b_j = \frac{2}{s}\sum_{t=1}^{s} f_t \sin\left(\frac{2j\pi}{s}t\right).$$

The above lemma states that every periodic function of period s defined on the integers can be expressed as the sum of its mean over s consecutive observations and $\lfloor s/2 \rfloor$ sinusoids at seasonal frequencies. As we did in Section 3.3.1, we can write the sinusoids in incremental form and then add a white noise.

So, if we call $\gamma_{j,t}$ the sinusoid at frequency $2j\pi/s$, using Lemma 3.2 we can express the mean-zero periodic function γ_t as $\gamma_t = \sum_{j=1}^{\lfloor s/2 \rfloor} \gamma_{j,t}$. If we represent $\gamma_{j,t}$ in incremental form as we did in Section 3.3.1, we can express the sinusoid at frequency $2j\pi/s$ as

$$\begin{bmatrix} \gamma_{j,t} \\ \gamma^*_{j,t} \end{bmatrix} = \begin{bmatrix} \cos(2j\pi/s) & \sin(2j\pi/s) \\ -\sin(2j\pi/s) & \cos(2j\pi/s) \end{bmatrix} \begin{bmatrix} \gamma_{j,t-1} \\ \gamma^*_{j,t-1} \end{bmatrix},$$

and by adding uncorrelated white noise to each equation we obtain the stochastic trigonometric form (3.13). This seasonal component is built using the trigonometric representation of a periodic function given by Lemma 3.2, where the deterministic sinusoids have been replaced by nonstationary stochastic sinusoids that allow the seasonal component to evolve over time.

3.4.2 Properties

Before analysing the properties of the two seasonal components it is useful to introduce two operators related to the backward and difference operators introduced in Chapter 2.

The *summation operator of order s* is defined by

$$\mathbb{S}_s = 1 + \mathbb{B} + \mathbb{B}^2 + \ldots + \mathbb{B}^{s-1}$$

and applied to the process $\{Y_t\}$ yields $\mathbb{S}_s Y_t = \sum_{j=0}^{s-1} Y_{t-j}$. By definition, if s is the seasonal period, then $\mathbb{S}_s$ applied to a deterministic seasonal component returns a zero constant.

The *trigonometric operator* is defined as

$$\mathbb{T}(\theta) = \begin{cases} 1 - \mathbb{B} & \text{if } \theta = 0, \\ 1 + \mathbb{B} & \text{if } \theta = \pi, \\ 1 - 2\cos(\theta)\mathbb{B} + \mathbb{B}^2 & \text{otherwise.} \end{cases}$$

and can be seen as an AR(2) operator with the two roots, $\cos(\theta) \mp \sin(\theta)$, on the unit circle. When $\theta = 0$ or $\theta = \pi$, the operator reduces to an AR(1) operator with root, respectively, 1 and -1.

We summarise the main properties of the new operators and of the two forms for the seasonal component in the following theorem.

Theorem 3.3. *The following results hold.*

- *Identities between the seasonal difference, simple difference, summation and trigonometric operators:*
  - $\Delta_s = \Delta \mathbb{S}_s$,
  - $\Delta_s = \prod_{j=0}^{\lfloor s/2 \rfloor} \mathbb{T}(2j\pi/s)$.
  - $\mathbb{S}_s = \prod_{j=1}^{\lfloor s/2 \rfloor} \mathbb{T}(2j\pi/s)$,
- *Reduced ARMA form of the seasonal components:*
  - *if γ_t is the stochastic dummy seasonal component (3.12), then $\mathbb{S}_s \gamma_t$ is the white noise process ω_t;*
  - *if γ_t is the stochastic trigonometric seasonal component (3.13), then $\mathbb{S}_s \gamma_t$ is a zero-mean MA$(s-2)$ process.*

- *Prediction of the seasonal components:*
  - *for the stochastic dummy seasonal form the optimal predictor based on its own past is*

$$\mathbb{E}[\gamma_{t+h}|\gamma_t, \gamma_{t-1}, \ldots, \gamma_{t-s+2}] = \begin{cases} -\sum_{j=1}^{s-1} \gamma_{t+1-j}, & \textit{for } h = 1, s+1, 2s+1, \ldots, \\ \gamma_{t+h-\lfloor 1+\frac{h-1}{s} \rfloor s}, & \textit{otherwise}, \end{cases}$$

  - *with MSE given by*

$$\begin{cases} \left(1 + 2\left\lfloor \frac{h-1}{s} \right\rfloor\right) \sigma_\omega^2, & \textit{for } h = 1, s+1, 2s+1, \ldots, \\ \left(2 + 2\left\lfloor \frac{h-1}{s} \right\rfloor\right) \sigma_\omega^2, & \textit{otherwise;} \end{cases}$$

  - *for the stochastic trigonometric seasonal form the optimal predictor based on the past of its sinusoids is*

$$\mathbb{E}[\gamma_{t+h}|\gamma_{1,t}, \gamma_{2,t}, \ldots, \gamma_{\lfloor s/2 \rfloor,t}] = \sum_{j=1}^{\lfloor s/2 \rfloor} \cos(2hj\pi/s)\gamma_{j,t} + \sin(2hj\pi/s)\gamma_{j,t}^*.$$

  - *with MSE given by* $h\lfloor s/2 \rfloor \sigma_\omega^2$.

Proof. Identities. The first identity is trivial: $(1 - \mathbb{B})(1 + \mathbb{B} + \ldots + \mathbb{B}^{s-1}) = 1 + \mathbb{B} + \ldots + \mathbb{B}^{s-1} - \mathbb{B} - \mathbb{B}^2 - \ldots - \mathbb{B}^s = 1 - \mathbb{B}^s$.

The second identity can quickly be proved using the algebra of complex numbers, but can also be obtained using trigonometric identities. The trigonometric way is lengthy and cumbersome and so we will follow the first approach (the reader not acquainted with complex numbers is invited to try the trigonometric proof for $s = 4$). The s solutions to the equation $x^s = 1$ are $\exp(-\mathrm{i}2j\pi/s)$ for $j = 0, 1, \ldots, s-1$, since $\exp(-\mathrm{i}2j\pi/s)^s = \exp(-\mathrm{i}2j\pi) = 1$. Thus, we can factorise $1 - \mathbb{B}^s$ as

$$1 - \mathbb{B}^s = \prod_{j=0}^{s-1} \left[1 - \exp(-\mathrm{i}2j\pi/s)\mathbb{B}\right],$$

and by noticing that, for $j = 1, 2, \ldots, s-1$, we have $\exp(-\mathrm{i}2(s-j)\pi/s) = \exp(\mathrm{i}2j\pi/s)$ we can write (the terms with an asterisk are present only if s is even)

$$
\begin{aligned}
\Delta_s &= 1 - \mathbb{B}^s \\
&= (1-\mathbb{B})(1+\mathbb{B})^* \prod_{j=1}^{\lfloor s/2 \rfloor - 1^*} \big[1 - \exp(-\mathrm{i}2j\pi/s)\mathbb{B}\big]\big[1 - \exp(\mathrm{i}2j\pi/s)\mathbb{B}\big] \\
&= (1-\mathbb{B})(1+\mathbb{B})^* \prod_{j=1}^{\lfloor s/2 \rfloor - 1^*} \big[1 - 2\big(\exp(-\mathrm{i}2j\pi/s) + \exp(\mathrm{i}2j\pi/s)\big)\mathbb{B} + \mathbb{B}^2\big] \\
&= (1-\mathbb{B})(1+\mathbb{B})^* \prod_{j=1}^{\lfloor s/2 \rfloor - 1^*} \big[1 - 2\cos(2j\pi/s)\mathbb{B} + \mathbb{B}^2\big] \\
&= \prod_{j=0}^{\lfloor s/2 \rfloor} \mathbb{T}(2j\pi/s).
\end{aligned}
$$

The third identity is an immediate consequence of the first two.

Reduced ARMA form. The reduced form of the sum of the stochastic dummy form of the seasonal component over s consecutive observations is, by definition, the white noise ω_t. For the trigonometric form we can exploit the representation (3.10), which we report here for $\gamma_{j,t}$:

$$
\gamma_{j,t} = \begin{cases} \frac{1}{\mathbb{T}(2j\pi/s)}\nu_{j,t} & \text{for } j = 1, 2, \ldots, \lfloor s/2 \rfloor \text{ and } j \neq n/2, \\ \frac{1}{1+\mathbb{B}}\omega_{j,t} & \text{for } j = n/2. \end{cases}
$$

where for notational compactness we have set $\nu_{j,t}$ equal to the MA(1) process $\big[\omega_{j,t} - \cos(2j\pi/s)\omega_{j,t-1} + \sin(2j\pi/s)\omega^*_{j,t-1}\big]$. When s is even, for notational coherence, let us also set $\nu_{n/2,t} = \omega_{n/2,t}$, which is a white noise process. Thus, the trigonometric seasonal component is given by

$$
\gamma_t = \sum_{j=1}^{\lfloor s/2 \rfloor} \frac{1}{\mathbb{T}(2j\pi/s)} \nu_{j,t},
$$

and by multiplying both sides of the equal sign times $\mathbb{S}_s$ we obtain

$$
\mathbb{S}\gamma_t = \sum_{j=1}^{\lfloor s/2 \rfloor} \chi_{j,t}
$$

where we have set

$$
\chi_{j,t} = \left(\prod_{k=1, k\neq j}^{\lfloor s/2 \rfloor} \mathbb{T}(2j\pi/s) \right) \nu_{j,t}.
$$

For $j \neq s/2$, the highest exponent of $\mathbb{B}$ in the operator in parenthesis is $s - 3$

and $\nu_{j,t}$ is a MA(1) process, thus, $\chi_{j,t}$ is a MA($s-2$) process: in fact, for opportunely chosen coefficients we can write

$$\chi_{j,t} = (1 + \psi_1 \mathbb{B} + \ldots + \psi_{s-2}\mathbb{B}^{s-3})(1 + \theta\mathbb{B})\varepsilon_t = (1 + \phi_1\mathbb{B} + \ldots + \phi_{s-2}\mathbb{B}^{s-2})\varepsilon_t.$$

When s is even, if $j = s/2$ the highest exponent of $\mathbb{B}$ in the operator in parenthesis is $s-2$ and $\nu_{n/2,t}$ equals $\omega_{n/2,t}$ which is a white noise, and so the maximum exponent is, again, $s-2$. So, we can conclude that $\mathbb{S}_s\gamma_t$ is a MA($s-2$) process.

Prediction. The derivation of the general prediction formula and its MSE for the stochastic dummy seasonal form is simple but very long (and tedious) to write down, so we leave it as an exercise for a fixed s (for example $s = 4$).

The same derivation for the trigonometric form can be easily derived by writing the $t+h$-th observation of the stochastic cycle as in equation (3.9)

$$\boldsymbol{\gamma}_{j,t+h} = \mathbf{R}(h \cdot 2j\pi/s)\boldsymbol{\gamma}_{j,t} + \sum_{k=0}^{h-1} \mathbf{R}(k \cdot 2j\pi/s)\boldsymbol{\omega}_{j,t-k},$$

and recalling that $\mathbf{R}(k \cdot 2j\pi/s)\mathbf{R}(k \cdot 2j\pi/s)^\top = \mathbf{I}$. Indeed, by taking the expectation conditionally on $\boldsymbol{\gamma}_{j,t}$ we obtain $\mathbf{R}(h \cdot 2j\pi/s)\boldsymbol{\gamma}_{j,t}$ and its MSE is

$$\mathbb{E}\left[\left(\sum_{k=0}^{h-1} \mathbf{R}(k \cdot 2j\pi/s)\boldsymbol{\omega}_{j,t-k}\right)\left(\sum_{k=0}^{h-1} \mathbf{R}(k \cdot 2j\pi/s)\boldsymbol{\omega}_{j,t-k}\right)^\top\right] = h\sigma_\omega^2\mathbf{I}.$$

The prediction and MSE formulae in the theorem derive from the fact that γ_{t+h} is the sum of the independent random quantities $\gamma_{j,t+h}$ (first elements of $\boldsymbol{\gamma}_{j,t+h}$) for $j = 1, \ldots, \lfloor s/2 \rfloor$.

□

Chapter 4

Regressors and Interventions

The UCM seen in Chapter 3 can be easily enriched to account for the effect of observable explanatory variables, or *regressors*. In particular, the regressors can be inserted in the equation that relates the dependent variable with the unobservable components (observation equation),

$$Y_t = \mu_t + \psi_t + \gamma_t + \boldsymbol{\delta}^\top \boldsymbol{X}_t + \varepsilon_t, \tag{4.1}$$

or they can be put in the equations defining the components, for example the trend,

$$\mu_t = \beta_{t-1} + \mu_{t-1} + \boldsymbol{\delta}^\top \boldsymbol{X}_t + \eta_t, \tag{4.2}$$

where in both equations $\boldsymbol{X}_t$ is the vector of explanatory variables at time t and $\boldsymbol{\delta}$ a vector of coefficients.

In the first case we speak of *static regression* since, for a given time point t, the regressors in $\boldsymbol{X}_t$ affect Y_t only at the same time point t. The second case, instead, is a particular declination of the *dynamic regression* model (also known as *transfer function* model) in which $\boldsymbol{X}_t$ affects $Y_t, Y_{t+1}, Y_{t+2}, \ldots$. In fact, μ_{t+1} depends on μ_t which is affected by $\boldsymbol{X}_t$, and, of course, the same reasoning can be carried on also for the time points $t+2$, $t+3$, etc., proving that the regressors at time t have an effect on all present and future values of μ and, thus, of Y.

In both equations (4.1) and (4.2) the regression coefficients can be made time-varying by defining their deterministic or stochastic evolution. Time varying regression coefficients are generally used in static regressions, but in principle they can be implemented without much effort also in dynamic regressions, in case a particular model should need them.

4.1 Static regression

It is often necessary in time series analysis to supplement the information contained in the past of the time series with that of explanatory variables (regressors) related to the variable under analysis. Sometimes these regressors are actual measurements carried out by companies (e.g., revenues), statistical institutions (e.g., GDP) or instruments (e.g., temperature), while sometimes they are built by the analyst to model outliers, temporary changes, structural breaks, etc.

Under this aspect working in the UCM framework is not too dissimilar from using a linear regression model, but some problems that arise in nonstationary time series regression (cf. Granger and Newbold, 1974, 1977; Phillips, 1986) can be easily solved by using nonstationary unobserved components such as the local linear trend. For instance, let us assume that our data are generated by the model

$$Y_t = \mu_t + \delta X_t,$$

where μ_t and X_t are independent random walks, but only X_t can be observed. In this case regressing Y_t on X_t using ordinary least squares (OLS) would lead to invalid inference because the regression error sequence (i.e., $Y_t - \delta X_t$) is a random walk and all the classical assumptions of the linear model are violated[1]. The correct approach to follow would be taking the difference of both Y_t and X_t, and estimating the coefficient δ by regressing ∇Y_t on ∇X_t. If in the model a mix of unobservable and observable stationary and non-stationary components is present the regression (+ ARMA) approach becomes even more problematic as it is easy to think of the relations among variables in levels, while it is much more involved to reconsider the same relations after a mix of simple and seasonal differences.

Example 4.1 (UCM vs. regression with ARMA errors).
We generated a sample path of 100 observations for the model

$$\begin{aligned} Y_t &= \mu_t + \delta X_t + \varepsilon_t \\ \mu_t &= \mu_{t-1} + \eta_t \\ X_t &= X_{t-1} + \zeta_t, \end{aligned}$$

with $\delta = 1$, and ε_t, η_t, ζ_t mutually uncorrelated standard normal white noises. We assumed that only Y_t and X_t were observable and estimated δ using an UCM and a regression + ARMA approach. Notice that by taking the first difference of both sides of the first equation above we can write

$$\nabla Y_t = \eta_t + \delta \nabla X_t + \nabla \varepsilon_t,$$

which is a regression model with MA(1) errors, in fact the sum of the white noise process η_t and of the (noninvertible) MA(1) process $\nabla \varepsilon_t$ is a (invertible) MA(1) process.

The estimates[2] using the two approaches are virtually the same, $\hat{\delta}_{UCM} = 0.952$, $\hat{\delta}_{REG} = 0.954$, and so are their standard errors (resp. 0.163 and 0.177). The UCM approach has some advantages: it does not

[1] Recall that statistical packages estimate regression coefficients, their standard errors and carry out tests on linear hypotheses under the assumption that regression errors are mean-zero, homoskedastic, serially uncorrelated and uncorrelated with the regressors.

[2] The random sample paths and the estimates have been computed using EViews 6.

require differentiations, the model is built by adding observable and unobservable components with direct interpretation, which is a very common way people tend to think of time series; it provides inference also on the unobservable components as illustrated in Figure 4.1.

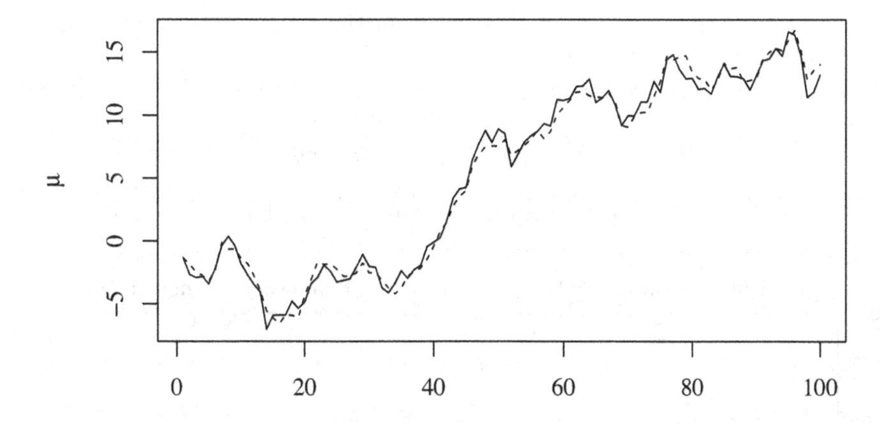

Figure 4.1 *Actual sample path of the unobservable component μ_t (solid line) and estimate (dashed line) by UCM.*

It is common in analysing time series, and particularly economic time series, to find abrupt changes in the data due to exogenous facts (e.g., strikes, regulatory changes, important sport events, moving holidays, etc.). These abnormal observations are generally named *outliers* and can have serious effects on the estimates and forecasts if not properly modelled. The static regressors used to adapt the model to fit outliers are defined below.

Additive outlier (AO) The AO is an abrupt change that involves only a single observation at time point t_0 and can be modelled using the variable

$$AO_t = \begin{cases} 1 & \text{if } t = t_0, \\ 0 & \text{otherwise.} \end{cases}$$

Temporary change (TC) A TC is an abrupt change lasting few (usually consecutive) time points, say $\Omega_0 = \{t_0, t_1, \ldots, t_m\}$ and can be modelled using the variable

$$TC_t = \begin{cases} 1 & \text{if } t \in \Omega_0, \\ 0 & \text{otherwise.} \end{cases}$$

Level shift (LS) A LS (sometimes also referred to as *structural break*) is an abrupt change that involves the time series starting from the time point t_0, and can be modelled using the variable

$$LS_t = \begin{cases} 1 & \text{if } t \geq t_0, \\ 0 & \text{otherwise.} \end{cases}$$

Slope shift (SS) An SS (sometimes referred to as *ramp*) is an abrupt change in the slope of a time series starting from the time point t_0, and can be modelled using the variable

$$SS_t = \begin{cases} t - t_0 + 1 & \text{if } t \geq t_0, \\ 0 & \text{otherwise.} \end{cases}$$

Figure 4.2 depicts the four variables defined above for 11 time points.

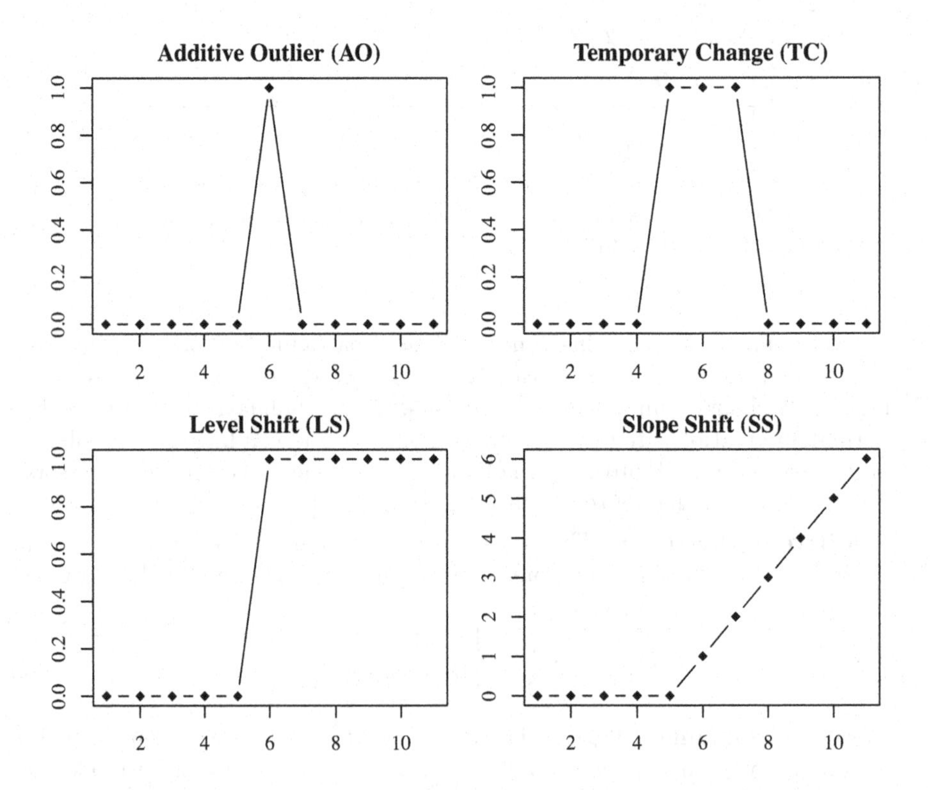

Figure 4.2 *Variables used to fit abrupt changes in time series.*

Example 4.2 (Shift in the Nile's flow).
The time series *Nile* contains yearly measurements of the annual flow of the river Nile at Ashwan in the period 1871-1970. We fitted two UCM: i) a local linear trend plus noise, ii) a local linear trend plus noise with

a levels shift starting with the year 1899. Indeed, as discussed by Cobb (1978), there are meteorological reasons that suggest a drastic drop in the river flow starting from about that year.

The estimated trend components under the two different models are depicted in Figure 4.3. In the first model the estimated variance of

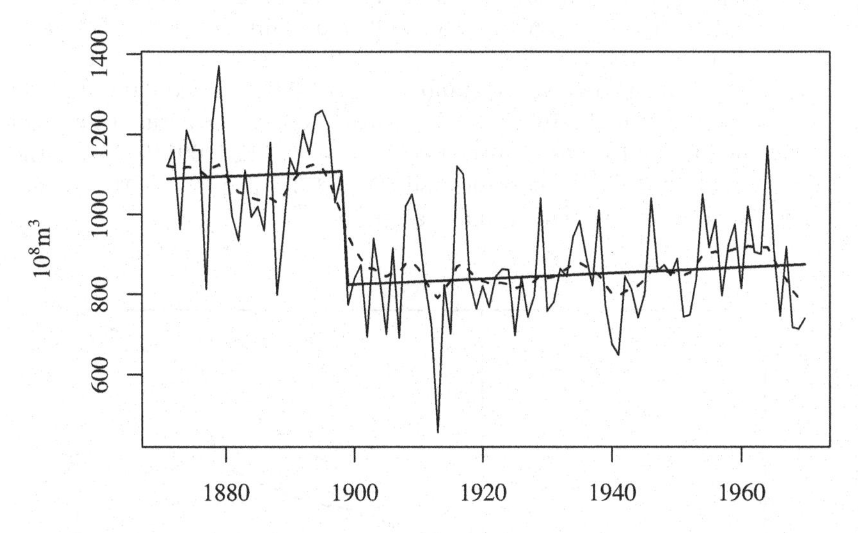

Figure 4.3 *Annual flow of the Nile river (thin line) with estimated trends under the hypotheses of no intervention (dashed line) and level shift starting from 1899 (thick line).*

the slope disturbance is zero ($\sigma_\zeta^2 = 0$) and, thus, the trend component reduces to a random walk with a slightly negative drift ($\beta = -3.4$). When we add the level shift both variances of the local linear trend component have zero estimates ($\sigma_\eta^2 = \sigma_\zeta^2 = 0$) and the trend reduces to a deterministic linear trend with a slightly positive drift ($\beta = 0.7$) and a negative shift of some $283 \cdot 10^8 m^3$ starting from 1899.

Example 4.3 (Shift in the inflation of Italian restaurants).
In the year 2002 many European countries switched from their national currencies to the common Euro currency. At that time many feared that, because of the need for price rounding, the level of prices would have suddenly increased and that this fact should have induced a higher level of inflation in the subsequent years.

Although the consumer price indexes (CPI) did not show any significant increase of the level and increment of prices in the year 2002, there are some sectors of the economy for which consumers tend to find noticeable increases in prices. In Italy, in particular, many consumers noted significant increases in the prices at restaurants.

We built a model for the Italian CPI of the sector *Restaurants* in which a level shift and a slope shift are superimposed to a UCM composed by local linear, seasonal component, stochastic cycle and measurement error:

$$Y_t = \mu_t + \gamma_t + \psi_t + \delta_1 LS_t + \delta_2 SS_t + \varepsilon_t,$$

with Y_t set equal to the natural log of the CPI for the restaurant industry, level shift date set to January 2002 and slope shift date set to February 2002. The estimates of the regression coefficients are $\delta_1 = 0.00165$ and $\delta_2 = 0.00110$ with respective standard errors 0.00135 and 0.00066: only δ_2 is significant at the 10% level. Figure 4.4 depicts the actual trend and the trend net of the effect of the two regressors. The 0.00110 shift in the slope of the log-CPI time series induces a 1.3% increase in the annual inflation of the "average" Italian restaurant.

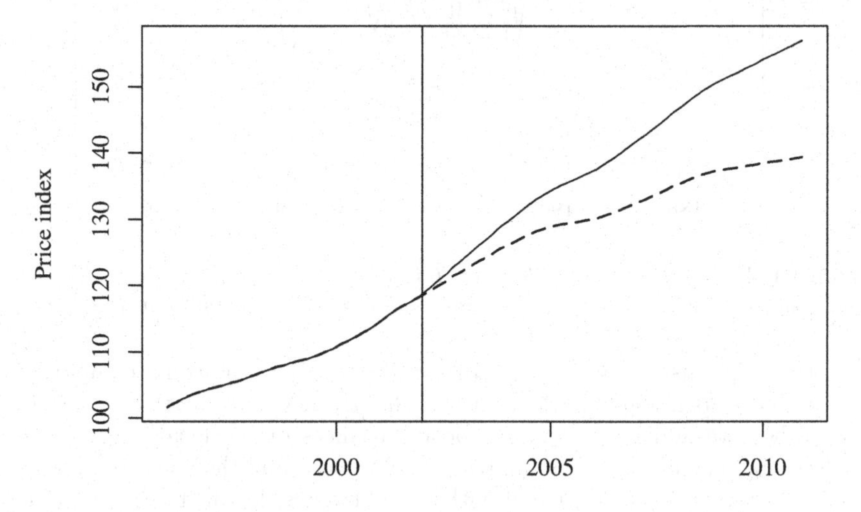

Figure 4.4 *Trend extracted from the Italian CPI of the restaurant industry with and without the effect of the introduction of the Euro currency in January 2002.*

Another set of artificial regressors often used in time series analysis is that of sinusoids. As already seen in Section 3.3, any periodic function of period s defined on integers can be written as linear combination of sinusoids: let $f : \mathbb{Z} \mapsto \mathbb{R}$ such that $f(t) = f(t + s)$, then

$$f(t) = a_0 + \sum_{j=1}^{\lfloor s/2 \rfloor} a_j \cos(2jt\pi/s) + b_j \sin(2jt\pi/s).$$

Furthermore, if the function f is smooth, then it can be well approximated with a reduced number of sinusoids and the summation in the equation can be stopped at $m \ll s/2$.

As already seen in Section 3.4 sinusoids are particularly suited to model smooth seasonal components.

Example 4.4 (Within-year periodicity in electricity demand).
The hourly demand for electricity is affected by different kinds of seasonal behaviours. There is a within-day periodic component (i.e., the period is 24 hours) due to the different needs for energy during the different times of the day and of the night. There is a within-week periodic component mainly due to the difference in the use of electricity on workdays with respect to Saturdays and Sundays. Finally, there is a within-year periodicity due to the different consumption of electricity in the four seasons.

Let us concentrate on the within-year seasonality and try to assess how many sinusoids are necessary for isolating and approximating this particular seasonal component. This is a preliminary analysis that can help determine which regressors are to be included in a model for forecasting the electricity demand.

In order to obtain a time series not affected by within-day periodicities we work on the total daily quantity (MWh). For each year in our sample, which includes the full years 2005–2013, we subtract the yearly mean and then average each of the 365 days of the year over the 9 years, obtaining the average spread of each day of the year with respect to the yearly mean. Since the so-obtained time series is still affected by the within-year periodicity, we apply a 7-day (centred) moving average and obtain a time series that can be considered a decent approximation of the smooth within-year periodic component of electricity demand.

We regressed this 365-day time series on 15 and 30 sinusoids obtaining the approximations depicted in Figure 4.5. The approximation with 30 sinusoids is virtually identical to the real component, but if abrupt changes such as Christmas, Easter and August vacations were properly modelled using temporary change variables, the number of sinusoids needed to approximate the within-year periodic component would certainly reduce significantly.

Very often, in real applications, a combination of sinusoids and *dummy variables* like those in Figure 4.2 has to be implemented. In fact, while a small number of sinusoids can easily approximate smooth periodic components, dummy variables are ideal for abrupt changes, which instead would require a large number of sinusoids to be modelled.

4.2 Regressors in components and dynamic regression

The regressors can also enter the equations that define the evolution of the unobserved components. Moreover, new unobserved components can be defined to model dynamic effects (i.e., effects that persist in time) that each regressor can have on the response variable.

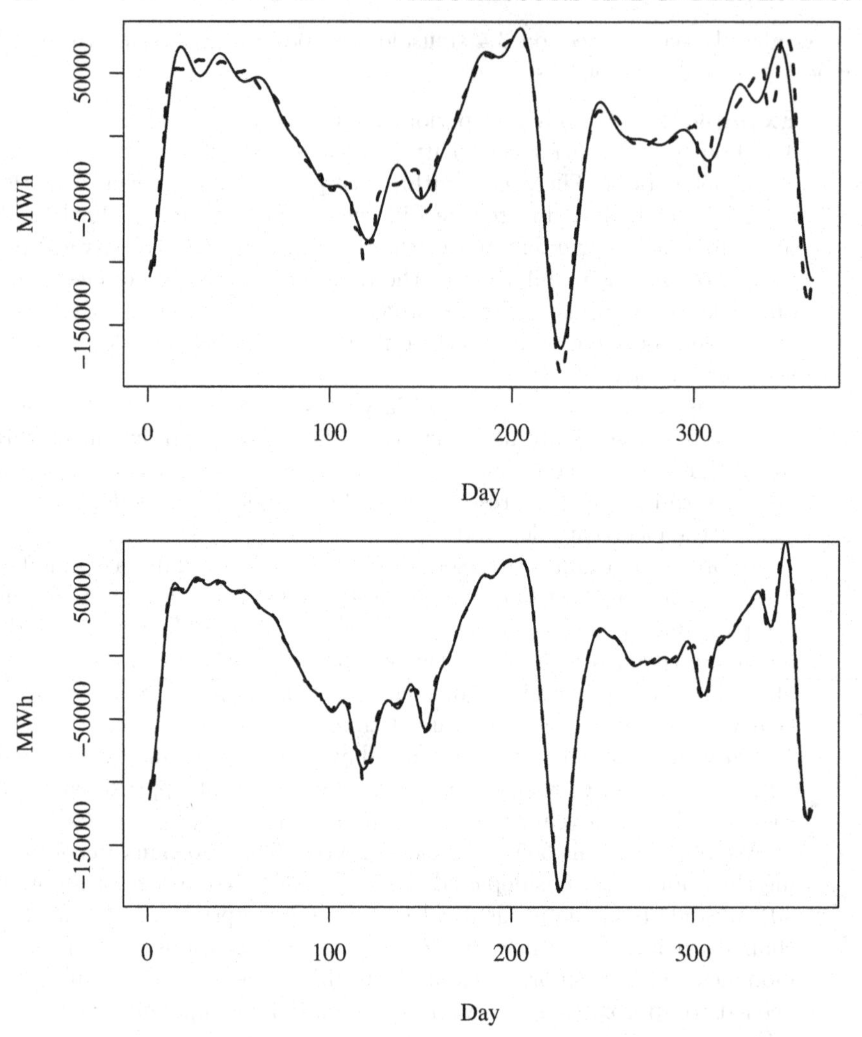

Figure 4.5 *Daily time series of the within-year periodic component of the mean-deviation of the quantity of electricity exchanged in the Italian Power Exchange (dashed line) and approximation with 15 sinusoids (line in top graph) and 30 sinusoids (line in bottom graph).*

In UCM with local linear trend the level shift and the slope shift can be obtained by introducing a dummy variable, respectively, in the level equation and in the slope equation. Let D_t be a dummy variable that takes the value one at time t_0 and zero otherwise. Then, by modifying the local linear trend equations (3.2) as follows

$$\mu_t = \mu_{t-1} + \beta_{t-1} + \delta D_t + \eta_t$$
$$\beta_t = \beta_{t-1} + \zeta_t,$$

the level μ_t undergoes an abrupt shift of δ units at time $t = t_0$. If the same dummy variable is put in the slope equation,

$$\mu_t = \mu_{t-1} + \beta_{t-1} + \eta_t$$
$$\beta_t = \beta_{t-1} + \delta D_t + \zeta_t,$$

the slope β_t is affected by a shift of δ units at time $t = t_0$ and, thus, the expected increment of μ_t increases by δ units starting at time $t = t_0 + 1$. Examples 4.2 and 4.3 can be replicated exactly using this approach.

There are cases in which an event causes an abrupt change in the shape of the seasonal pattern. In a UCM this situation can be easily modelled whatever type of seasonal component has been chosen (i.e., stochastic dummies or sinusoids). If s is the seasonal period (e.g., $s = 4$ for quarterly time series and $s = 12$ for monthly data), then $s - 1$ coefficients have to be estimated. In the stochastic sinusoid case the abrupt change can be modelled by adapting equation (3.13) as

$$\gamma_t = \sum_{j=2}^{\lfloor s/2 \rfloor} \gamma_{j,t},$$

$$\begin{bmatrix} \gamma_{j,t} \\ \gamma^*_{j,t} \end{bmatrix} = \begin{bmatrix} \cos(2j\pi/s) & \sin(2j\pi/s) \\ -\sin(2j\pi/s) & \cos(2j\pi/s) \end{bmatrix} \begin{bmatrix} \gamma_{j,t-1} \\ \gamma^*_{j,t-1} \end{bmatrix} + \begin{bmatrix} \delta_j \\ \delta^*_j \end{bmatrix} D_t + \begin{bmatrix} \omega_{j,t} \\ \omega^*_{j,t} \end{bmatrix},$$

for $j = 1, \ldots, \lfloor s/2 \rfloor$, where D_t is a dummy variable which takes the value one on the event date and zero otherwise, and δ_j and δ^*_j are coefficients to be estimated, which capture the jumps in amplitude and angle of the sinusoids.

In the stochastic dummy case, equation (3.12) can be adapted as

$$\gamma_t = \sum_{j=1}^{s-1} (-\gamma_{t-j} + \delta_j D_{j,t}) + \omega_t,$$

where the $s - 1$ dummies $D_{j,t}$ are defined as

$$D_{j,t} = \begin{cases} 1 & \text{if } t = t_0 + j - 1 \\ 0 & \text{otherwise.} \end{cases}$$

In this case a sequence of dummy variables is needed for letting each "season" update after the abrupt change.

An alternative and more parsimonious way to obtain the same results for both types of seasonal models can be obtained by letting dummy variables modify the values of the white noise variances. In fact, an abrupt change in the seasonal component can be seen as the result of an increased variance of the error term driving the component. For the trigonometric seasonal model we can use

$$\gamma_t = \sum_{j=2}^{\lfloor s/2 \rfloor} \gamma_{j,t},$$

$$\begin{bmatrix} \gamma_{j,t} \\ \gamma_{j,t}^* \end{bmatrix} = \begin{bmatrix} \cos(2j\pi/s) & \sin(2j\pi/s) \\ -\sin(2j\pi/s) & \cos(2j\pi/s) \end{bmatrix} \begin{bmatrix} \gamma_{j,t-1} \\ \gamma_{j,t-1}^* \end{bmatrix} + \begin{bmatrix} \omega_{j,t} \\ \omega_{j,t}^* \end{bmatrix},$$

$$\begin{bmatrix} \omega_{j,t} \\ \omega_{j,t}^* \end{bmatrix} \sim \mathrm{WN}\Big(\mathbf{0}, (\sigma_\omega^2 + \tau^2 D_t)\boldsymbol{I}_2\Big),$$

where τ is a parameter to estimate and D_t takes value one at time t_0 and zero otherwise. For the stochastic dummies model we can implement the following modification

$$\gamma_t = -\sum_{j=1}^{s-1} \gamma_{t-j} + \omega_t,$$

$$\omega_t \sim \mathrm{WN}(0, \sigma_\omega^2 + \tau^2 \tilde{D}_t),$$

where τ is a parameter to estimate and $\tilde{D}_t$ takes value one at times $\{t_0, t_0 + 1, \ldots, t_0 + s - 1\}$ and zero otherwise.

Example 4.5 (English weddings).
Until the end of 1968 in the United Kingdom for fiscal reasons it was financially convenient to get married in the first quarter of the year, while starting with 1969 a new law, already announced in 1968, made the taxation scheme neutral to the choice of the season of the wedding.

The change in the seasonal pattern is evident in Figure 4.6, where the peak of the first quarter of each year disappears starting with 1969.

We fitted a simple model with random walk plus seasonal component plus white noise on the time series over the period 1947–2011. We used both types of seasonal components (dummy and sinusoids) and implemented all the abrupt change models shown above. The seasonal components extracted using the four different models are virtually identical and so Figure 4.7 depicts just one of them for the restricted period 1965–1972.

The abrupt change in the seasonal pattern has been well picked up by the adapted seasonal components (see Figure 4.7). If a standard seasonal component without interventions were fit to the same time series, the change in seasonal pattern would eventually be "learned" by the model, but this would take a longer time and the variance of the disturbance sequence $\{\omega_t\}$ would be overestimated causing an overfit of the evolving seasonal pattern.

Regressors can also be made dynamic by defining state variables that react

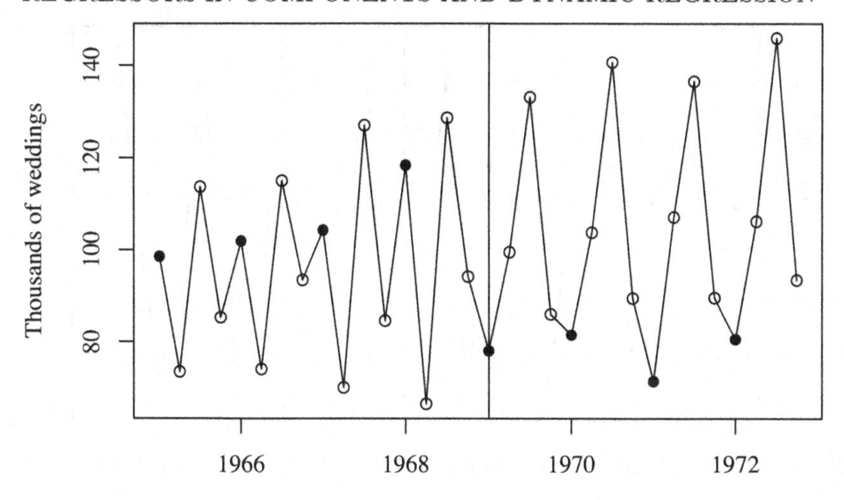

Figure 4.6 *Number of weddings in England and Wales per quarter. The black points denote the first quarter of each year.*

to them dynamically using ARMA-like filters. Let $\{X_t\}$ be the regressor time series, then in the state variable defined by

$$W_t = \delta_1 W_{t-1} + \ldots + \delta_r W_{t-r} + \omega_0 X_t + \omega_1 X_{t-1} + \ldots \omega_{t-s} X_{t-s} \qquad (4.3)$$

the regressor at time t, X_t, affects the values $W_t, W_{t+1}, W_{t+2}, \ldots$, and the type of response is determined by the values of the δ and ω coefficients. Equation (4.3) is generally referred to as *transfer function* (Box and Jenkins, 1976) or *dynamic regression* (Pankratz, 1991) model, and using polynomials in the lag operator can be synthetically rewritten as

$$\delta_r(\mathbb{B}) W_t = \omega_s(\mathbb{B}) X_t \quad \text{or} \quad W_t = \frac{\omega_s(\mathbb{B})}{\delta_r(\mathbb{B})} X_t,$$

with $\delta_r(\mathbb{B}) = 1 - \delta_1 \mathbb{B}^1 - \ldots - \delta_r \mathbb{B}^r$ and $\omega_s(\mathbb{B}) = \omega_0 + \omega_1 \mathbb{B}^1 + \ldots + \omega_s \mathbb{B}^s$.

The coefficient ω_0 measures the immediate impact that X_t has on W_t, while the dynamic effect of X_t on the values of $W_{t+1}, W_{t+2}, \ldots$ depends on all other ω- and δ-coefficients. In particular, if the δ-coefficients are zero, the coefficient ω_j is the value by which the variable X_t is multiplied to affect W_{t+j}, and this is usually referred to as *distributed lag model*. Much more interesting is the effect of the operator $\delta_r(\mathbb{B})$, which is responsible for long-run effects of the series $\{X_t\}$ on $\{W_t\}$. The best way to visualise these effects is by means of the *impulse response function* (IRF): assume that for $t < 0$ the series $\{X_t\}$ and $\{W_t\}$ are constantly equal to zero, while $X_0 = 1$ and $X_t = 0$ for $t > 0$; the impulse response function is just the set of values taken by $\{W_t\}$ as a function of its indexes $t = 0, 1, 2, \ldots$.

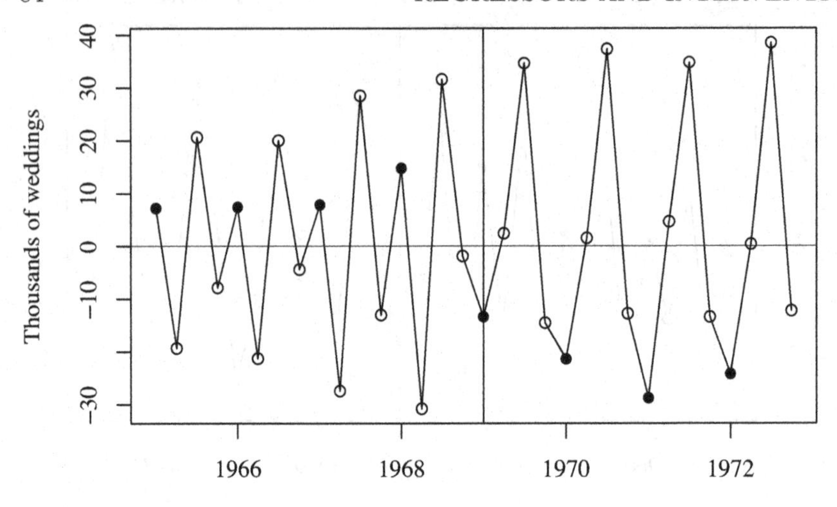

Figure 4.7 *Seasonal component for the quarterly series representing the number of weddings in England and Wales. The black points denote the first quarter of each year.*

The long-run behaviour of the IRF and, thus, of the effect of $\{X_t\}$ on $\{W_t\}$ depends on the solutions of the r-th order equation in z, $\delta_r(z) = 0$. If all the roots of this equation are outside the unit circle (i.e., $|z_j| > 1$ for $j = 1, \ldots, r$ such that $\delta_r(z_j) = 0$), then the IRF converge to zero as t diverges. If one or more roots is on the unit circle (i.e., $|Z_j| = 1$ for some z_j) and all the other are outside, then the IRF converges to a value different than zero or oscillates within a finite range. If at least one root is inside the unit circle (i.e., $|z_j| < 1$ for some z_j), then the IRF diverges. Moreover, if some of the roots are complex (with non-zero imaginary parts), then the IRF shows a sinusoidal behaviour.

Generally, most real-world situations can be coped with via δ polynomials of order $r = 1$ or $r = 2$. Figure 4.8 depicts the IRFs for two transfer functions models of orders $r = 1, s = 0$ and $r = 2, s = 0$. In both cases the δ-polynomial roots are outside the unit circle and in the second model the two roots are complex conjugate. It appears clear that the first model is good for fitting an effect that dies out monotonically, while the second one matches an effect that dies out sinusoidally. While there can be found hundreds of examples for the first IRF pattern (e.g., effect of advertisement on sales, see also Example 4.6), the second IRF shape is encountered in situations such as government incentives to scrap old cars and buy new less polluting vehicles, or sales promotions on products that after a certain time have to be replaced, like toothpastes, etc. Indeed, in these specific cases the incentives/promotions concentrate sales in a given period of time causing a subsequent drop in sales in the following years (vehicles) or weeks (toothpaste). At the same time it is likely that another (lower) positive peak of sales will be recorded after a pe-

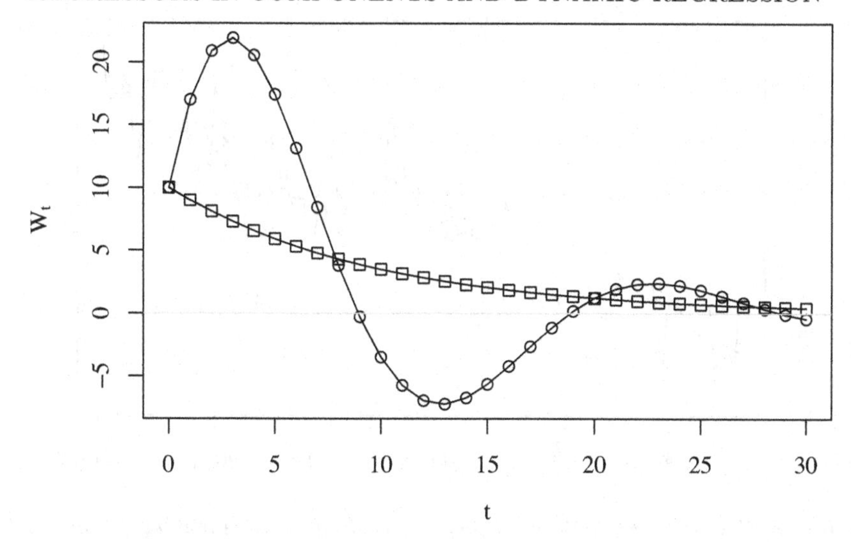

Figure 4.8 *Impulse response functions of the transfer function models* $W_t = 0.9W_{t-1} + 10X_t$ *(squares) and* $W_t = 1.7W_{t-1} - 0.8W_{t-2} + 10X_t$ *(circles).*

riod as long as the average life of the product (years for vehicles and weeks for toothpastes).

Example 4.6 (Effect on 9/11 and Iraq war outbreak on airline passengers from and to Australia).
At the beginning of the new millennium two events had a negative effect on the airline market: the terrorist attack of September 11th, 2001, and the beginning of the Iraq war on March 20th, 2003.

In this example we want to assess the effect of these two events on the number of passengers flying to and from Australia[3]. We fit an UCM to the log of the number of passengers with local linear trend, stochastic dummy seasonal component and two transfer function models of the type

$$W_t = \delta W_{t-1} + \omega_0 X_t,$$

where X_t is a dummy variable that takes 1 in the month following the event and zero otherwise.

The estimated values of the transfer function coefficients are $\omega_0 = -0.10, \delta = 0.25$ for the 9/11 events and $\omega_0 = -0.15, \delta = 0.90$ for the Iraq war start. Figure 4.9 depicts (part of) the passenger time series together with the estimated local linear trend component and the trend compo-

[3]The monthly time series starts on Jan 1985 and ends on Dec 2013 and was published on the site of the Department of Infrastructure and Regional Development of the Australian government (www.bitre.gov.au/publications/).

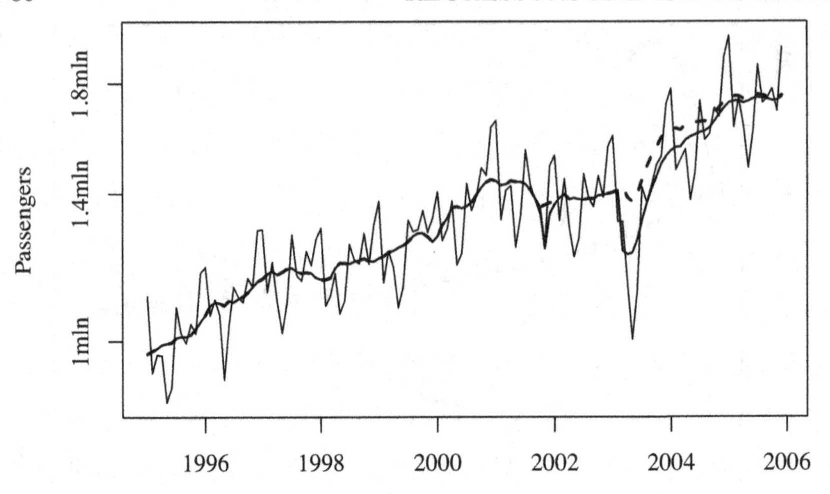

Figure 4.9 *Number of monthly passengers flying from or to Australia; actual level (thick line) and level in absence of the 9/11 events and Iraq war (thick dashed line).*

nent combined with the transfer function effects. The difference between these two trend curves represents the cumulative effect of the events on the time series. This effect amounts to -175 thousand passengers for 9/11 and to -2 million passengers for the Iraq war start.

Figure 4.10 depicts the IRFs of the two events. Since we modelled the log of the time series, the figures on the y-axis approximate relative changes (i.e., -0.15 means 15% fewer passengers per month). The Iraq war outbreak has a stronger immediate impact, which is also much more persistent than the effect of 9/11.

When the input variable, X_t, of model (4.3) is a dummy variable, then the transfer function analysis is usually referred to as *intervention analysis*. The reader interested in a higher level of details on transfer function or intervention analysis should refer to the classical books by Box and Jenkins (1976, Part 3) and Pankratz (1991) or to the more recent Yaffee and McGee (2000, Chapters 8-9).

4.3 Regression with time-varying coefficients

In a UCM model with regressors, such as that in equation (4.1) for instance, the regression coefficients can also be let vary over time as unobserved components:

$$Y_t = \mu_t + \psi_t + \gamma_t + \delta_t X_t + \varepsilon_t,$$

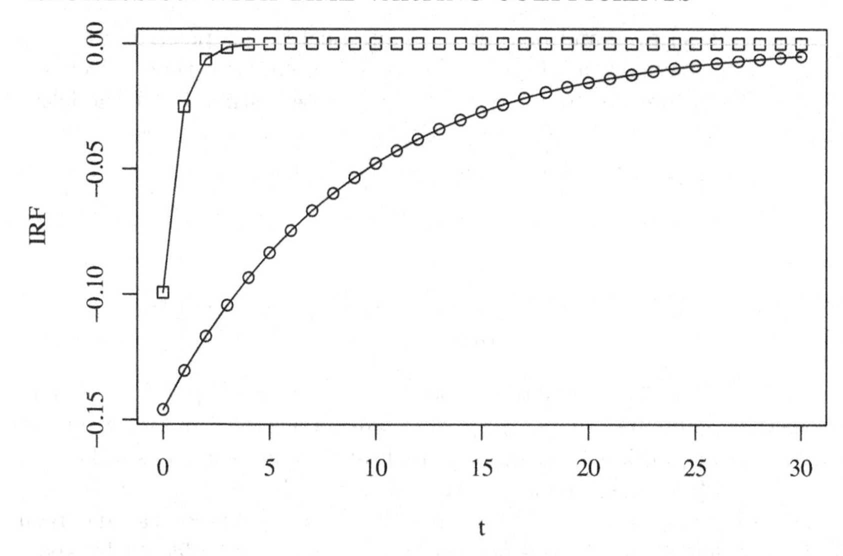

Figure 4.10 *Impulse response functions of the two transfer function models estimated on the log of the Australia air passengers time series: 9/11 events effect (squares), Iraq war outbreak effect (circles).*

where for simplicity just one regressor coefficient (δ_t) is present. If there are good reasons to believe that the regression coefficients have a fixed mean, then they can be modelled as low-order AR process – typically AR(1) – with roots outside the unit circle. Otherwise, the coefficients are generally modelled as random walks or integrated random walks, which make the coefficients evolve very smoothly over time. Of course, if for some reason some regression coefficients should follow the business cycle or, possibly, the seasonal periodicity, then they could be modelled using the respective processes seen in sections 3.3 and 3.4.

Example 4.7 (Okun's law).
Okun's original statement of his law (actually empirical evidence rather than law) affirms that 2% increase in output (real GDP) corresponds to a 1% decline in the rate of cyclical unemployment (Okun, 1962). The typical way applied economists verify this statement is through the regression:

$$\nabla Y_t = \alpha + \beta \frac{\nabla X_t}{X_{t-1}} + \varepsilon_t,$$

where Y_t is the unemployment rate and X_t the real GDP. If we fit this regression on U.S. quarterly time series spanning from 1950Q1 to 2014Q2 by ordinary least squares, we obtain an estimate of $\beta = -0.38$, which is not that far away from the value implied by Okun's law (i.e., -0.5).

However, there are good reasons to believe that this relationship is not stable in time and possibly asymmetrical (i.e., a positive growth of the GDP may have a different effect on unemployment than a decline in the GPD). In order to answer these doubts, we estimate two models:

$$\nabla_4 Y_t = \alpha + (\mu_t + \psi_t)\frac{\nabla_4 X_t}{X_{t-4}} + \varepsilon_t$$

and

$$\nabla_4 Y_t = \alpha + (\mu_{+,t} + \psi_t)\left[\frac{\nabla_4 X_t}{X_{t-4}}\right]^+ + (\mu_{-,t} + \psi_t)\left[\frac{\nabla_4 X_t}{X_{t-4}}\right]^- + \varepsilon_t$$

where $[\cdot]^+$ and $[\cdot]^-$ denote the positive and negative part[4] of their respective contents and μ_t, $\mu_{+,t}$, $\mu_{-,t}$ are integrated random walk processes (see Section 3.2), ψ_t is a stochastic cycle (see Section 3.3) and ε_t is a white noise sequence. We chose to split the β coefficient in a smooth component, μ, affected mostly by the technological and regulatory changes and a business cycle component, ψ, affected by short- and mid-term shocks. Figure 4.11 depicts the smooth component of the

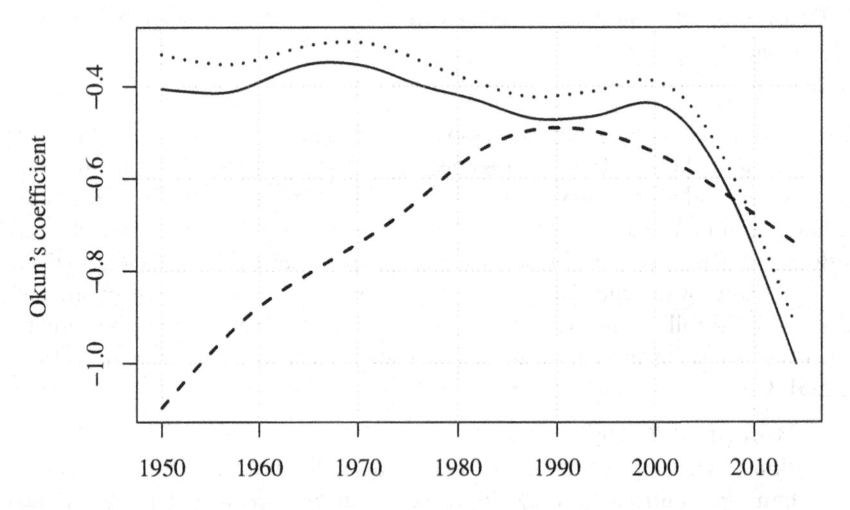

Figure 4.11 *Smooth component of time varying Okun's coefficients for U.S. in the symmetric model (line), and in the asymmetric model, when the GDP increases (dotted line) or declines (dashed line).*

time-varying beta coefficients for the symmetric and for the asymmetric model. The symmetric and positive-growth Okun's coefficients stay around the value −0.4 from 1950 to the end of the '90s and then decline quickly toward −1.0. On the contrary, the Okun's coefficient relative to declines in the GDP grows from −1.0 to −0.5 in the period 1950–1990

[4]i.e., $[x]^+ = \max(x, 0)$ and $[x]^- = \min(x, 0)$.

and then starts to decrease to some -0.7. The fact that the symmetric coefficients and the positive growth coefficient are so similar should not surprise since $\nabla_4 X_t$ is positive 88% of the time. From the asymmetric model coefficients we can conclude that from 1950 to 1990 an output decline had a stronger effect on unemployment than a positive growth, but this spread has gradually reduced in time. In the last 14–15 years in the sample the two coefficients are very close and at the very end the reaction to positive growths become more important than the effect of output declines.

Chapter 5

Estimation

All the models introduced so far and many other linear time series models can be represented in a form called *state space form*. The great advantage of representing time series models in the state space form is the availability of a set of general algorithms, the *Kalman filter* and its relatives, for the estimation by linear prediction of the unobservable components and for the computation of the Gaussian likelihood, which can be numerically maximised to obtain maximum likelihood (ML) estimate of the model's parameters.

It is important that time series practitioners acquire a good familiarity with the state space form, since it is the tool by which all the ingredients of our model (unobserved components, dynamic regressors, etc.) can be transformed into the final dish (the fitted model).

5.1 The state space form

There are different versions and notations of the state space form; the one adopted here is very common in the econometric literature and can be easily translated into most statistical software packages and languages.

The state space form is a system of equations in which one or more observable time series are linearly related to a set of unobservable state variables which evolve as a (possibly time-varying) VAR(1) process.

Definition 5.1 (State space form). Let $\{\boldsymbol{Y}_1, \boldsymbol{Y}_2, \ldots, \boldsymbol{Y}_n\}$ be an observable time series of possibly vector-valued observations, then, the state space form is defined by the following system of equations,

Observation or measurement equation

$$\boldsymbol{Y}_t = \boldsymbol{c}_t + \mathbf{Z}_t \boldsymbol{\alpha}_t + \boldsymbol{\epsilon}_t, \tag{5.1}$$

where $\{\boldsymbol{\epsilon}_t\}$ is a sequence of serially uncorrelated random variables with zero mean vector and covariance matrix $\mathbf{H}_t$.

State or transition equation

$$\boldsymbol{\alpha}_{t+1} = \boldsymbol{d}_t + \mathbf{T}_t \boldsymbol{\alpha}_t + \boldsymbol{\nu}_t, \tag{5.2}$$

where $\{\boldsymbol{\nu}_t\}$ is a sequence of serially uncorrelated random variables with zero mean vector and covariance matrix $\mathbf{Q}_t$, and the following notation and conditions hold:

- $\mathbb{E}[\boldsymbol{\alpha}_1] = \boldsymbol{a}_{1|0}$ and $\mathbb{E}[\boldsymbol{\alpha}_1\boldsymbol{\alpha}_1^\top] = \mathbf{P}_{1|0}$;
- $\mathbb{E}[\boldsymbol{\epsilon}_t\boldsymbol{\nu}_t^\top] = \mathbf{G}_t$ and $\mathbb{E}[\boldsymbol{\epsilon}_s\boldsymbol{\nu}_t^\top] = \mathbf{0}$ for $s \neq t$;
- $\mathbb{E}[\boldsymbol{\alpha}_1\boldsymbol{\epsilon}_t^\top] = \mathbf{0}$ and $\mathbb{E}[\boldsymbol{\alpha}_1\boldsymbol{\nu}_t^\top] = \mathbf{0}$, for $t = 1, 2, \ldots$;
- the system vectors and matrices $\boldsymbol{c}_t$, $\boldsymbol{d}_t$, $\mathbf{Z}_t$, $\mathbf{T}_t$, $\mathbf{H}_t$, $\mathbf{Q}_t$ and $\mathbf{G}_t$ are allowed to vary over time in a deterministic fashion.

If $\boldsymbol{\alpha}_1$ and the random sequences $\{\boldsymbol{\epsilon}_t\}$ and $\{\boldsymbol{\nu}_t\}$ (disturbances) are normally distributed, then so is the time series $\{\boldsymbol{Y}_t\}$, and the system is referred to as *Gaussian state space*.

When all the system vectors and matrices are not time-varying, the state space form is named *time-homogeneous* or *time-invariant* and the time index t is dropped from the constant vectors and matrices. Time-homogeneity is a necessary condition for stationarity, but not sufficient. Indeed, as the next pages will show, UCM with stochastic trends and/or seasonal components have time-invariant state space representations but are clearly non-stationary.

All UCM found in this book and used in practice have measurement errors, $\boldsymbol{\epsilon}_t$, that are uncorrelated with the state disturbances, $\boldsymbol{\nu}_t$, but for the sake of generality we will provide results for general covariance matrices $\mathbf{G}_t$.

So far we have only discussed UCM for univariate time series, for which $\boldsymbol{Y}_t$ reduces to a scalar and the matrix $\mathbf{Z}_t$ to a row vector. Chapter 7 deals with multivariate UCM for which we need the general definition of state space given above.

Our definition of state space form is not the only one used in the literature and implemented in software packages, even though it is the most common in recent works in the fields of econometrics and statistics. Very often the vectors $\boldsymbol{c}_t$ and $\boldsymbol{d}_t$ are dropped since they can be easily absorbed into additional elements in the *state vector* $\boldsymbol{\alpha}_t$. Furthermore, the *future* form of the transition equation given in (5.2) in some works (for instance Harvey, 1989; West and Harrison, 1989) is substituted by the *contemporaneous* form $\boldsymbol{\alpha}_t = \boldsymbol{d}_t + \mathbf{T}_t\boldsymbol{\alpha}_{t-1} + \boldsymbol{\nu}_t$. If the measurement errors are uncorrelated with the state disturbances, and the system matrices time-invariant, then the two forms are equivalent, otherwise they can be made equivalent by opportunely redesigning the system matrices (see Harvey and Proietti, 2005, pages 3-4). Other definitions found in the literature retouch errors and disturbances. For instance, the state disturbance in Harvey (1989) is set equal to $\mathbf{R}_t\boldsymbol{\nu}_t$ so that its covariance matrix changes to $\mathbf{R}_t\mathbf{Q}_t\mathbf{R}_t^\top$. Durbin and Koopman (2001), instead, use just one vector of uncorrelated disturbances $\boldsymbol{\varepsilon}_t$ whose components are then related to the observation and state equations using opportune matrices, $\boldsymbol{\epsilon}_t = \tilde{\mathbf{H}}_t \boldsymbol{\varepsilon}_t$, $\boldsymbol{\nu}_t = \tilde{\mathbf{G}}_t \boldsymbol{\varepsilon}_t$, and the equivalence with our covariance matrices is $\mathbf{H}_t = \tilde{\mathbf{H}}_t \tilde{\mathbf{H}}_t^\top$, $\mathbf{Q}_t = \tilde{\mathbf{G}}_t \tilde{\mathbf{G}}_t^\top$, $\mathbf{G}_t = \tilde{\mathbf{H}}_t \tilde{\mathbf{G}}_t^\top$.

5.2 Models in state space form

We now illustrate how the models presented in this book so far can be stated in state space form.

5.2.1 ARIMA processes in state space form

Even though there are specific algorithms to estimate and forecast ARIMA processes, many software packages use the state space form and the Kalman filter to achieve this goal. Another reason for learning to write ARIMA process in state space form is being able to create hybrid models that mix different kinds of processes.

Let us begin with simple examples and then generalise to any ARIMA model. Notice that state space representations do not need to be unique (i.e., the same correlation structure can be implied by different state space forms). We will provide the simplest form to implement, but alternative ones are possible. A state space representation is said to be *minimal* if there is no equivalent representation with a smaller number of state variables.

Example 5.1 (AR(1) in state space form).
The AR(1) model can be easily set in state space form as follows.

$$\begin{aligned} &\text{Transition equation:} \quad && \alpha_{t+1} = \phi_1 \alpha_t + \nu_t, \\ &\text{Measurement equation:} \quad && Y_t = c + \alpha_t. \end{aligned}$$

Thus, the system matrices and vectors reduce to $\boldsymbol{c}_t = c$, $\boldsymbol{d}_t = 0$, $\mathbf{Z}_t = 1$, $\mathbf{T}_t = \phi$ $\mathbf{H}_t = 0$, $\mathbf{Q}_t = \sigma$ and $\mathbf{G}_t = 0$.

The transition equation defines a zero-mean AR(1) process and the measurement equation just adds a constant (the mean) to the process. No measurement error is present.

The fact that the white noise process ν_t has index t and not $t+1$, as one would expect from the usual definition of the AR(1) process, should not upset the reader. Indeed, without loss of generality, we could define a "new" white noise process as $\varepsilon_t = \nu_{t-1}$ and the equation above would read exactly $\alpha_{t+1} = \phi_1 \alpha_t + \varepsilon_{t+1}$ as in the usual definition of the AR(1) recursion.

The natural choice for the first two moments of the distribution of the initial state α_1, if no better information is available and $|\phi| < 1$, consists in the marginal moments of the AR(1) process: $\mathbb{E}[\alpha_1] = 0$ and $\mathbb{E}[\alpha_1^2] = \sigma^2/(1-\phi^2)$.

Example 5.2 (AR(2) in state space form).
While the state space representation of the AR(1) was trivial, the AR(2)

representation needs a trick with which the reader has to become confident.

$$\text{Transition equation:} \qquad \begin{bmatrix}\alpha_{1,t+1}\\ \alpha_{2,t+1}\end{bmatrix} = \begin{bmatrix}\phi_1 & \phi_2\\ 1 & 0\end{bmatrix}\begin{bmatrix}\alpha_{1,t}\\ \alpha_{2,t}\end{bmatrix} + \begin{bmatrix}\nu_t\\ 0\end{bmatrix}$$

$$\text{Measurement equation:} \qquad Y_t = c + \begin{bmatrix}1 & 0\end{bmatrix}\begin{bmatrix}\alpha_{1,t}\\ \alpha_{2,t}\end{bmatrix},$$

with $\nu_t \sim \text{WN}(0, \sigma^2)$. Thus, the system matrices used in this form are

$$\mathbf{T}_t = \begin{bmatrix}\phi_1 & \phi_2\\ 1 & 0\end{bmatrix}, \quad \mathbf{Z}_t = \begin{bmatrix}1 & 0\end{bmatrix}, \quad \mathbf{Q}_t = \begin{bmatrix}\sigma^2 & 0\\ 0 & 0\end{bmatrix},$$

$\mathbf{H}_t = 0$, $\mathbf{G}_t = \mathbf{0}$.

The trick mentioned above is in the second line of the transition equation that reads $\alpha_{2,t+1} = \alpha_{1,t}$, from which $\alpha_{2,t} = \alpha_{1,t-1}$. Substituting this in the first line of the transition equation we obtain the AR(2) process $\alpha_{1,t+1} = \phi_1\alpha_{1,t} + \phi_2\alpha_{1,t-1} + \nu_t$. In the observation equation, the matrix $\mathbf{Z}_t$ selects the first element of the state vector, which is a zero-mean AR(2) process, and the mean c is then added.

The state space has to be completed with the first two moments of $\boldsymbol{\alpha}_1$, which, if the AR(2) is casual stationary, are given by

$$\mathbb{E}[\boldsymbol{\alpha}_1] = \begin{bmatrix}0\\ 0\end{bmatrix}, \qquad \mathbb{E}[\boldsymbol{\alpha}_1\boldsymbol{\alpha}_1^\top] = \begin{bmatrix}\gamma(0) & \gamma(1)\\ \gamma(1) & \gamma(0)\end{bmatrix},$$

where $\gamma(k)$ is the autocovariance function of the AR(2) process. Indeed, the first element of the state vector $\boldsymbol{\alpha}_1$ is the AR(2) process at time $t = 1$ and the second element is the same process at time $t = 0$.

In the AR(2) case the autocovariance is a not too complicated function of the autoregressive coefficients, but for a general stationary ARMA model $\gamma(k)$ can become rather intricate. Later in this chapter it will be shown how to compute the mean vector and covariance matrix of $\boldsymbol{\alpha}_1$ for any stationary state vector.

Example 5.3 (MA(2) in state space form).
We mentioned above that state space representations are not necessarily unique. Here we show two alternative representations of the same MA(2) process.

In the first representation the transition equation is only used to generate a white noise sequence with two lags and then in the measurement equation the lagged white noises are multiplied by their respective coefficients.

$$\text{Transition equation:} \quad \begin{bmatrix} \alpha_{1,t+1} \\ \alpha_{2,t+1} \\ \alpha_{3,t+1} \end{bmatrix} = \begin{bmatrix} 0 & 0 & 0 \\ 1 & 0 & 0 \\ 0 & 1 & 0 \end{bmatrix} \begin{bmatrix} \alpha_{1,t} \\ \alpha_{2,t} \\ \alpha_{3,t} \end{bmatrix} + \begin{bmatrix} \nu_t \\ 0 \\ 0 \end{bmatrix}$$

$$\text{Measurement equation:} \quad Y_t = c + \begin{bmatrix} 1 & \theta_1 & \theta_2 \end{bmatrix} \begin{bmatrix} \alpha_{1,t} \\ \alpha_{2,t} \\ \alpha_{3,t} \end{bmatrix},$$

with $\nu_t \sim \mathrm{WN}(0, \sigma^2)$. From the first line of the transition equation, $\alpha_{1,t+1} = \nu_t$, it appears clear that $\alpha_{1,t+1}$ is just a white noise sequence. The second and the third lines are used to lag it up to two periods: $\alpha_{2,t+1} = \alpha_{1,t} = \nu_{t-1}$ and $\alpha_{3,t+1} = \alpha_{2,t} = \alpha_{1,t-1} = \nu_{t-2}$. Thus, by substituting in the measurement equation we obtain: $Y_t = \nu_{t-1} + \theta_1 \nu_{t-2} + \theta_2 \nu_{t-3}$, which is a MA(2) process. Again, do not let the time index of the white noise sequence ν_t confuse you, as we could just rename it as $\varepsilon_t = \nu_{t-1}$ and read $Y_t = \varepsilon_t + \theta_1 \varepsilon_{t-1} + \theta_2 \varepsilon_{t-2}$. The covariance matrices of this system are clearly given by

$$\mathbf{H}_t = 0, \qquad \mathbf{G}_t = \mathbf{0}, \qquad \mathbf{Q}_t = \begin{bmatrix} \sigma^2 & 0 & 0 \\ 0 & 0 & 0 \\ 0 & 0 & 0 \end{bmatrix}.$$

The second representation is slightly less intuitive but has one advantage over the first: the MA(2) process is completely defined in the transition equation.

$$\text{Transition equation:} \quad \begin{bmatrix} \alpha_{1,t+1} \\ \alpha_{2,t+1} \\ \alpha_{3,t+1} \end{bmatrix} = \begin{bmatrix} 0 & 1 & 0 \\ 0 & 0 & 1 \\ 0 & 0 & 0 \end{bmatrix} \begin{bmatrix} \alpha_{1,t} \\ \alpha_{2,t} \\ \alpha_{3,t} \end{bmatrix} + \begin{bmatrix} 1 \\ \theta_1 \\ \theta_2 \end{bmatrix} \nu_t$$

$$\text{Measurement equation:} \quad Y_t = c + \begin{bmatrix} 1 & 0 & 0 \end{bmatrix} \begin{bmatrix} \alpha_{1,t} \\ \alpha_{2,t} \\ \alpha_{3,t} \end{bmatrix},$$

with $\nu_t \sim \mathrm{WN}(0, \sigma^2)$. To see why $\alpha_{1,t}$ (and thus Y_t) is a MA(2) process let us substitute the identities in the transition equations starting from the last line up to the first:

$$\begin{aligned} \alpha_{3,t+1} &= \theta_2 \nu_t, \\ \alpha_{2,t+1} &= \alpha_{3,t} + \theta_1 \nu_t = \theta_2 \nu_{t-1} + \theta_1 \nu_t \\ \alpha_{1,t+1} &= \alpha_{2,t} + \nu_t = \theta_2 \nu_{t-2} + \theta_1 \nu_{t-1} + \nu_t. \end{aligned}$$

Thus, $\alpha_{1,t}$ is a MA(2) process and the measurement equation just sets Y_t equal to $\alpha_{1,t}$. The covariance matrices of the system are given by $\mathbf{H}_t = 0$, $\mathbf{G}_t = \mathbf{0}$,

$$\mathbf{Q}_t = \mathbb{E}\left(\nu_t^2 \begin{bmatrix} 1 \\ \theta_1 \\ \theta_2 \end{bmatrix} \begin{bmatrix} 1 & \theta_1 & \theta_2 \end{bmatrix} \right) = \sigma^2 \begin{bmatrix} 1 & \theta_1 & \theta_2 \\ \theta_1 & \theta_1^2 & \theta_1\theta_2 \\ \theta_2 & \theta_1\theta_2 & \theta_2^2 \end{bmatrix}.$$

In both cases the state space form has to be completed with the mean vector and covariance matrix of $\boldsymbol{\alpha}_1$, but, as already mentioned, in the next pages we will provide a general formula to compute the marginal mean and covariance matrix of any stationary state vector process that should be used as initial values if no better information is available to the modeller.

Before giving the general formulation of ARMA processes in state space form, let us examine one last example for a mixed model: the ARMA(1,1).

Example 5.4 (ARMA(1,1) in state space form).
Let us examine two different state space forms for the ARMA(1,1). The first form exploits the fact that if we pass an AR(q) process though a MA(q) filter we obtain an ARMA(p, q) process (the same result is obtained if we pass a MA process though an AR filter):

$$\text{Transition equation:} \qquad \begin{bmatrix} \alpha_{1,t+1} \\ \alpha_{2,t+1} \end{bmatrix} = \begin{bmatrix} \phi_1 & 0 \\ 1 & 0 \end{bmatrix} \begin{bmatrix} \alpha_{1,t} \\ \alpha_{2,t} \end{bmatrix} + \begin{bmatrix} \nu_t \\ 0 \end{bmatrix},$$

$$\text{Measurement equation:} \qquad Y_t = c + \begin{bmatrix} 1 & \theta_1 \end{bmatrix} \begin{bmatrix} \alpha_{1,t} \\ \alpha_{2,t} \end{bmatrix},$$

with $\nu_t \sim \text{WN}(0, \sigma^2)$.

The first line of the transition equation clearly defines $\alpha_{1,t}$ as an AR(1) process, while $\alpha_{2,t} = \alpha_{1,t-1}$ is the lag of that process. Using the lag polynomial, we can write the AR(1) process as $\alpha_{1,t} = (1-\phi_1\mathbb{B})^{-1}\nu_{t-1}$ and, thus, $\alpha_{2,t} = \mathbb{B}(1 - \phi_1\mathbb{B})^{-1}\nu_{t-1}$. Substituting in the measurement equation we obtain

$$\begin{aligned} Y_t &= c + (1 - \phi_1\mathbb{B})^{-1}\nu_{t-1} + \theta_1\mathbb{B}(1 - \phi_1\mathbb{B})^{-1}\nu_{t-1} \\ &= c + \frac{(1 + \theta_1\mathbb{B})}{(1 - \phi_1\mathbb{B})}\nu_{t-1}, \end{aligned}$$

which is an ARMA(1,1) process.

The other form is defined as follows:

$$\text{Transition equation:} \qquad \begin{bmatrix} \alpha_{1,t+1} \\ \alpha_{2,t+1} \end{bmatrix} = \begin{bmatrix} \phi_1 & 1 \\ 0 & 0 \end{bmatrix} \begin{bmatrix} \alpha_{1,t} \\ \alpha_{2,t} \end{bmatrix} + \begin{bmatrix} 1 \\ \theta_1 \end{bmatrix} \nu_t,$$

$$\text{Measurement equation:} \qquad Y_t = c + \begin{bmatrix} 1 & 0 \end{bmatrix} \begin{bmatrix} \alpha_{1,t} \\ \alpha_{2,t} \end{bmatrix},$$

with $\nu_t \sim \text{WN}(0, \sigma^2)$. Examining the transition equation from the bottom to the top we have

$$\begin{aligned} \alpha_{2,t+1} &= \theta_1\nu_t \\ \alpha_{1,t+1} &= \phi_1\alpha_{1,t} + \alpha_{2,t} + \nu_t = \phi_1\alpha_{1,t} + \theta_1\nu_{t-1} + \nu_t, \end{aligned}$$

and $\alpha_{1,t+1}$ turns out to be the ARMA(1,1) process.

The reader can easily derive the covariance matrices $\mathbf{H}_t$, $\mathbf{G}_t$ and $\mathbf{Q}_t$ for both forms.

Let us give two of the most common state space representations of a general ARMA process.

Theorem 5.1 (ARMA in state space form). *The following two state space forms define the same ARMA(p, q) process. Let* $r = \max(p, q+1)$, *then the ARMA process can be written as*

$$\underbrace{\begin{bmatrix} \alpha_{1,t+1} \\ \alpha_{2,t+1} \\ \vdots \\ \alpha_{r-1,t+1} \\ \alpha_{r,t+1} \end{bmatrix}}_{\boldsymbol{\alpha}_{t+1}} = \underbrace{\begin{bmatrix} \phi_1 & \phi_2 & \dots & \phi_{r-1} & \phi_r \\ 1 & 0 & \dots & 0 & 0 \\ \vdots & \ddots & \dots & \vdots & \vdots \\ 0 & 0 & \ddots & 0 & 0 \\ 0 & 0 & \dots & 1 & 0 \end{bmatrix}}_{\mathbf{T}} \underbrace{\begin{bmatrix} \alpha_{1,t} \\ \alpha_{2,t} \\ \vdots \\ \alpha_{r-1,t} \\ \alpha_{r,t} \end{bmatrix}}_{\boldsymbol{\alpha}_t} + \underbrace{\begin{bmatrix} \nu_t \\ 0 \\ \vdots \\ 0 \\ 0 \end{bmatrix}}_{\boldsymbol{\nu}_t}$$

$$Y_t = \underbrace{\begin{bmatrix} 1 & \theta_1 \dots & \theta_{r-1} \end{bmatrix}}_{\mathbf{Z}} \boldsymbol{\alpha}_t,$$

where the covariance matrices are given by

$$\mathbf{H} = 0, \qquad \mathbf{G} = \mathbf{0}, \qquad \mathbf{Q} = \begin{bmatrix} \sigma^2 & 0 & \dots & 0 \\ 0 & 0 & \dots & 0 \\ \vdots & \vdots & \vdots & \vdots \\ 0 & 0 & \dots & 0 \end{bmatrix}.$$

The same process can also be written as

$$\underbrace{\begin{bmatrix} \alpha_{1,t+1} \\ \alpha_{2,t+1} \\ \vdots \\ \alpha_{r-1,t+1} \\ \alpha_{r,t+1} \end{bmatrix}}_{\boldsymbol{\alpha}_{t+1}} = \underbrace{\begin{bmatrix} \phi_1 & 1 & \dots & 0 & 0 \\ \phi_2 & 0 & \ddots & 0 & 0 \\ \vdots & \vdots & \dots & \ddots & \vdots \\ \phi_{r-1} & 0 & \dots & 0 & 1 \\ \phi_r & 0 & \dots & 0 & 0 \end{bmatrix}}_{\mathbf{T}} \underbrace{\begin{bmatrix} \alpha_{1,t} \\ \alpha_{2,t} \\ \vdots \\ \alpha_{r-1,t} \\ \alpha_{r,t} \end{bmatrix}}_{\boldsymbol{\alpha}_t} + \underbrace{\begin{bmatrix} 1 \\ \theta_1 \\ \vdots \\ \theta_{r-1} \\ \theta_r \end{bmatrix} \nu_t}_{\boldsymbol{\nu}_t}$$

$$Y_t = \underbrace{\begin{bmatrix} 1 & 0 \dots & 0 & 0 \end{bmatrix}}_{\mathbf{Z}} \boldsymbol{\alpha}_t,$$

where the covariance matrices are given by

$$\mathbf{H} = 0, \qquad \mathbf{G} = \mathbf{0}, \qquad \mathbf{Q} = \sigma^2 \begin{bmatrix} 1 & \theta_1 & \dots & \theta_{r-1} \\ \theta_1 & \theta_1^2 & \dots & \theta_1\theta_{r-1} \\ \vdots & \vdots & \vdots & \vdots \\ \theta_{r-1} & \theta_1\theta_{r-1} & \dots & \theta_{r-1}^2 \end{bmatrix}.$$

In both cases, if $r > p$ *then* $\phi_{p+1} = \ldots = \phi_r = 0$, *if* $r > q+1$ *then* $\theta_{q+1} = \ldots = \theta_{r-1} = 0$.

The reader can easily prove the above statements using the same arguments seen in examples 5.1–5.4. Notice that other representations are possible, but these are the most frequently used in time series books and software packages. In the rest of the volume, we will refer to these two state space representations of the ARMA process respectively as *horizontal* and *vertical*.

If the time series is integrated or seasonally integrated, it is possible to use the above state space forms after taking the opportune differences, as customary in Box–Jenkins ARIMA modelling, or it is also possible to include the integration operation in the state space form.

Example 5.5 (ARIMA(1, 1, 1) in state space form).
For implementing integration into the state space form, the second representation of the ARMA model in Theorem 5.1 is more convenient. The trick consists in adding a state variable that is equal to its own lagged value plus the ARMA model:

$$\begin{bmatrix}\alpha_{1,t+1}\\ \alpha_{2,t+1}\\ \alpha_{3,t+1}\end{bmatrix} = \begin{bmatrix}1 & 1 & 0\\ 0 & \phi_1 & 1\\ 0 & 0 & 0\end{bmatrix}\begin{bmatrix}\alpha_{1,t}\\ \alpha_{2,t}\\ \alpha_{3,t}\end{bmatrix} + \begin{bmatrix}0\\ 1\\ \theta_1\end{bmatrix}\nu_t,$$
$$Y_t = c + \begin{bmatrix}1 & 0 & 0\end{bmatrix}\boldsymbol{\alpha}_t,$$

with $\nu_t \sim \mathrm{WN}(0, \sigma^2)$. In this system $\alpha_{2,t}$ is the ARMA(1, 1) process and $\alpha_{1,t+1} = \alpha_{1,t} + \alpha_{2,t}$ is integrating $\alpha_{2,t}$ to obtain the ARIMA(1, 1, 1). The covariance matrices of this system are

$$\mathbf{H}_t = 0, \qquad \mathbf{G}_t = \mathbf{0}, \qquad \mathbf{Q}_t = \sigma^2\begin{bmatrix}0 & 0 & 0\\ 0 & 1 & \theta_1\\ 0 & \theta_1 & \theta_1^2\end{bmatrix}.$$

An alternative representation with a smaller state vector can be built by noticing that the ARI(1,1) operator $(1-\phi_1\mathbb{B})(1-\mathbb{B}) = 1-(1+\phi_1)\mathbb{B}+\phi_1\mathbb{B}^2$ can be seen as a non-stationary AR(2) operator:

$$\begin{bmatrix}\alpha_{1,t+1}\\ \alpha_{2,t+1}\end{bmatrix} = \begin{bmatrix}-(1+\phi_1) & 1\\ \phi_1 & 0\end{bmatrix}\begin{bmatrix}\alpha_{1,t}\\ \alpha_{2,t}\end{bmatrix} + \begin{bmatrix}1\\ \theta_1\end{bmatrix}\nu_t,$$
$$Y_t = c + \begin{bmatrix}1 & 0\end{bmatrix}\boldsymbol{\alpha}_t,$$

with $\nu_t \sim \mathrm{WN}(0, \sigma^2)$.

This second representation has the computational advantage of having smaller system matrices, but in the former representation the stationary and non-stationary components are well separated and this makes it easier to assign the values to the mean and covariance matrix of the initial state $\boldsymbol{\alpha}_1$ as explained later in this chapter.

A very popular ARIMA model is the so-called *Airline model*, which is the ARIMA$(0,1,1)(0,1,1)$ process that Box and Jenkins (1976) fitted to the international U.S. airline passengers time series depicted in Figure 2.1. This model has become popular among practitioners because it fits well many real-world time series. As Harvey (1989, Section 2.5.5) shows, the autocorrelation structure of the Airline model is very similar to that of an UCM composed by a local linear trend and a stochastic seasonal component.

Example 5.6 (ARIMA$(0,1,1)(0,1,1)_4$).
Let us build the state space form of the Airline model for quarterly data, by exploiting the fact that the combined difference operators $(1-\mathbb{B})(1-\mathbb{B}^4) = (1-\mathbb{B}-\mathbb{B}^4+\mathbb{B}^5)$ can be seen as a non-stationary AR(5) operator. Moreover, the product of the MA operators yields $(1+\theta_1\mathbb{B})(1+\Theta_1\mathbb{B}^4) = (1+\theta_1\mathbb{B}+\Theta_1\mathbb{B}^4+\theta_1\Theta_1\mathbb{B}^5)$. The state space is

$$\begin{bmatrix}\alpha_{1,t+1}\\ \alpha_{2,t+1}\\ \alpha_{3,t+1}\\ \alpha_{4,t+1}\\ \alpha_{5,t+1}\\ \alpha_{6,t+1}\end{bmatrix} = \begin{bmatrix}1 & 1 & 0 & 0 & 0 & 0\\ -1 & 0 & 1 & 0 & 0 & 0\\ 0 & 0 & 0 & 1 & 0 & 0\\ -1 & 0 & 0 & 0 & 1 & 0\\ 1 & 0 & 0 & 0 & 0 & 1\\ 0 & 0 & 0 & 0 & 0 & 0\end{bmatrix} \begin{bmatrix}\alpha_{1,t}\\ \alpha_{2,t}\\ \alpha_{3,t}\\ \alpha_{4,t}\\ \alpha_{5,t}\\ \alpha_{6,t}\end{bmatrix} + \begin{bmatrix}1\\ \theta_1\\ 0\\ 0\\ \Theta_1\\ \theta_1\Theta_1\end{bmatrix} \nu_t$$
$$Y_t = \begin{bmatrix}1 & 0 & 0 & 0 & 0 & 0\end{bmatrix} \boldsymbol{\alpha}_t,$$

with $\nu_t \sim \mathrm{WN}(0,\sigma^2)$. The reader can easily obtain the system correlation matrices.

5.2.2 *UCM in state space form*

Putting UCM in state space form is relatively easy since most components are already stated in this form. Let us see what each component looks like in state space form, and then we will provide a simple rule to build the complete UCM.

5.2.2.1 *Local linear trend*

The local linear trend (LLT, see equation 3.2) is defined by two equations driven by two uncorrelated shocks:

$$\begin{bmatrix}\mu_{t+1}\\ \beta_{t+1}\end{bmatrix} = \begin{bmatrix}1 & 1\\ 0 & 1\end{bmatrix} \begin{bmatrix}\mu_t\\ \beta_t\end{bmatrix} + \begin{bmatrix}\eta_t\\ \zeta_t\end{bmatrix}, \tag{5.3}$$

so that the transition equation is characterised by the matrices

$$\mathbf{T}_\mu = \begin{bmatrix}1 & 1\\ 0 & 1\end{bmatrix}, \qquad \mathbf{Q}_\mu = \begin{bmatrix}\sigma_\eta^2 & 0\\ 0 & \sigma_\zeta^2\end{bmatrix}.$$

If the LLT is the only component present in the system, then we have $\mathbf{Z}_\mu = [1 \;\; 0]$.

If we have no information on the distribution of the initial level μ_1 and slope β_1, we can state our ignorance using a *diffuse distribution*, that is a distribution with arbitrary mean (generally 0) and an infinite variance. The infinite variance suggests that we have no idea about the range of values that μ_1 and β_1 can assume.

5.2.2.2 Stochastic cycle

The stochastic cycle was defined in equation (3.5) and it is already in state space form. We report it here again as a reference using the notation we have chosen to define the general state space form:

$$\begin{bmatrix} \psi_{t+1} \\ \psi^*_{t+1} \end{bmatrix} = \rho \begin{bmatrix} \cos(\lambda) & \sin(\lambda) \\ -\sin(\lambda) & \cos(\lambda) \end{bmatrix} \begin{bmatrix} \psi_t \\ \psi^*_t \end{bmatrix} + \begin{bmatrix} \kappa_t \\ \kappa^*_t \end{bmatrix}. \tag{5.4}$$

The transition equation matrices are

$$\mathbf{T}_\psi = \rho \begin{bmatrix} \cos(\lambda) & \sin(\lambda) \\ -\sin(\lambda) & \cos(\lambda) \end{bmatrix}, \qquad \mathbf{Q}_\psi = \sigma^2_\kappa \mathbf{I}_2;$$

and if only the cycle is present in the UCM, then $\mathbf{Z}_\psi = [1 \;\; 0]$. If $|\rho| < 1$, then the cycle is stationary and the distribution of the initial value can be set equal to the marginal distribution, whose first two moments are

$$\mathbb{E}\begin{bmatrix} \psi_1 \\ \psi^*_1 \end{bmatrix} = \begin{bmatrix} 0 \\ 0 \end{bmatrix}, \qquad \mathbb{E}\left\{ \begin{bmatrix} \psi_1 \\ \psi^*_1 \end{bmatrix} [\psi_1 \;\; \psi^*_1] \right\} = \frac{\sigma^2_\kappa}{1-\rho^2} \mathbf{I}_2.$$

For a second-order cycle the 4×4 matrices of the transition equation become

$$\mathbf{T}_\psi = \begin{bmatrix} \rho\mathbf{R}(\lambda) & \mathbf{I}_2 \\ \mathbf{0} & \rho\mathbf{R}(\lambda) \end{bmatrix}, \qquad \mathbf{Q}_\psi = \begin{bmatrix} \mathbf{0} & \mathbf{0} \\ \mathbf{0} & \sigma^2_\kappa \mathbf{I}_2 \end{bmatrix},$$

where $\mathbf{R}(\lambda)$ is the rotation matrix defined in equation (3.6). If the second-order cycle is the only component in the UCM, then $\mathbf{Z}_\psi = [1 \;\; 0 \;\; 0 \;\; 0]$.

In the third-order cycle the 6×6 transition equation matrices are

$$\mathbf{T}_\psi = \begin{bmatrix} \rho\mathbf{R}(\lambda) & \mathbf{I}_2 & \mathbf{0} \\ \mathbf{0} & \rho\mathbf{R}(\lambda) & \mathbf{I}_2 \\ \mathbf{0} & \mathbf{0} & \rho\mathbf{R}(\lambda) \end{bmatrix}, \qquad \mathbf{Q}_\psi = \begin{bmatrix} \mathbf{0} & \mathbf{0} & \mathbf{0} \\ \mathbf{0} & \mathbf{0} & \mathbf{0} \\ \mathbf{0} & \mathbf{0} & \sigma^2_\kappa \mathbf{I}_2 \end{bmatrix},$$

and if this cycle is the only component of the UCM, then $\mathbf{Z}_\psi = [1 \;\; 0 \;\; 0 \;\; 0 \;\; 0 \;\; 0]$.

We are sure the reader is now able to generalise to any m-order cycle. If $|\rho| < 1$ the marginal mean of any higher-order cycle vector is $\mathbf{0}$ while the covariance matrix has the following structure, derived by Trimbur (2005):

$$\begin{bmatrix} \mathbf{\Sigma}_{1,1} & \cdots & \mathbf{\Sigma}_{1,m} \\ \vdots & \ddots & \vdots \\ \mathbf{\Sigma}_{m,1} & \cdots & \mathbf{\Sigma}_{m,m} \end{bmatrix}$$

where $\mathbf{\Sigma}_{i,j}$ is the 2×2 matrix

$$\mathbf{\Sigma}_{i,j} = \frac{\sigma_\kappa^2 \sum_{r=0}^{i-1} \binom{i-1}{r} \binom{j-1}{r+j-1} \rho^{2r+j-i}}{(1-\rho^2)^{i+j-1}} \mathbf{R}(\lambda(j-i)).$$

5.2.2.3 *Trigonometric seasonal component*

The trigonometric seasonal component for time series with periodicity s is given by the sum of $\lfloor s/2 \rfloor$ stochastic cycles with frequencies respectively equal to

$$\lambda_j = \frac{2\pi j}{s}, \qquad j = 1, 2, \ldots, \lfloor s/2 \rfloor.$$

Thus, the transition equation matrices for this component are

$$\mathbf{T}_\gamma = \begin{bmatrix} \mathbf{R}(\lambda_1) & \mathbf{0} & \ldots & \mathbf{0} \\ \mathbf{0} & \mathbf{R}(\lambda_2) & \ldots & \mathbf{0} \\ \vdots & \vdots & \ddots & \vdots \\ \mathbf{0} & \mathbf{0} & \ldots & \mathbf{R}(\lambda_{\lfloor s/2 \rfloor}) \end{bmatrix}, \qquad \mathbf{Q}_\gamma = \sigma_\omega^2 \mathbf{I}_{s-1}.$$

Notice that when s is an even number, then the second row of $\mathbf{R}(\lambda_{\lfloor s/2 \rfloor})$ can be dropped (cf. discussion in Section 3.4) so that $\mathbf{T}_\gamma$ and $\mathbf{Q}_\gamma$ are always $(s-1) \times (s-1)$ matrices.

Since only the first component of each sinusoid-pair has to be selected for the observation equation, we have $\mathbf{Z}_\gamma = [1 \; 0 \; 1 \; 0 \; \ldots \; 1 \; 0]$ if s is odd and $\mathbf{Z}_\gamma = [1 \; 0 \; 1 \; 0 \; \ldots \; 1]$ when s is even. Regardless of s being odd or even, $\mathbf{Z}_\gamma$ is always a row vector with $s-1$ elements.

As for the initial distribution of the state vector, if nothing better is known, it should be diffuse (i.e., zero mean and infinite variance for each component) as all the components are non-stationary.

5.2.2.4 *Stochastic dummy seasonal component*

The stochastic dummy seasonal component is based on the recursive definition

$$\gamma_t = -\gamma_{t-1} - \gamma_{t-2} - \ldots - \gamma_{t-s+1} + \omega_t,$$

with $\omega_t \sim \mathrm{WN}(0, \sigma_\omega^2)$. This simple equation can be stated in state space form by using the same trick already seen for ARMA models in state space form; that is, the second state variable is equal to the lagged first variable, the third state variable is the lagged second variable and so on:

$$\begin{bmatrix} \alpha_{1,t+1} \\ \alpha_{2,t+1} \\ \ldots \\ \alpha_{s-2,t+1} \\ \alpha_{s-1,t+1} \end{bmatrix} = \begin{bmatrix} -1 & -1 & \ldots & -1 & -1 \\ 1 & 0 & \ldots & 0 & 0 \\ \vdots & \ddots & \vdots & \vdots & \vdots \\ 0 & 0 & \ddots & 0 & 0 \\ 0 & 0 & \ldots & 1 & 0 \end{bmatrix} \begin{bmatrix} \alpha_{1,t} \\ \alpha_{2,t} \\ \ldots \\ \alpha_{s-2,t} \\ \alpha_{s-1,t} \end{bmatrix} + \begin{bmatrix} \omega_t \\ 0 \\ \ldots \\ 0 \\ 0 \end{bmatrix}.$$

The second line reads $\alpha_{2,t} = \alpha_{1,t-1}$, and substituting this in the third line we get $\alpha_{3,t} = \alpha_{2,t-1} = \alpha_{1,t-2}$. Going ahead in the same fashion we obtain $\alpha_{m,t} = \alpha_{1,t-m+1}$ for $m = 2, 3, \ldots, s-1$. Finally, substituting these identities into the first line we obtain

$$\alpha_{1,t+1} = -\alpha_{1,t} - \alpha_{1,t-1} - \ldots - \alpha_{1,t-s+2} + \omega_t,$$

which implies that $\alpha_{1,t+1}$ is equal to the γ_t defined above.

So, the state vector has $s-1$ elements and the matrices of the transition equations are

$$\mathbf{T}_\gamma = \begin{bmatrix} -\mathbf{1}_{s-2}^\top & -1 \\ \mathbf{I}_{s-2} & \mathbf{0} \end{bmatrix}, \qquad \mathbf{Q}_\gamma = \begin{bmatrix} \sigma_\omega^2 & 0 & \ldots & 0 \\ 0 & 0 & \ldots & 0 \\ \vdots & \vdots & \vdots & \vdots \\ 0 & 0 & \ldots & 0 \end{bmatrix},$$

where $\mathbf{1}_{s-2}$ is a column vector with $s-2$ ones. In the observation equation only the first element of $\boldsymbol{\alpha}_t$ is to be selected, and so $\mathbf{Z}_\gamma = [1 \; 0 \; \ldots \; 0]$. As in the trigonometric seasonal component the initial conditions of the state vector are diffuse, if no better information is available.

5.2.2.5 *Building UCM in state space form*

A UCM is composed by a selection of components such as trend, cycle, seasonality, but possibly also by static or dynamic regressors and, sometimes, even ARMA processes.

Building the state space form of a UCM using the desired components is really straightforward. First of all, notice that all the covariances between the disturbances within and across the components are zero and also the correlations between the observation error and the component disturbances are zero.

If we stack the state vector of the different components in one long state vector, then the transition matrix of the UCM, $\mathbf{T}$, is obtained by *diagonal concatenation* of the transition matrices of the components, and the same holds also for the covariance matrix $\mathbf{Q}$. For example if we wish to model a time series with LLT, stochastic cycle and trigonometric seasonal component, the transition equation matrices become

$$\mathbf{T} = \begin{bmatrix} \mathbf{T}_\mu & \mathbf{0} & \mathbf{0} \\ \mathbf{0} & \mathbf{T}_\psi & \mathbf{0} \\ \mathbf{0} & \mathbf{0} & \mathbf{T}_\gamma \end{bmatrix}, \qquad \mathbf{Q} = \begin{bmatrix} \mathbf{Q}_\mu & \mathbf{0} & \mathbf{0} \\ \mathbf{0} & \mathbf{Q}_\psi & \mathbf{0} \\ \mathbf{0} & \mathbf{0} & \mathbf{Q}_\gamma \end{bmatrix}.$$

The matrix $\mathbf{Z}$ of the observation equation, which in the univariate case is just a row vector, is obtained by staking horizontally all the corresponding matrices (vectors) of the components, in the same order used to build the composite state vector:

$$\mathbf{Z} = \begin{bmatrix} \mathbf{Z}_\mu & \mathbf{Z}_\psi & \mathbf{Z}_\gamma \end{bmatrix}.$$

An analogous treatment holds also for the mean vector and covariance matrix of the initial state $\boldsymbol{\alpha}_1$. So, if we call $\boldsymbol{a}$ and $\mathbf{P}$ these two quantities, we have

$$\boldsymbol{a} = \begin{bmatrix} \boldsymbol{a}_\mu \\ \boldsymbol{a}_\psi \\ \boldsymbol{a}_\gamma \end{bmatrix}, \qquad \mathbf{P} = \begin{bmatrix} \mathbf{P}_\mu & \mathbf{0} & \mathbf{0} \\ \mathbf{0} & \mathbf{P}_\psi & \mathbf{0} \\ \mathbf{0} & \mathbf{0} & \mathbf{P}_\gamma \end{bmatrix}.$$

Example 5.7 (Basic structural models (BSM)).
The UCM with LLT and stochastic dummy seasonal component is generally referred to as *basic structural model*, or just BSM. This model fits well many different kinds of time series especially of economic content. The state space form of this model for quarterly time series ($s = 4$) is

$$\underbrace{\begin{bmatrix} \mu_{t+1} \\ \beta_{t+1} \\ \hline \gamma_{t+1} \\ \gamma_t \\ \gamma_{t-1} \end{bmatrix}}_{\boldsymbol{\alpha}_{t+1}} = \underbrace{\left[\begin{array}{cc|ccc} 1 & 1 & 0 & 0 & 0 \\ 0 & 1 & 0 & 0 & 0 \\ \hline 0 & 0 & -1 & -1 & -1 \\ 0 & 0 & 1 & 0 & 0 \\ 0 & 0 & 0 & 1 & 0 \end{array}\right]}_{\mathbf{T}} \underbrace{\begin{bmatrix} \mu_t \\ \beta_t \\ \hline \gamma_t \\ \gamma_{t-1} \\ \gamma_{t-2} \end{bmatrix}}_{\boldsymbol{\alpha}_t} + \underbrace{\begin{bmatrix} \eta_t \\ \zeta_t \\ \hline \omega_t \\ 0 \\ 0 \end{bmatrix}}_{\boldsymbol{\nu}_t},$$

where the covariance matrix of $\boldsymbol{\nu}_t$ is clearly

$$\mathbf{Q} = \left[\begin{array}{cc|ccc} \sigma^2_\eta & 0 & 0 & 0 & 0 \\ 0 & \sigma^2_\zeta & 0 & 0 & 0 \\ \hline 0 & 0 & \sigma^2_\omega & 0 & 0 \\ 0 & 0 & 0 & 0 & 0 \\ 0 & 0 & 0 & 0 & 0 \end{array}\right].$$

Thus, the observation equation should select μ_t and γ_t, take their sum and add a white noise:

$$Y_t = \underbrace{\left[\begin{array}{cc|ccc} 1 & 0 & 1 & 0 & 0 \end{array}\right]}_{\mathbf{Z}} \boldsymbol{\alpha}_t + \epsilon_t.$$

The remaining covariance matrices are $\mathbf{H} = \sigma^2_\epsilon$ and $\mathbf{G} = \mathbf{0}$.

If one prefers to use the trigonometric seasonal component instead of stochastic dummy form then, recalling that $\cos(\pi/2) = 0$, $\sin(\pi/2) = 1$ and $\cos(\pi) = -1$, the transition equation becomes

$$\underbrace{\begin{bmatrix} \mu_{t+1} \\ \beta_{t+1} \\ \hline \gamma_{1,t+1} \\ \gamma^*_{1,t+1} \\ \gamma_{2,t+1} \end{bmatrix}}_{\boldsymbol{\alpha}_{t+1}} = \underbrace{\left[\begin{array}{cc|ccc} 1 & 1 & 0 & 0 & 0 \\ 0 & 1 & 0 & 0 & 0 \\ \hline 0 & 0 & 0 & 1 & 0 \\ 0 & 0 & -1 & 0 & 0 \\ 0 & 0 & 0 & 0 & -1 \end{array}\right]}_{\mathbf{T}} \underbrace{\begin{bmatrix} \mu_t \\ \beta_t \\ \hline \gamma_{1,t} \\ \gamma^*_{1,t} \\ \gamma_{2,t} \end{bmatrix}}_{\boldsymbol{\alpha}_t} + \underbrace{\begin{bmatrix} \eta_t \\ \zeta_t \\ \hline \omega_{1,t} \\ \omega^*_{1,t} \\ \omega_{2,t} \end{bmatrix}}_{\boldsymbol{\nu}_t},$$

where the covariance matrix of $\boldsymbol{\nu}_t$ is now

$$\mathbf{Q} = \left[\begin{array}{cc|ccc} \sigma_\eta^2 & 0 & 0 & 0 & 0 \\ 0 & \sigma_\zeta^2 & 0 & 0 & 0 \\ \hline 0 & 0 & \sigma_\omega^2 & 0 & 0 \\ 0 & 0 & 0 & \sigma_\omega^2 & 0 \\ 0 & 0 & 0 & 0 & \sigma_\omega^2 \end{array}\right];$$

and the observation equation becomes

$$Y_t = \underbrace{\left[\begin{array}{cc|ccc} 1 & 0 & 1 & 0 & 1 \end{array}\right]}_{\mathbf{Z}} \boldsymbol{\alpha}_t + \epsilon_t,$$

In both forms, all the components are non-stationary and, therefore, a diffuse initialisation of the state vector process is the natural choice in absence of better information: $\boldsymbol{a} = \mathbf{0}$, $\mathbf{P} = \tau \mathbf{I}_5$ with $\tau \to \infty$.

From the point of view of estimation, none of the two forms is to be preferred to the other: in both cases the state vectors have the same dimension and there are four unknown parameters (i.e., σ_η, σ_ζ, σ_ω, σ_ϵ) to estimate.

Example 5.8 (Integrated random walk and stochastic cycle).
For reasons that will be discussed in Chapter 6, when the UCM contains a stochastic cycle, the trend is often modelled using the integrated random walk (IRW) instead of the LLT. The IRW is a LLT with the variance of the first disturbance set to zero. The transition equation is

$$\underbrace{\left[\begin{array}{c} \mu_{t+1} \\ \beta_{t+1} \\ \hline \psi_{t+1} \\ \psi^*_{t+1} \end{array}\right]}_{\boldsymbol{\alpha}_{t+1}} = \underbrace{\left[\begin{array}{cc|cc} 1 & 1 & 0 & 0 \\ 0 & 1 & 0 & 0 \\ \hline 0 & 0 & \rho\cos(\lambda) & \rho\sin(\lambda) \\ 0 & 0 & -\rho\sin(\lambda) & \rho\cos(\lambda) \end{array}\right]}_{\mathbf{T}} \underbrace{\left[\begin{array}{c} \mu_t \\ \beta_t \\ \hline \psi_t \\ \psi^*_t \end{array}\right]}_{\boldsymbol{\alpha}_t} + \underbrace{\left[\begin{array}{c} 0 \\ \zeta_t \\ \hline \kappa_t \\ \kappa^*_t \end{array}\right]}_{\boldsymbol{\nu}_t},$$

where the covariance matrix of $\boldsymbol{\nu}_t$ is

$$\mathbf{Q} = \left[\begin{array}{cc|cc} 0 & 0 & 0 & 0 \\ 0 & \sigma_\zeta^2 & 0 & 0 \\ \hline 0 & 0 & \sigma_\kappa^2 & 0 \\ 0 & 0 & 0 & \sigma_\kappa^2 \end{array}\right].$$

In the observation equation the first and third state variable are selected and summed to a white noise:

$$Y_t = \underbrace{\left[\begin{array}{cc|cc} 1 & 0 & 1 & 0 \end{array}\right]}_{\mathbf{Z}} \boldsymbol{\alpha}_t + \epsilon_t,$$

with $\mathbf{H} = \sigma_\epsilon^2$ and $\mathbf{G} = 0$.

Since the two state variables of the LLT are non-stationary, while the two stochastic cycle variables are zero-mean stationary, the distribution of the state vector at time $t = 1$ has the first two moments

$$\boldsymbol{a} = \mathbf{0}, \qquad \mathbf{P} = \begin{bmatrix} \infty & 0 & 0 & 0 \\ 0 & \infty & 0 & 0 \\ 0 & 0 & \sigma_\psi^2 & 0 \\ 0 & 0 & 0 & \sigma_\psi^2 \end{bmatrix},$$

with $\sigma_\psi^2 = \sigma_\kappa^2/(1-\rho^2)$.

The parameters to be estimated are the variances σ_η^2, σ_κ^2 and the coefficients ρ and λ, even though sometimes the latter two are fixed to fit a cyclical component consistent with the economists' definition of business cycle.

5.2.3 *Regression models in state space form*

As seen in Chapter 4 real and artificial regressors can enter an UCM in different forms.

A classical linear model can be put in state space form in two alternative ways: one considers the regression coefficients as unobserved components that are constant over time, in the other one the regression coefficients are to be estimated as any other unknown parameter in the model.

Let $\boldsymbol{X}_t$ be the vector of regressors at time t, then the first representation is given by

$$\begin{aligned} Y_t &= \boldsymbol{X}_t^\top \boldsymbol{\alpha}_t + \epsilon_t \\ \boldsymbol{\alpha}_{t+1} &= \mathbf{I}\boldsymbol{\alpha}_t, \end{aligned} \tag{5.5}$$

with system matrices $\mathbf{Z}_t = \boldsymbol{X}_t$, $\mathbf{T} = \mathbf{I}$, $\mathbf{H} = \sigma_\epsilon^2$, $\mathbf{Q} = \mathbf{0}$ and $\mathbf{G} = \mathbf{0}$. If no better information on the coefficients is available, the distribution of $\boldsymbol{\alpha}_1$ should be diffuse.

The second representation does not need a transition equation and the regressors multiplied by the respective coefficients enter the state space form through the system vector $\boldsymbol{c}_t$:

$$Y_t = \boldsymbol{X}_t^\top \boldsymbol{\beta} + \epsilon_t, \tag{5.6}$$

with all system matrices equal to zero except $\boldsymbol{c}_t = \boldsymbol{X}_t^\top \boldsymbol{\beta}$ and $\mathbf{H} = \sigma_\epsilon^2$. The first representation is preferable if in the UCM there are not so many other state variables but many parameters to estimate by maximum likelihood, while the second representation is more convenient when in the UCM there are already many state variables but not so many parameters to estimate.

If the regression is time-varying and the coefficients evolve over time, we can use the first form and substitute the transition equation in (5.5) with one of the following alternatives. If the coefficients do not tend to return toward fixed values, then the coefficients can be modelled by random walks,

$$\boldsymbol{\alpha}_{t+1} = \mathbf{I}\boldsymbol{\alpha}_t + \boldsymbol{\zeta}_t,$$

where the white noise vector ζ_t can have uncorrelated or correlated elements (generally, in order to limit the number of parameters to be estimated, the components are kept uncorrelated). If the time-evolution of the regression coefficients has to be smoother, then they can be modelled as integrated random walks. A simple way to achieve this is by using the transition equation matrices

$$\mathbf{T} = \begin{bmatrix} \mathbf{I}_k & \mathbf{I}_k \\ \mathbf{0} & \mathbf{I}_k \end{bmatrix}, \qquad \mathbf{Q} = \begin{bmatrix} 0 & 0 \\ 0 & \mathbf{D} \end{bmatrix},$$

where k is the number of regressors (including the constant if one is included) and $\mathbf{D}$ is a $k \times k$ diagonal matrix with variances on the main diagonal and the state vector has $2k$ elements. The observation equation is

$$Y_t = \begin{bmatrix} \boldsymbol{X}_t^\top & \mathbf{0}^\top \end{bmatrix} \boldsymbol{\alpha}_t + \varepsilon_t,$$

where $\mathbf{0}$ is a column vector with k zeros.

If the regression coefficients are "attracted" by fixed values, then they could be modelled as stationary AR(1) processes. In this case the transition equation becomes

$$\boldsymbol{\alpha}_{t+1} = \boldsymbol{d} + \boldsymbol{\Phi}\boldsymbol{\alpha}_t + \boldsymbol{\zeta}_t,$$

where $\boldsymbol{\Phi}$ is a diagonal matrix with the AR(1) coefficients on the main diagonal. If it is sensible to assume that all coefficients revert to the mean at the same rate, then we can set $\boldsymbol{\Phi} = \phi \mathbf{I}_k$. Again, the covariance matrix $\mathbf{Q}$ can be diagonal or not.

Example 5.9 (Regression with integrated errors).
In Example 4.1 we presented the case

$$\begin{aligned} Y_t &= \mu_t + \delta X_t + \varepsilon_t \\ \mu_t &= \mu_{t-1} + \eta_t \\ X_t &= X_{t-1} + \zeta_t, \end{aligned}$$

where Y_t and X_t are observable while μ_t is not and we are interested in estimating the regression coefficient δ. Since X_t is observable, it is used as a regressor and does not need a state variable. So, the state space representation, when δ is treated as a parameter to be estimated by maximum likelihood is straightforward (we only provide the system matrices):

$$\mathbf{T} = 1,\ \boldsymbol{c}_t = \delta \boldsymbol{X}_t,\ \mathbf{Q} = \sigma_\eta^2,\ \mathbf{Z} = 1,\ \boldsymbol{d} = 0,\ \mathbf{H} = \sigma_\varepsilon^2,\ \mathbf{G} = 0,$$

with diffuse initial conditions, $\mathbb{E}[\alpha_1] = 0$, $\mathbb{E}[\alpha_1^2] = \infty$.

In the other representation we presented in the text, δ is treated as an additional state variable which is constant over time:

$$\begin{aligned} \boldsymbol{\alpha}_{t+1} &= \begin{bmatrix} 1 & 0 \\ 0 & 1 \end{bmatrix} \boldsymbol{\alpha}_t + \begin{bmatrix} \eta_t \\ 0 \end{bmatrix} \\ Y_t &= \begin{bmatrix} 1 & X_t \end{bmatrix} \boldsymbol{\alpha}_t + \epsilon_t, \end{aligned}$$

where the elements in the state vector are to interpreted as $\boldsymbol{\alpha}_t^\top = [\mu_t \; \delta_t]$. We are sure the reader can build the remaining system vectors and matrices, that should include diffuse distribution moments for both state variables at time $t = 1$.

The latter representation has two parameters (i.e., σ_η^2, σ_ϵ^2) to be estimated by maximum likelihood, while the former had three (i.e., the two variances of the latter representation plus the regression coefficient δ).

If we need the regression coefficients to vary over time, then the transition equation can be changed into

$$\boldsymbol{\alpha}_{t+1} = \begin{bmatrix} 1 & 0 \\ 0 & 1 \end{bmatrix} \boldsymbol{\alpha}_t + \begin{bmatrix} \eta_t \\ \zeta_t \end{bmatrix},$$

if the coefficient is modelled as a random walk, or into

$$\boldsymbol{\alpha}_{t+1} = \begin{bmatrix} 1 & 0 \\ 0 & \phi \end{bmatrix} \boldsymbol{\alpha}_t + \begin{bmatrix} 0 \\ \delta \end{bmatrix} + \begin{bmatrix} \eta_t \\ \zeta_t \end{bmatrix},$$

if there is a mean value (in this case $\delta/(1-\phi)$) toward which the coefficient always reverts. The observation equation remains unchanged and the two white noise sequences, η_t and ζ_t are generally assumed uncorrelated.

If we need to write in state space form transfer function (or dynamic regression) models, we can adapt from the ARMA representation seen in Theorem 5.1. Indeed, a transfer function model is just an ARMA model whose (unobservable) driving white noise is substituted with an observable regressor or, otherwise stated, it is an ARMA filter applied to an observable input variable:

$$W_t = \delta_1 W_{t-1} + \ldots \delta_p W_{t-r} + \omega_0 X_t + \ldots + \omega_s X_{t-s},$$

where the only difference with respect to the ARMA filter is that here there is the ω_0 coefficient which does not correspond to any θ_0 coefficient in the ARMA filter and which represents the contemporaneous impact of X_t on W_t.

The transfer function model comprises also some noise, as in

$$Y_t = W_t + \varepsilon_t,$$

where ε_t can be white noise or some other process.

Theorem 5.2 (Transfer function in state space form). *The following two state space forms define the same transfer function process of order* (r, s). *Let* $v = \max(r, s+1)$, *then the transfer function model with white noise* *error can be written as*

$$\underbrace{\begin{bmatrix} \alpha_{1,t+1} \\ \alpha_{2,t+1} \\ \vdots \\ \alpha_{r-1,t+1} \\ \alpha_{r,t+1} \end{bmatrix}}_{\boldsymbol{\alpha}_{t+1}} = \underbrace{\begin{bmatrix} \delta_1 & \delta_2 & \dots & \delta_{v-1} & \delta_v \\ 1 & 0 & \dots & 0 & 0 \\ \vdots & \ddots & \dots & \vdots & \vdots \\ 0 & 0 & \ddots & 0 & 0 \\ 0 & 0 & \dots & 1 & 0 \end{bmatrix}}_{\mathbf{T}} \underbrace{\begin{bmatrix} \alpha_{1,t} \\ \alpha_{2,t} \\ \vdots \\ \alpha_{v-1,t} \\ \alpha_{v,t} \end{bmatrix}}_{\boldsymbol{\alpha}_t} + \underbrace{\begin{bmatrix} X_t \\ 0 \\ \vdots \\ 0 \\ 0 \end{bmatrix}}_{\mathbf{d}_t}$$

$$Y_t = \underbrace{\begin{bmatrix} \omega_0 & \omega_1 \dots & \omega_{v-1} \end{bmatrix}}_{\mathbf{Z}} \boldsymbol{\alpha}_t + \epsilon_t,$$

where the covariance matrices are given by $\mathbf{H} = \sigma_\varepsilon^2$, $\mathbf{G} = \mathbf{0}$, $\mathbf{Q} = \mathbf{0}$.

The same process can also be written as

$$\underbrace{\begin{bmatrix} \alpha_{1,t+1} \\ \alpha_{2,t+1} \\ \vdots \\ \alpha_{r-1,t+1} \\ \alpha_{r,t+1} \end{bmatrix}}_{\boldsymbol{\alpha}_{t+1}} = \underbrace{\begin{bmatrix} \delta_1 & 1 & \dots & 0 & 0 \\ \delta_2 & 0 & \ddots & 0 & 0 \\ \vdots & \vdots & \dots & \ddots & \vdots \\ \delta_{v-1} & 0 & \dots & 0 & 1 \\ \delta_v & 0 & \dots & 0 & 0 \end{bmatrix}}_{\mathbf{T}} \underbrace{\begin{bmatrix} \alpha_{1,t} \\ \alpha_{2,t} \\ \vdots \\ \alpha_{v-1,t} \\ \alpha_{v,t} \end{bmatrix}}_{\boldsymbol{\alpha}_t} + \underbrace{\begin{bmatrix} \omega_0 X_t \\ \omega_1 X_t \\ \vdots \\ \omega_{v-1} X_t \\ \omega_v X_t \end{bmatrix}}_{\mathbf{d}_t}$$

$$Y_t = \underbrace{\begin{bmatrix} 1 & 0 \dots & 0 & 0 \end{bmatrix}}_{\mathbf{Z}} \boldsymbol{\alpha}_t + \epsilon_t,$$

where the covariance matrices are given by $\mathbf{H} = \sigma_\varepsilon^2$, $\mathbf{G} = \mathbf{0}$, $\mathbf{Q} = \mathbf{0}$.

In both cases, if $v > r$ *then* $\delta_{r+1} = \ldots = \delta_v = 0$, *if* $v > s+1$ *then* $\omega_{s+1} = \ldots = \omega_v = 0$.

As we did for the ARMA family, we will refer to the two state space representations of the transfer function model respectively as *horizontal* and *vertical.*

As for the initial distribution of the state vector, we can think of two different cases. When X_t is an intervention variable (a dummy variable) and W_t is zero before the intervention takes place, then the initial state vector $\boldsymbol{\alpha}_1$ is just a non-stochastic vector of zeros, $\boldsymbol{a} = 0$, and its covariance matrix is, therefore, a matrix of zeros $\mathbf{P} = \mathbf{0}$. Otherwise, if X_t is a real regressor the distribution of the state vector should be set diffuse unless X_t is generated by a stationary process and the roots of the equations $\delta_r(z) = 0$ are outside the unit root. In this case the state vector is stationary and its marginal distribution can be used. Later in this chapter we will show how to compute the first two moments of the marginal distribution of any stationary state vector.

Example 5.10 (Effect of advertisement on sales).
Suppose that Y_t represents the weekly value of sales of some product and X_t the amount of the investment in advertisement for that product. If the sales are well represented by a random walk plus an advertisement effect that dies out geometrically, that is,

$$\begin{aligned} Y_t &= \mu_t + W_t + \varepsilon_t, \\ \mu_t &= \mu_{t-1} + \eta_t, \\ W_t &= \delta_1 W_{t-1} + \omega_0 X_t, \end{aligned} \tag{5.7}$$

with $0 < \delta < 1$, then the state space form is given by

$$\boldsymbol{\alpha}_{t+1} = \underbrace{\begin{bmatrix} 1 & 0 \\ 0 & \delta_1 \end{bmatrix}}_{\mathbf{T}} \boldsymbol{\alpha}_t + \underbrace{\begin{bmatrix} 0 \\ \omega_0 X_{t+1} \end{bmatrix}}_{\boldsymbol{d}_t} + \underbrace{\begin{bmatrix} \eta_t \\ 0 \end{bmatrix}}_{\boldsymbol{\nu}_t},$$

$$Y_t = \underbrace{\begin{bmatrix} 1 & 1 \end{bmatrix}}_{\mathbf{Z}} \boldsymbol{\alpha}_t + \varepsilon_t.$$

Here, the distribution of the state vector $\boldsymbol{\alpha}_1$ should be diffuse if no better information is available. Notice that since Y_t and X_t have to be synchronised the rhs of the second line of the transition equation contains X_{t+1} and not X_t.

If, instead of the future state space form, the contemporaneous form were available, then equation (5.7) would be already in the right form.

5.2.4 Putting the pieces together

Modelling real data implies the use of a mix of unobserved components, (dynamic, static, real, artifical) regressors and, sometimes, ARMA processes. One of the advantages of the UCM approach with respect to alternatives (for example, the Box–Jenkins approach) is its modularity. The analyst can concentrate on every single feature of the data and build a specific component (in wide sense) for that feature. When all aspects of the time series have been dealt with their distinct components, all components are put together in a complete model, which will be easy to interpret and explain to non-experts in statistics.

As already seen in Section 5.2.2.5, if the (wide sense) components to be put in an UCM are independent from each other, then the transition equation vectors have to be stacked and the matrices of the complete transition equation are to be built by diagonal concatenation of the same matrices of the submodels. Namely, if the system vectors and matrices of the state space form of the i-th (wide sense) component are $\mathbf{T}_t^{(i)}$, $\boldsymbol{d}_t^{(i)}$, $\mathbf{Q}_t^{(i)}$, $\mathbf{Z}_t^{(i)}$, $\boldsymbol{c}_t^{(i)}$, $\mathbf{H}_t^{(i)}$, $\mathbf{G}_t^{(i)}$, $\boldsymbol{\alpha}_t^{(i)}$, $\boldsymbol{a}^{(i)}$, $\mathbf{P}^{(i)}$, then the system matrices of the state space form of the complete model are:

$$\boldsymbol{\alpha}_t = \begin{bmatrix} \boldsymbol{\alpha}_t^{(1)} \\ \boldsymbol{\alpha}_t^{(2)} \\ \vdots \end{bmatrix}, \quad \boldsymbol{d}_t = \begin{bmatrix} \boldsymbol{d}_t^{(1)} \\ \boldsymbol{d}_t^{(2)} \\ \vdots \end{bmatrix}, \quad \boldsymbol{c}_t = \boldsymbol{c}_t^{(1)} + \boldsymbol{c}_t^{(2)} + \ldots,$$

$$\mathbf{T}_t = \begin{bmatrix} \mathbf{T}_t^{(1)} & \mathbf{0} & \ldots \\ \mathbf{0} & \mathbf{T}_t^{(2)} & \ldots \\ \vdots & \vdots & \ddots \end{bmatrix}, \quad \mathbf{Q}_t = \begin{bmatrix} \mathbf{Q}_t^{(1)} & \mathbf{0} & \ldots \\ \mathbf{0} & \mathbf{Q}_t^{(2)} & \ldots \\ \vdots & \vdots & \ddots \end{bmatrix},$$

$$\mathbf{Z}_t = \begin{bmatrix} \mathbf{Z}_t^{(1)} & \mathbf{Z}_t^{(2)} & \ldots \end{bmatrix}, \quad \mathbf{G}_t = \begin{bmatrix} \mathbf{G}_t^{(1)} & \mathbf{G}_t^{(2)} & \ldots \end{bmatrix},$$

$$\boldsymbol{a} = \begin{bmatrix} \boldsymbol{a}^{(1)} \\ \boldsymbol{a}^{(2)} \\ \vdots \end{bmatrix}, \quad \mathbf{P} = \begin{bmatrix} \mathbf{P}^{(1)} & \mathbf{0} & \ldots \\ \mathbf{0} & \mathbf{P}^{(2)} & \ldots \\ \vdots & \vdots & \ddots \end{bmatrix},$$

and $\mathbf{H}_t$ will be equal to some variance σ^2 if at least one (scalar) matrix $\mathbf{H}_t^{(i)}$ is not zero.

Example 5.11 (Effect on 9/11 and the Iraq war outbreak on airline passengers from and to Australia continued).
In Example 4.6 we modelled the monthly time series of international airline passengers from and to Australia using a model with LLT, stochastic seasonal dummies and two intervention variables with order $(r = 1, s = 0)$ transfer functions.

Suppose we want also to add a static regressor that counts the number of holidays (including weekends) in each month and an AR(1) component to fit local persistent deviations from the trend. To keep the size of the matrices small enough to fit the page size of this book, we will suppose that the time series is quarterly and not monthly and that only one intervention takes place.

Let us write the system matrices when the components follow the order, LLT, seasonal dummies, intervention, static regressor and AR(1), and the dummy variable for the intervention is D_t and the holiday count variable is X_t.

$$\mathbf{T} = \left[\begin{array}{cc|ccc|c|c|c} 1 & 1 & 0 & 0 & 0 & 0 & 0 & 0 \\ 0 & 1 & 0 & 0 & 0 & 0 & 0 & 0 \\ \hline 0 & 0 & -1 & -1 & -1 & 0 & 0 & 0 \\ 0 & 0 & 1 & 0 & 0 & 0 & 0 & 0 \\ 0 & 0 & 0 & 1 & 0 & 0 & 0 & 0 \\ \hline 0 & 0 & 0 & 0 & 0 & \delta_1 & 0 & 0 \\ \hline 0 & 0 & 0 & 0 & 0 & 0 & 1 & 0 \\ \hline 0 & 0 & 0 & 0 & 0 & 0 & 0 & \phi \end{array}\right], \quad \boldsymbol{d}_t = \left[\begin{array}{c} 0 \\ 0 \\ \hline 0 \\ 0 \\ 0 \\ \hline \omega_0 D_{t+1} \\ \hline 0 \\ \hline 0 \end{array}\right],$$

$$
\mathbf{Q} = \left[\begin{array}{cc|ccc|c|c|c}
\sigma^2_\eta & 0 & 0 & 0 & 0 & 0 & 0 & 0 \\
0 & \sigma^2_\zeta & 0 & 0 & 0 & 0 & 0 & 0 \\
\hline
0 & 0 & \sigma^2_\omega & 0 & 0 & 0 & 0 & 0 \\
0 & 0 & 0 & 0 & 0 & 0 & 0 & 0 \\
0 & 0 & 0 & 0 & 0 & 0 & 0 & 0 \\
\hline
0 & 0 & 0 & 0 & 0 & 0 & 0 & 0 \\
\hline
0 & 0 & 0 & 0 & 0 & 0 & 0 & 0 \\
\hline
0 & 0 & 0 & 0 & 0 & 0 & 0 & \sigma^2_\varepsilon
\end{array}\right], \quad
\mathbf{Z}_t^\top = \left[\begin{array}{c}
1 \\ 0 \\ \hline 1 \\ 0 \\ 0 \\ \hline 1 \\ \hline X_t \\ \hline 1
\end{array}\right],
$$

$$
\mathbf{P} = \left[\begin{array}{cc|ccc|c|c|c}
\infty & 0 & 0 & 0 & 0 & 0 & 0 & 0 \\
0 & \infty & 0 & 0 & 0 & 0 & 0 & 0 \\
\hline
0 & 0 & \infty & 0 & 0 & 0 & 0 & 0 \\
0 & 0 & 0 & \infty & 0 & 0 & 0 & 0 \\
0 & 0 & 0 & 0 & \infty & 0 & 0 & 0 \\
\hline
0 & 0 & 0 & 0 & 0 & 0 & 0 & 0 \\
\hline
0 & 0 & 0 & 0 & 0 & 0 & \infty & 0 \\
\hline
0 & 0 & 0 & 0 & 0 & 0 & 0 & \frac{\sigma^2_\varepsilon}{(1-\phi^2)}
\end{array}\right], \quad
\boldsymbol{a} = \left[\begin{array}{c}
0 \\ 0 \\ \hline 0 \\ 0 \\ 0 \\ \hline 0 \\ \hline 0 \\ \hline 0
\end{array}\right],
$$

and if an observation error is added, we have $\mathbf{H} = \sigma^2$. The covariance matrix of the observation error with the state disturbances is a row vector of zeros. The parameters to be estimated by maximum likelihood are seven: the coefficients δ_1, ω_0, ϕ and the variances $\sigma^2_\eta, \sigma^2_\zeta, \sigma^2_\omega, \sigma^2$.

5.3 Inference for the unobserved components

Suppose for now that all the parameters in the system matrices of the state space form are known. In this case the only unknown quantities are the unobservable components. Since the components are (generally) specified as random variables (i.e., the stochastic process $\{\boldsymbol{\alpha}_t\}$) to be guessed through the realisation of other random variables (i.e., the stochastic process $\{\boldsymbol{Y}_t\}$), then the inference to carry out is the *statistical prediction* seen in Chapter 1.

Let us name the observable time series up to observation t as $\mathcal{Y}_t = \{\boldsymbol{Y}_1, \ldots, \boldsymbol{Y}_t\}$. The statistical predictions we are usually interested in are the following:

forecasting: statistical prediction of $\boldsymbol{\alpha}_t$ based on $\mathcal{Y}_s$ with $s < t$;

filtering: statistical prediction of $\boldsymbol{\alpha}_t$ based on $\mathcal{Y}_t$;

smoothing: statistical prediction of $\boldsymbol{\alpha}_t$ based on $\mathcal{Y}_s$ with $s > t$.

Since the state space form defines linear recursions, and stochastic properties of the system are defined up to the second moments (means and covariance matrices), we can build optimal linear predictors of $\boldsymbol{\alpha}_t$ based on $\mathcal{Y}_s$ without imposing further assumptions on the system (see Section 1.2). Of course, optimal linear predictors will be better than any other predictor if we can assume that the white noise sequences $\{\boldsymbol{\epsilon}_t\}$ and $\{\boldsymbol{\nu}_t\}$, and the initial state $\boldsymbol{\alpha}_1$ (and, thus the observable time series $\{\boldsymbol{Y}_t\}$) are jointly Gaussian.

In the next sections we will use the following notation for the linear prediction of the state vector based on the observable time series and the relative prediction error covariance matrix:

$$\boldsymbol{a}_{t|s} = \mathbb{P}[\boldsymbol{\alpha}_t|\mathcal{Y}_s], \qquad \mathbf{P}_{t|s} = \mathbb{E}[\boldsymbol{\alpha}_t - \boldsymbol{a}_{t|s}][\boldsymbol{\alpha}_t - \boldsymbol{a}_{t|s}]^\top.$$

Under the hypothesis of normality of the state space form, the following identities hold

$$\boldsymbol{a}_{t|s} = \mathbb{E}[\boldsymbol{\alpha}_t|\mathcal{Y}_s], \qquad \mathbf{P}_{t|s} = \mathbb{V}\mathrm{ar}[\boldsymbol{\alpha}_t|\mathcal{Y}_s],$$

where $\mathbb{V}\mathrm{ar}[\boldsymbol{\alpha}_t|\mathcal{Y}_s]$ denotes the conditional variance-covariance matrix of $\boldsymbol{\alpha}_t$ given $\mathcal{Y}_s$.

5.3.1 *The Kalman filter*

The *Kalman filter* is an algorithm for computing the pair $\{\boldsymbol{a}_{t|t-1}, \mathbf{P}_{t|t-1}\}$ starting from $\{\boldsymbol{a}_{t-1|t-1}, \mathbf{P}_{t-1|t-1}\}$, and $\{\boldsymbol{a}_{t|t}, \mathbf{P}_{t|t}\}$ from $\{\boldsymbol{a}_{t|t-1}, \mathbf{P}_{t|t-1}\}$. This way, by providing an initial value for $\{\boldsymbol{a}_{1|0}, \mathbf{P}_{1|0}\}$ or $\{\boldsymbol{a}_{0|0}, \mathbf{P}_{0|0}\}$, the two equations can be iterated to compute $\{\boldsymbol{a}_{t|t}, \mathbf{P}_{t|t}\}$ and $\{\boldsymbol{a}_{t+1|t}, \mathbf{P}_{t+1|t}\}$ for $t = 1, 2, \ldots, n$:

$$\{\boldsymbol{a}_{1|0}, \mathbf{P}_{1|0}\} \mapsto \{\boldsymbol{a}_{1|1}, \mathbf{P}_{1|1}\} \mapsto \{\boldsymbol{a}_{2|1}, \mathbf{P}_{2|1}\} \mapsto \{\boldsymbol{a}_{2|2}, \mathbf{P}_{2|2}\} \mapsto \ldots$$

As a byproduct the Kalman filter also provides the sequence of innovations (i.e., the one-step-ahead prediction errors) with the relative covariance matrix, which can be used to compute the Gaussian likelihood of the model in state space form (see Section 5.4 below).

Theorem 5.3 (Kalman filter recursions). *Assume the time series $\{\mathbf{Y}_1, \ldots, \mathbf{Y}_n\}$ is generated by a model in state space form as in Definition 5.1, then*

Prediction step

$$\begin{aligned} \boldsymbol{a}_{t|t-1} &= \mathbf{T}_{t-1}\boldsymbol{a}_{t-1|t-1} + \boldsymbol{d}_{t-1} + \mathbf{G}_{t-1}^\top \mathbf{F}_{t-1}^{-1} \boldsymbol{\iota}_{t-1} \\ \mathbf{P}_{t|t-1} &= \mathbf{T}_{t-1}\mathbf{P}_{t-1|t-1}\mathbf{T}_{t-1}^\top + \mathbf{Q}_{t-1} - \mathbf{G}_{t-1}^\top \mathbf{F}_{t-1}^{-1}\mathbf{G}_{t-1} \end{aligned} \tag{5.8}$$

One-step-ahead forecast and innovation

$$\begin{aligned} \hat{\mathbf{Y}}_{t|t-1} &= \mathbb{P}[\mathbf{Y}_t|\mathcal{Y}_t] = \mathbf{Z}_t\boldsymbol{a}_{t|t-1} + \boldsymbol{c}_t \\ \boldsymbol{\iota}_t &= \mathbf{Y}_t - \hat{\mathbf{Y}}_{t|t-1} \\ \mathbf{F}_t &= \mathbb{E}[\boldsymbol{\iota}_t\boldsymbol{\iota}_t^\top] = \mathbf{Z}_t\mathbf{P}_{t|t-1}\mathbf{Z}_t^\top + \mathbf{H}_t \end{aligned} \tag{5.9}$$

Updating step

$$\begin{aligned} \boldsymbol{a}_{t|t} &= \boldsymbol{a}_{t|t-1} + \mathbf{P}_{t|t-1}\mathbf{Z}_t^\top \mathbf{F}_t^{-1}\boldsymbol{\iota}_t \\ \mathbf{P}_{t|t} &= \mathbf{P}_{t|t-1} - \mathbf{P}_{t|t-1}\mathbf{Z}_t^\top \mathbf{F}_t^{-1}\mathbf{Z}_t\mathbf{P}_{t|t-1} \end{aligned} \tag{5.10}$$

Proof. One-step-ahead forecast and innovation. We start by proving equation (5.9) since it is very elementary and the results appear also in equation (5.8). Applying the unbiasedness and linearity properties of the optimal linear predictor (see Theorem 1.4) we obtain:

$$\begin{aligned}
\hat{\boldsymbol{Y}}_{t|t-1} &= \mathbb{P}[\boldsymbol{Y}_t|\mathcal{Y}_{t-1}] = \mathbb{P}[\mathbf{Z}_t\boldsymbol{\alpha}_t + \boldsymbol{d}_t + \boldsymbol{\epsilon}_t|\mathcal{Y}_{t-1}] \\
&= \mathbf{Z}_t\mathbb{P}[\boldsymbol{\alpha}_t|\mathcal{Y}_{t-1}] + \boldsymbol{d}_t + \mathbb{P}[\boldsymbol{\epsilon}_t|\mathcal{Y}_{t-1}] = \mathbf{Z}_t\boldsymbol{a}_{t|t-1} + \boldsymbol{d}_t; \\
\mathbf{F}_t &= \mathbb{E}[\boldsymbol{Y}_t - \hat{\boldsymbol{Y}}_{t|t-1}][\boldsymbol{Y}_t - \hat{\boldsymbol{Y}}_{t|t-1}]^\top \\
&= \mathbb{E}[\mathbf{Z}_t\boldsymbol{\alpha}_t + \boldsymbol{d}_t + \boldsymbol{\epsilon}_t - \mathbf{Z}_t\boldsymbol{a}_{t|t-1} - \boldsymbol{d}_t][\mathbf{Z}_t\boldsymbol{\alpha}_t + \boldsymbol{d}_t + \boldsymbol{\epsilon}_t - \mathbf{Z}_t\boldsymbol{a}_{t|t-1} - \boldsymbol{d}_t]^\top \\
&= \mathbb{E}[\mathbf{Z}_t(\boldsymbol{\alpha}_t - \boldsymbol{a}_{t|t-1})][\mathbf{Z}_t(\boldsymbol{\alpha}_t - \boldsymbol{a}_{t|t-1})]^\top + \mathbb{E}[\boldsymbol{\epsilon}_t\boldsymbol{\epsilon}_t^\top] \\
&= \mathbf{Z}_t\mathbf{P}_{t|t-1}\mathbf{Z}_t^\top + \mathbf{H}_t,
\end{aligned}$$

where the cross products disappeared because the assumptions on the state space model imply that $\mathbb{E}[(\boldsymbol{\alpha}_t - \boldsymbol{a}_{t|t-1})\boldsymbol{\epsilon}_t^\top] = \mathbf{0}$.

Prediction step. Applying again the unbiasedness and linearity properties of the optimal linear predictor to the (lagged) transition equation, we get

$$\begin{aligned}
\boldsymbol{a}_{t|t-1} = \mathbb{P}[\boldsymbol{\alpha}_t|\mathcal{Y}_{t-1}] &= \mathbb{P}[\mathbf{T}_{t-1}\boldsymbol{\alpha}_{t-1} + \boldsymbol{d}_{t-1} + \boldsymbol{\nu}_{t-1}|\mathcal{Y}_{t-1}] \\
&= \mathbf{T}_{t-1}\boldsymbol{a}_{t-1|t-1} + \boldsymbol{d}_{t-1} + \mathbb{P}[\boldsymbol{\nu}_{t-1}|\mathcal{Y}_{t-1}].
\end{aligned} \tag{5.11}$$

In the common case in which the measurement error and the transition equation disturbances are uncorrelated (i.e., $\mathbf{G}_{t-1} = \mathbf{0}$), then $\mathbb{P}[\boldsymbol{\nu}_{t-1}|\mathcal{Y}_{t-1}] = \mathbf{0}$. If, instead, $\mathbf{G}_{t-1} \neq \mathbf{0}$, then we can first notice that $\mathbb{P}[\boldsymbol{\nu}_{t-1}|\mathcal{Y}_{t-1}] = \mathbb{P}[\boldsymbol{\nu}_{t-1}|\boldsymbol{Y}_{t-1}]$ as by assumption $\boldsymbol{\nu}_{t-1}$ is uncorrelated with $\mathcal{Y}_{t-2}$. Then, using the orthogonal decomposition of $\boldsymbol{Y}_t$ as

$$\boldsymbol{Y}_{t-1} = \mathbb{P}[\boldsymbol{Y}_{t-1}|\mathcal{Y}_{t-2}] + \big(\boldsymbol{Y}_{t-1} - \mathbb{P}[\boldsymbol{Y}_{t-1}|\mathcal{Y}_{t-2}]\big) = \hat{\boldsymbol{Y}}_{t-1|t-2} + \boldsymbol{\iota}_{t-1},$$

where we used the same notation as in equation (5.9), we can exploit the uncorrelatedness of the two addends to can compute the linear projection as

$$\mathbb{P}[\boldsymbol{\nu}_{t-1}|\boldsymbol{Y}_{t-1}] = \mathbb{P}[\boldsymbol{\nu}_{t-1}|\hat{\boldsymbol{Y}}_{t-1|t-2} + \boldsymbol{\iota}_{t-1}] = \mathbb{P}[\boldsymbol{\nu}_{t-1}|\hat{\boldsymbol{Y}}_{t-1|t-2}] + \mathbb{P}[\boldsymbol{\nu}_{t-1}|\boldsymbol{\iota}_{t-1}],$$

where $\mathbb{P}[\boldsymbol{\nu}_{t-1}|\hat{\boldsymbol{Y}}_{t-1|t-2}] = \mathbf{0}$ as $\boldsymbol{\nu}_{t-1}$ is uncorrelated with $\mathcal{Y}_{t-2}$. Now we can compute

$$\mathbb{P}[\boldsymbol{\nu}_{t-1}|\boldsymbol{\iota}_{t-1}] = \mathbb{E}[\boldsymbol{\nu}_{t-1}, \boldsymbol{\iota}_{t-1}]\mathbb{E}[\boldsymbol{\iota}_{t-1}\boldsymbol{\iota}_{t-1}^\top]^{-1}\boldsymbol{\iota}_{t-1} = \mathbf{G}_{t-1}^\top\mathbf{F}_{t-1}^{-1}\boldsymbol{\iota}_{t-1},$$

and substituting in equation (5.11) we obtain the result in the first line of equation (5.8).

The covariance matrix of this prediction error is given by

$$\begin{aligned}
\mathbf{P}_{t|t-1} &= \mathbb{E}[\boldsymbol{\alpha}_t - \boldsymbol{a}_{t|t-1}][\boldsymbol{\alpha}_t - \boldsymbol{a}_{t|t-1}]^\top \\
&= \mathbb{E}[\mathbf{T}_{t-1}(\boldsymbol{\alpha}_{t-1} - \boldsymbol{a}_{t-1|t-1}) + \boldsymbol{\nu}_{t-1} - \mathbf{G}_{t-1}^\top \mathbf{F}_{t-1}^{-1}\iota_{t-1}][\ldots = \ldots]^\top \\
&= \mathbf{T}_{t-1}\mathbf{P}_{t-1|t-1}\mathbf{T}_{t-1}^\top + \mathbf{Q}_{t-1} + \\
&\quad + \mathbf{G}_{t-1}^\top \mathbf{F}_{t-1}^{-1}\mathbf{F}_{t-1}\mathbf{F}_{t-1}^{-1}\mathbf{G}_{t-1} - 2\mathbf{G}_{t-1}^\top \mathbf{F}_{t-1}^{-1}\mathbf{G}_{t-1} \\
&= \mathbf{T}_{t-1}\mathbf{P}_{t-1|t-1}\mathbf{T}_{t-1}^\top + \mathbf{Q}_{t-1} - \mathbf{G}_{t-1}^\top \mathbf{F}_{t-1}^{-1}\mathbf{G}_{t-1},
\end{aligned}$$

since by assumption $\boldsymbol{\nu}_{t-1}$ is uncorrelated with $(\boldsymbol{\alpha}_{t-1} - \boldsymbol{a}_{t-1|t-1})$.

Updating step. Adapting the formula to update a linear prediction (Theorem 1.4, property 7.) we can write

$$\mathbb{P}[\boldsymbol{\alpha}_t|\boldsymbol{Y}_t, \mathcal{Y}_{t-1}] = \mathbb{P}[\boldsymbol{\alpha}_t|\mathcal{Y}_{t-1}] + \boldsymbol{\Sigma}_{\boldsymbol{\alpha}_t\boldsymbol{Y}_t|\mathcal{Y}_{t-1}}\boldsymbol{\Sigma}_{\boldsymbol{Y}_t\boldsymbol{Y}_t|\mathcal{Y}_{t-1}}^{-1}(\boldsymbol{Y}_t - \mathbb{P}[\boldsymbol{Y}_t|\mathcal{Y}_{t-1}]),$$

where

$$\begin{aligned}
\boldsymbol{\Sigma}_{\boldsymbol{\alpha}_t\boldsymbol{Y}_t|\mathcal{Y}_{t-1}} &= \mathbb{E}[\boldsymbol{\alpha}_t - \boldsymbol{a}_{t|t-1}][\boldsymbol{Y}_t - \hat{\boldsymbol{Y}}_{t|t-1}]^\top \\
&= \mathbb{E}[\boldsymbol{\alpha}_t - \boldsymbol{a}_{t|t-1}][\mathbf{Z}_t(\boldsymbol{\alpha}_t - \boldsymbol{a}_{t|t-1}) + \boldsymbol{\epsilon}_t]^\top \\
&\mathbf{P}_{t|t-1}\mathbf{Z}_t^\top,
\end{aligned}$$

as $\boldsymbol{\epsilon}_t$ is uncorrelated with $(\boldsymbol{\alpha}_t - \boldsymbol{a}_{t|t-1})$ by assumption, and

$$\boldsymbol{\Sigma}_{\boldsymbol{Y}_t\boldsymbol{Y}_t|\mathcal{Y}_{t-1}} = \mathbb{E}[\boldsymbol{Y}_t - \hat{\boldsymbol{Y}}_{t|t-1}][\boldsymbol{Y}_t - \hat{\boldsymbol{Y}}_{t|t-1}]^\top = \mathbf{F}_t.$$

Substituting in the formula to update the prediction we obtain the first line of equation (5.10). The second line follows from the application of the MSE formula for updating a linear prediction (Theorem 1.4, property 7.) to our case:

$$\begin{aligned}
MSE_{\boldsymbol{\alpha}_t|\mathcal{Y}_t} &= MSE_{\boldsymbol{\alpha}_t|\mathcal{Y}_{t-1}} - \boldsymbol{\Sigma}_{\boldsymbol{\alpha}_t\boldsymbol{Y}_t|\mathcal{Y}_{t-1}}\boldsymbol{\Sigma}_{\boldsymbol{Y}_t\boldsymbol{Y}_t|\mathcal{Y}_{t-1}}^{-1}\boldsymbol{\Sigma}_{\boldsymbol{\alpha}_t\boldsymbol{Y}_t|\mathcal{Y}_{t-1}}^\top, \\
\mathbf{P}_{t|t} &= \mathbf{P}_{t|t-1} - \mathbf{P}_{t|t-1}\mathbf{Z}_t^\top \mathbf{F}_t^{-1}\mathbf{Z}_t\mathbf{P}_{t|t-1}.
\end{aligned}$$

□

If the system matrices $\mathbf{T}_t$, $\mathbf{Q}_t$, $\mathbf{Z}_t$, $\mathbf{H}_t$, $\mathbf{G}_t$ are time-invariant (the vectors $\boldsymbol{c}_t$ and $\boldsymbol{d}_t$ can be time-varying), under mild regularity conditions (see Harvey, 1989, Section 3.3) always holding in a well-specified UCM, the covariance matrices $\mathbf{P}_{t|t-1}$ and $\mathbf{P}_{t|t}$ eventually converge to two fixed matrices. When this happens, one says that the Kalman filter has reached the *steady state*.

Example 5.12 (Kalman filter for the local level model).
Let us see how the Kalman filter recursions work on the *local level model*, that is, the simple model

$$Y_t = \mu_t + \epsilon_t, \qquad \mu_{t+1} = \mu_t + \nu_t,$$

with the white noise sequences $\{\epsilon_t\}$ and $\{\nu_t\}$ being uncorrelated (i.e., $\mathbf{G} = 0$).

For this model the Kalman filter formulae specialise to

$$a_{t|t-1} = a_{t-1|t-1}, \qquad P_{t|t-1} = P_{t-1|t-1} + \sigma^2_\nu$$
$$Y_{t|t-1} = a_{t|t-1}, \qquad \iota_t = Y_t - a_{t|t-1}, \qquad F_t = P_{t|t-1} + \sigma^2_\epsilon$$
$$a_{t|t} = a_{t|t-1} + \frac{P_{t|t-1}}{P_{t|t-1} + \sigma^2_\epsilon}(Y_t - a_{t|t-1}),$$
$$P_{t|t} = P_{t|t-1} - \frac{P^2_{t|t-1}}{P_{t|t-1} + \sigma^2_\epsilon} = \frac{P_{t|t-1}\sigma^2_\epsilon}{P_{t|t-1} + \sigma^2_\epsilon}.$$

The component μ_t is nonstationary and, thus, a diffuse distribution should be used for μ_1. Thus, let us set

$$a_{1|0} = 0, \qquad P_{1|0} = \tau,$$

where τ is a large positive number that we will let diverge to infinity. The one-step-ahead forecast, the innovation and its variance at time $t = 1$ are (equation 5.9):

$$\hat{Y}_{1|0} = 0, \qquad \iota_1 = Y_1, \qquad F_1 = \tau + \sigma^2_\epsilon.$$

The first iteration concludes with the updating step (equation 5.10):

$$a_{1|1} = \frac{\tau}{\tau + \sigma^2_\epsilon} Y_1, \qquad P_{1|1} = \frac{\tau\sigma^2_\epsilon}{\tau + \sigma^2_\epsilon}.$$

Now it is interesting to see what happens to the above quantities when we let $\tau \to \infty$ so that the distribution of μ_1 is diffuse:

$$\lim_{\tau\to\infty} a_{1|1} = Y_1, \qquad \lim_{\tau\to\infty} P_{1|1} = \sigma^2_\epsilon.$$

Thus, even though the MSE is infinite when predicting μ_1 with no information, after the first observation Y_1 is available, the MSE of the prediction (filter) of μ_1 based on Y_1 is finite. In general if the initial state vector has d elements with diffuse distributions, then the variance of the prediction becomes finite when at least d observations are available.

We mentioned that in a time-invariant state space the matrices $\mathbf{P}_{t|t-1}$ and $\mathbf{P}_{t|t}$ converge. Let us find the values to which they converge in the local level model. First, let us substitute the definition of $P_{t|t}$ into the rhs of the definition of $P_{t+1|t}$ so that we obtain the recursive equation

$$P_{t+1|t} = \frac{P_{t|t-1}\sigma^2_\epsilon}{P_{t|t-1} + \sigma^2_\epsilon} + \sigma^2_\nu.$$

When $P_{t+1|t}$ converges to some P the above equation continues to hold and so also P must satisfy

$$P = \frac{P\sigma_\epsilon^2}{P+\sigma_\epsilon^2} + \sigma_\nu^2,$$

which can be rewritten

$$P^2 - \sigma_\nu^2 P - \sigma_\nu^2\sigma_\epsilon^2 = 0,$$

whose solutions are given by

$$P = \frac{\sigma_\nu^2 \pm \sqrt{\sigma_\nu^4 + 4\sigma_\nu^2\sigma_\epsilon^2}}{2}.$$

The quantity under square root is certainly not smaller than σ_ν^2 and so we can only take the solution with "+" as P must be non-negative (it is a MSE). Given P, the steady state solution for $P_{t|t}$, say $\tilde{P}$, is easily derived from its definition:

$$\tilde{P} = \frac{P\sigma_\epsilon^2}{P+\sigma_\epsilon^2}.$$

Figure 5.1 depicts the time series of a simulated local level model with the Kalman filter inference for the level variable μ_t (top graph), and the MSE of both the one-step-ahead predictor and filter (bottom graph). The variances were set to $\sigma_\epsilon^2 = 1$, $\sigma_\nu^2 = 0.1$ and $\tau = 10^{10}$, and so the steady state values of the MSE can be easily computed with the formulae above: $P = 0.37$ and $\tilde{P} = 0.27$.

Notice that when the Kalman filter reaches the steady state the filtering recursion can be written as (recall that for this model $a_{t|t-1} = a_{t-1|t-1}$)

$$a_{t|t} = a_{t-1|t-1} + \lambda(Y_t - a_{t-1|t-1}) = \lambda Y_t + (1-\lambda)a_{t-1|t-1},$$

with $0 < \lambda = P/(P+\sigma_\epsilon^2) \leq 1$, which is known as *exponential smoothing* or *exponentially weighted moving average* (EWMA). This class of forecasting methods is treated, expanded and related to the state space form in the book by Hyndman et al. (2008).

Often the two steps of the Kalman filter are merged into just one recursion for the one-step-ahead predictor[1] $\boldsymbol{a}_{t+1|t}$:

$$\begin{aligned}\boldsymbol{a}_{t+1|t} &= \mathbf{T}_t\boldsymbol{a}_{t|t-1} + \boldsymbol{d}_t + \mathbf{K}\boldsymbol{\iota}_t,\\ \mathbf{P}_{t+1|t} &= \mathbf{T}_t\mathbf{P}_{t|t-1}\mathbf{T}_t^\top + \mathbf{Q}_t - \mathbf{K}_t\mathbf{F}_t\mathbf{K}_t^\top,\end{aligned} \tag{5.12}$$

[1]The reader is invited to derive these equations by substituting equation (5.10) into (5.8).

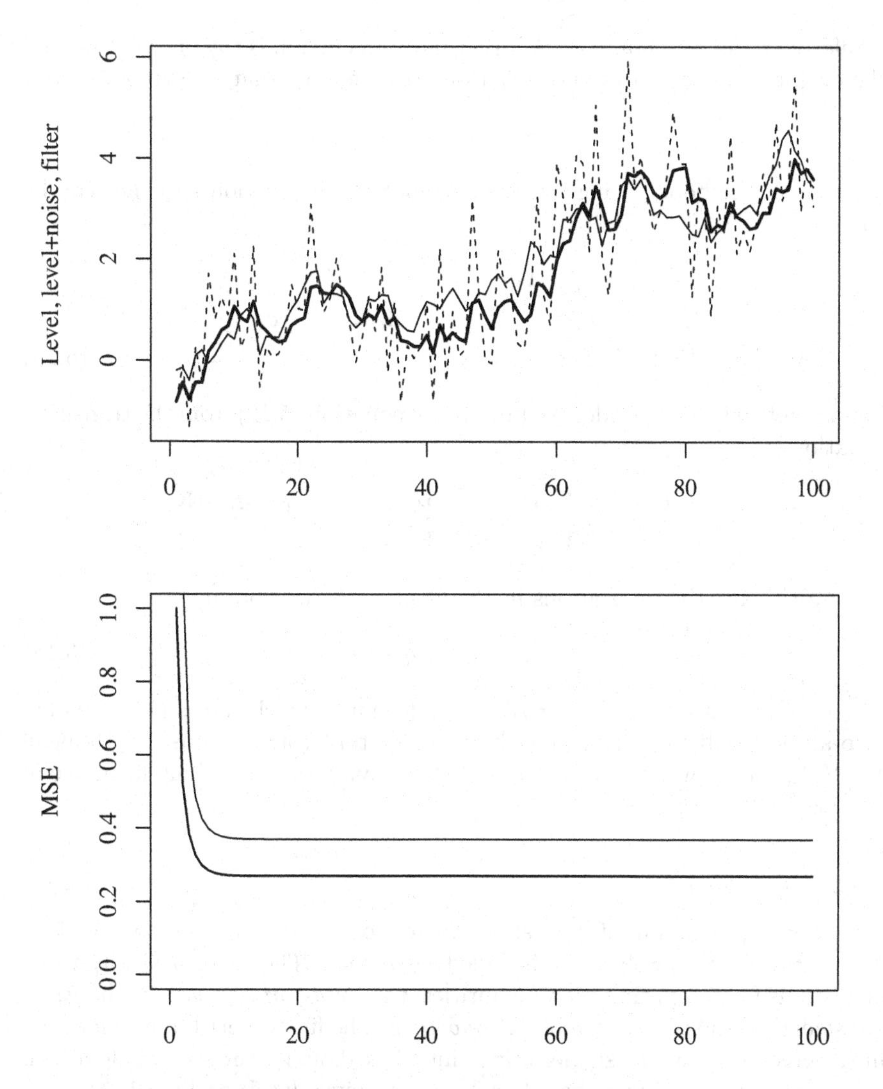

Figure 5.1 *Top graph: observed time series Y_t (dashed line), real unobservable level μ_t (thin line) and its filter $a_{t|t}$ (thick line). Bottom graph: MSE of the one-step-ahead predictor, $P_{t|t-1}$ (thinner line), and of the filter, $P_{t|t}$ (thicker line). The data were simulated from the local level model with $\sigma_\varepsilon^2 = 1$ and $\sigma_\nu^2 = 0.1$. The initial variance for the state variable was $\tau = 10^{10}$.*

where the matrix

$$\mathbf{K}_t = \left(\mathbf{T}_t\mathbf{P}_{t|t-1}\mathbf{Z}_t^\top\mathbf{F}_t^{-1} + \mathbf{G}_t^\top\mathbf{F}_t^{-1}\right)$$

is referred to as *Kalman gain*. Another useful recursion is the one that defines the evolution of the one-step-prediction errors for $\boldsymbol{\alpha}_t$, that we symbolise as

$$\boldsymbol{\chi}_t = \boldsymbol{\alpha}_t - \boldsymbol{a}_{t|t-1}$$

throughout the book. Using this new symbol, the innovation sequence can be written as

$$\begin{aligned}\boldsymbol{\iota}_t &= \boldsymbol{Y}_t - \hat{\boldsymbol{Y}}_{t|t-1}\\ &= \mathbf{Z}_t\boldsymbol{\alpha}_t + \boldsymbol{c}_t + \boldsymbol{\epsilon}_t - \mathbf{Z}_t\boldsymbol{a}_{t|t-1} - \boldsymbol{c}_t\\ &= \mathbf{Z}_t\boldsymbol{\chi}_t + \boldsymbol{\epsilon}_t. \end{aligned} \tag{5.13}$$

Let us subtract, side by side, the first line of equation (5.12) from the transition equation (5.2):

$$\begin{aligned}\boldsymbol{\alpha}_{t+1} - \boldsymbol{a}_{t+1|t} &= \mathbf{T}_t\boldsymbol{\alpha}_t + \boldsymbol{d}_t + \boldsymbol{\nu}_t - \mathbf{T}_t\boldsymbol{a}_{t|t-1} - \boldsymbol{d}_t - \mathbf{K}\boldsymbol{\iota}_t,\\ \boldsymbol{\chi}_{t+1} &= \mathbf{T}_t\boldsymbol{\chi}_t + \boldsymbol{\nu}_t - \mathbf{K}_t\boldsymbol{\iota}_t.\end{aligned}$$

By substituting the innovations in the form (5.13) we obtain

$$\boldsymbol{\chi}_{t+1} = \mathbf{L}_t\boldsymbol{\chi}_t + \boldsymbol{\nu}_t - \mathbf{K}_t\boldsymbol{\epsilon}_t, \tag{5.14}$$

where we have set $\mathbf{L}_t = \mathbf{T}_t - \mathbf{K}_t\mathbf{Z}_t$. This formula, which shows how the one-step-ahead prediction errors for the state vector evolve also as a zero-mean VAR(1) process, will prove very useful for deriving the smoothing formulae in Section 5.3.2.

5.3.2 *Smoothing*

The linear prediction of the state vector $\boldsymbol{\alpha}_t$ based on the observations $\mathcal{Y}_s = \{\boldsymbol{Y}_1, \ldots, \boldsymbol{Y}_s\}$ with $s > t$ is called *smoothing*. There are basically three algorithms for computing these quantities: the *fixed-point smoother*, the *fixed-lag smoother* and the *fixed-interval smoother*. The first algorithm provides the linear prediction for the single time point t based on $\mathcal{Y}_s$; the second algorithm provides predictions for $\boldsymbol{\alpha}_t$ based on $\mathcal{Y}_{t+k}$ for a given $k \in \mathbb{N}$ and $t = 1, 2, \ldots, n$; the third algorithm provides predictions for $\boldsymbol{\alpha}_t$, $t = 1, 2, \ldots, n$, based on the whole time series $\mathcal{Y}_n$. Here, we treat only the latter algorithm which is the one typically needed in statistics and econometrics and found in the relative software packages.

Consistent with the previous sections, the smoother and the covariance matrix of its prediction error are named

$$\boldsymbol{a}_{t|n} = \mathbb{P}[\boldsymbol{\alpha}_t|\mathcal{Y}_n], \qquad \mathbf{P}_{t|n} = \mathbb{E}[\boldsymbol{\alpha}_t - \boldsymbol{a}_{t|n}][\boldsymbol{\alpha}_t - \boldsymbol{a}_{t|n}]^\top.$$

Theorem 5.4 (Fixed-interval smoother). *Assume the time series $\{\mathbf{Y}_1, \ldots, \mathbf{Y}_n\}$ is generated by a model in state space form as in Definition 5.1 and that the Kalman filter presented in Theorem 5.3 has run and $\boldsymbol{r}_n = \mathbf{0}$, $\mathbf{N}_n = 0$; then, for $t = n, n-1, \ldots, 1$,*

$$\begin{aligned}
\boldsymbol{r}_{t-1} &= \mathbf{Z}_t^\top \mathbf{F}_t^{-1} \boldsymbol{\iota}_t + \mathbf{L}_t^\top \boldsymbol{r}_t, \\
\mathbf{N}_{t-1} &= \mathbf{Z}_t^\top \mathbf{F}_t^{-1} \mathbf{Z}_t + \mathbf{L}_t^\top \mathbf{N}_t \mathbf{L}_t, \\
\boldsymbol{a}_{t|n} &= \boldsymbol{a}_{t|t-1} + \mathbf{P}_{t|t-1} \boldsymbol{r}_{t-1}, \\
\mathbf{P}_{t|n} &= \mathbf{P}_{t|t-1} - \mathbf{P}_{t|t-1} \mathbf{N}_{t-1} \mathbf{P}_{t|t-1}.
\end{aligned} \tag{5.15}$$

Proof. Before computing the smoother, we need the following result: for $k = 0, 1, \ldots$

$$\mathbb{E}\left[\boldsymbol{\chi}_t \, \boldsymbol{\chi}_{t+k}^\top\right] = \mathbf{P}_{t|t-1} \mathbf{L}_t^\top \mathbf{L}_{t+1}^\top \cdots \mathbf{L}_{t+k-1}^\top,$$

which is easy to derive being the cross-covariance function of the time-varying VAR(1) process (5.14).

Furthermore, we exploit the identity

$$\mathbb{P}[\boldsymbol{\alpha}_t | \mathcal{Y}_n] = \mathbb{P}[\boldsymbol{\alpha}_t | \mathcal{Y}_{t-1}, \boldsymbol{\iota}_t, \boldsymbol{\iota}_{t+1}, \ldots, \boldsymbol{\iota}_n],$$

which holds since the observations $\{\mathbf{Y}_t, \mathbf{Y}_{t+1}, \ldots, \mathbf{Y}_n\}$ can be recovered as linear transformations of the innovations $\{\boldsymbol{\iota}_t, \boldsymbol{\iota}_{t+1}, \ldots, \boldsymbol{\iota}_n\}$. The advantage of predicting using the innovations is that they are serially uncorrelated and also uncorrelated with past observations and so we can write the formula for updating a linear projection as (see Theorem 1.4 Properties 6 and 7)

$$\begin{aligned}
\mathbb{P}[\boldsymbol{\alpha}_t | \mathcal{Y}_{t-1}, \boldsymbol{\iota}_t, \boldsymbol{\iota}_{t+1}, \ldots, \boldsymbol{\iota}_n] &= \boldsymbol{a}_{t|t-1} + \sum_{j=t}^{n} \mathbb{E}\left[(\boldsymbol{\alpha}_t - \boldsymbol{a}_{t|t-1}) \boldsymbol{\iota}_j^\top\right] \mathbf{F}_j^{-1} \boldsymbol{\iota}_j \\
\boldsymbol{a}_{t|n} &= \boldsymbol{a}_{t|t-1} + \sum_{j=t}^{n} \mathbb{E}\left[\boldsymbol{\chi}_t (\mathbf{Z}_j \boldsymbol{\chi}_j + \boldsymbol{\epsilon}_j)^\top\right] \mathbf{F}_j^{-1} \boldsymbol{\iota}_j \\
\boldsymbol{a}_{t|n} &= \boldsymbol{a}_{t|t-1} + \sum_{j=t}^{n} \mathbb{E}\left[\boldsymbol{\chi}_t \, \boldsymbol{\chi}_j^\top\right] \mathbf{Z}_j^\top \mathbf{F}_j^{-1} \boldsymbol{\iota}_j, \\
\boldsymbol{a}_{t|n} &= \boldsymbol{a}_{t|t-1} + \mathbf{P}_{t|t-1} \sum_{j=t}^{n} \mathbf{L}_t^\top \mathbf{L}_{t+1}^\top \cdots \mathbf{L}_{j-1}^\top \mathbf{Z}_j^\top \mathbf{F}_j^{-1} \boldsymbol{\iota}_j,
\end{aligned}$$

where we used the identity (5.13) and the fact that $\boldsymbol{\epsilon}_j$ is uncorrelated with $\boldsymbol{\chi}_t$ for $j \geq t$, and for $j = t$ the element in the summation is to be read as as $\mathbf{Z}_j^\top \mathbf{F}_j^{-1} \boldsymbol{\iota}_j$. Now, let us rewrite the last expression letting t take the values

$n, n-1, \ldots, 1$:

$$\begin{aligned}
a_{n|n} &= a_{t|t-1} + \mathbf{P}_{n|n-1}\mathbf{Z}_n^\top \mathbf{F}_n^{-1}\iota_n,\\
a_{n-1|n} &= a_{n-1|n-2} + \mathbf{P}_{n-1|n-2}\left[\mathbf{Z}_{n-1}^\top \mathbf{F}_{n-1}^{-1}\iota_{n-1} + \mathbf{L}_{n-1}^\top \mathbf{Z}_n^\top \mathbf{F}_n^{-1}\iota_n\right],\\
a_{n-2|n} &= a_{n-2|n-3} + \mathbf{P}_{n-2|n-3}\times\\
&\quad \times\left[\mathbf{Z}_{n-2}^\top \mathbf{F}_{n-2}^{-1}\iota_{n-2} + \mathbf{L}_{n-2}^\top \mathbf{Z}_{n-1}^\top \mathbf{F}_{n-1}^{-1}\iota_{n-1} + \mathbf{L}_{n-2}^\top \mathbf{L}_{n-1}^\top \mathbf{Z}_n^\top \mathbf{F}_n^{-1}\iota_n\right],\\
\ldots &= \ldots
\end{aligned}$$

Notice that we can reproduce the sequence in square brackets using the recursion: $r_n = 0$, and for $t = n-1, n-2, \ldots, 1$,

$$r_{t-1} = \mathbf{Z}_t^\top \mathbf{F}_t^{-1}\iota_t + \mathbf{L}_t^\top r_t.$$

Indeed, by applying the recursion for $t = n, n-1, n-3, \ldots$ we obtain

$$\begin{aligned}
r_{n-1} &= \mathbf{Z}_n^\top \mathbf{F}_n^{-1}\iota_n,\\
r_{n-2} &= \mathbf{Z}_{n-1}^\top \mathbf{F}_{n-1}^{-1}\iota_{n-1} + \mathbf{L}_{n-1}^\top \mathbf{Z}_n^\top \mathbf{F}_n^{-1}\iota_n,\\
r_{n-3} &= \mathbf{Z}_{n-2}^\top \mathbf{F}_{n-2}^{-1}\iota_{n-2} + \mathbf{L}_{n-2}^\top \mathbf{Z}_{n-1}^\top \mathbf{F}_{n-1}^{-1}\iota_{n-1} + \mathbf{L}_{n-2}^\top \mathbf{L}_{n-1}^\top \mathbf{Z}_n^\top \mathbf{F}_n^{-1}\iota_n,\\
\ldots &= \ldots
\end{aligned}$$

and so, for $t = n-1, n-2, \ldots, 1$, we can write

$$a_{t|n} = a_{t|t-1} + \mathbf{P}_{t|t-1}r_{t-1}.$$

Now, let us compute the covariance sequence of r_t:

$$\begin{aligned}
\mathbf{N}_{t-1} = \mathbb{E}[r_{t-1}r_{t-1}^\top] &= \mathbb{E}\left[(\mathbf{Z}_t^\top \mathbf{F}_t^{-1}\iota_t + \mathbf{L}_t^\top r_t)(\mathbf{Z}_t^\top \mathbf{F}_t^{-1}\iota_t + \mathbf{L}_t^\top r_t)^\top\right]\\
&= \mathbf{Z}_t^\top \mathbf{F}_t^{-1}\mathbf{Z}_t + \mathbf{L}_t^\top \mathbf{N}_t \mathbf{L}_t,
\end{aligned}$$

where the cross-products are zero because the innovation sequence is serially uncorrelated and r_t is a linear function only of the future innovations $\iota_{t+1}, \iota_{t+2}\ldots, \iota_n$. Of course, since $r_n = 0$, then $\mathbf{N}_n = \mathbf{0}$. Let us see how $\mathbf{N}_t$ evolves for $t = n-1, n-2, \ldots$:

$$\begin{aligned}
\mathbf{N}_{n-1} =& \mathbf{Z}_n^\top \mathbf{F}_n^{-1}\mathbf{Z}_n,\\
\mathbf{N}_{n-2} =& \mathbf{Z}_{n-1}^\top \mathbf{F}_{n-1}^{-1}\mathbf{Z}_{n-1} + \mathbf{L}_{n-1}^\top \mathbf{Z}_n^\top \mathbf{F}_n^{-1}\mathbf{Z}_n\mathbf{L}_{n-1},\\
\mathbf{N}_{n-3} =& \mathbf{Z}_{n-2}^\top \mathbf{F}_{n-2}^{-1}\mathbf{Z}_{n-2} +\\
&+ \mathbf{L}_{n-2}^\top \mathbf{Z}_{n-1}^\top \mathbf{F}_{n-1}^{-1}\mathbf{Z}_{n-1}\mathbf{L}_{n-2} + \mathbf{L}_{n-2}^\top \mathbf{L}_{n-1}^\top \mathbf{Z}_n^\top \mathbf{F}_n^{-1}\mathbf{Z}_n\mathbf{L}_{n-1}\mathbf{L}_{n-2},\\
\ldots =& \ldots
\end{aligned}$$

Again, using Properties 6 and 7 of Theorem 1.4, we can derive the covariance matrix of the smoothing error:

$$\mathbf{P}_{t|n} = \mathbf{P}_{t|t-1} - \mathbf{P}_{t|t-1}\left[\sum_{j=t}^{n} \mathbf{L}_t^\top \mathbf{L}_{t+1}^\top \cdots \mathbf{L}_{j-1}^\top \mathbf{Z}_j^\top \mathbf{F}_j^{-1}\mathbf{Z}_j\mathbf{L}_{j-1}\cdots\mathbf{L}_{t+1}\mathbf{L}_t\right]\mathbf{P}_{t|t-1}.$$

It is straightforward to verify that the quantity in square brackets is exactly equal to $\mathbf{N}_{t-1}$ defined above. □

The following alternative fixed-interval smoothing algorithm can be found in many books such as Anderson and Moore (1979), Harvey (1989), Hamilton (1994), Lütkepohl (2007): for $t = n-1, n-2, \ldots, 1$,

$$\begin{aligned} \boldsymbol{a}_{t|n} &= \boldsymbol{a}_{t|t} + \mathbf{O}_t(\boldsymbol{a}_{t+1|n} - \boldsymbol{a}_{t+1|t}), \\ \mathbf{P}_{t|n} &= \mathbf{P}_{t|t} + \mathbf{O}_t(\mathbf{P}_{t+1|n} - \mathbf{P}_{t+1|t})\mathbf{O}_t^\top, \end{aligned} \tag{5.16}$$

with $\mathbf{O}_t = \mathbf{P}_{t|t}\mathbf{T}_t^\top \mathbf{P}_{t+1|t}^{-1}$.[2] The compact formulae may seem appealing, but the computational burden of this algorithm is generally much higher than the one in Theorem 5.4 because of the inversion of the sequence of matrices $\{\mathbf{P}_{t+1|t}\}$. In contrast, the algorithm in Theorem 5.4 requires the inversion of the sequence of matrices $\{\mathbf{F}_t\}$ which is already needed to compute the Kalman filter and whose dimensions are generally much smaller than those of $\mathbf{P}_{t+1|t}$. For example, in models for scalar time series $\mathbf{F}_t$ is just 1×1.

Example 5.13 (Fixed-interval smoothing for the local level model). In Example 5.12 we considered a random walk plus noise (local level model) and saw that, when the Kalman filter has reached the steady state, the one-step-ahead prediction recursion is given by

$$a_{t+1|t} = \lambda Y_t + (1-\lambda)a_{t|t-1},$$

with $\lambda = P/(P + \sigma_\epsilon^2)$.

Let us now derive the smoothing recursion assuming the Kalman filter is in the steady state. We will use equation (5.16) because it easily provides the result we want to obtain. Notice that in steady state the matrix $\mathbf{O}_t$ is also constant and equals $\sigma_\epsilon^2/(P+\sigma_\epsilon^2) = (1-\gamma)$. So, recalling that for this model $a_{t+1|t} = a_{t|t}$, we can write the smoother as

$$\begin{aligned} a_{t|n} &= a_{t|t} + O(a_{t+1|n} - a_{t+1|t}) \\ &= a_{t+1|t} + (1-\lambda)(a_{t+1|n} - a_{t+1|t}) \\ &= \lambda a_{t+1|t} + (1-\lambda)a_{t+1|n}. \end{aligned}$$

By observing the recursions for $a_{t+1|t}$ and for $a_{t|n}$ we note that the first one has the structure of an EWMA applied to the original time series $\{Y_t\}$, and the second one has the same structure but it is applied backwards to the previously computed sequence $a_{t+1|t}$.

The top graph of Figure 5.2 depicts the time series, Y_t, the random walk component, μ_t, and its prediction by filtering, $a_{t|t}$, and smoothing, $P_{t|n}$. On the average, the smoother rests closer to the real component than the filter. The bottom graph compares the MSE of the filter with

[2] For a proof using methods similar to the ones used in this book refer to Hamilton (1994, Section 13.6).

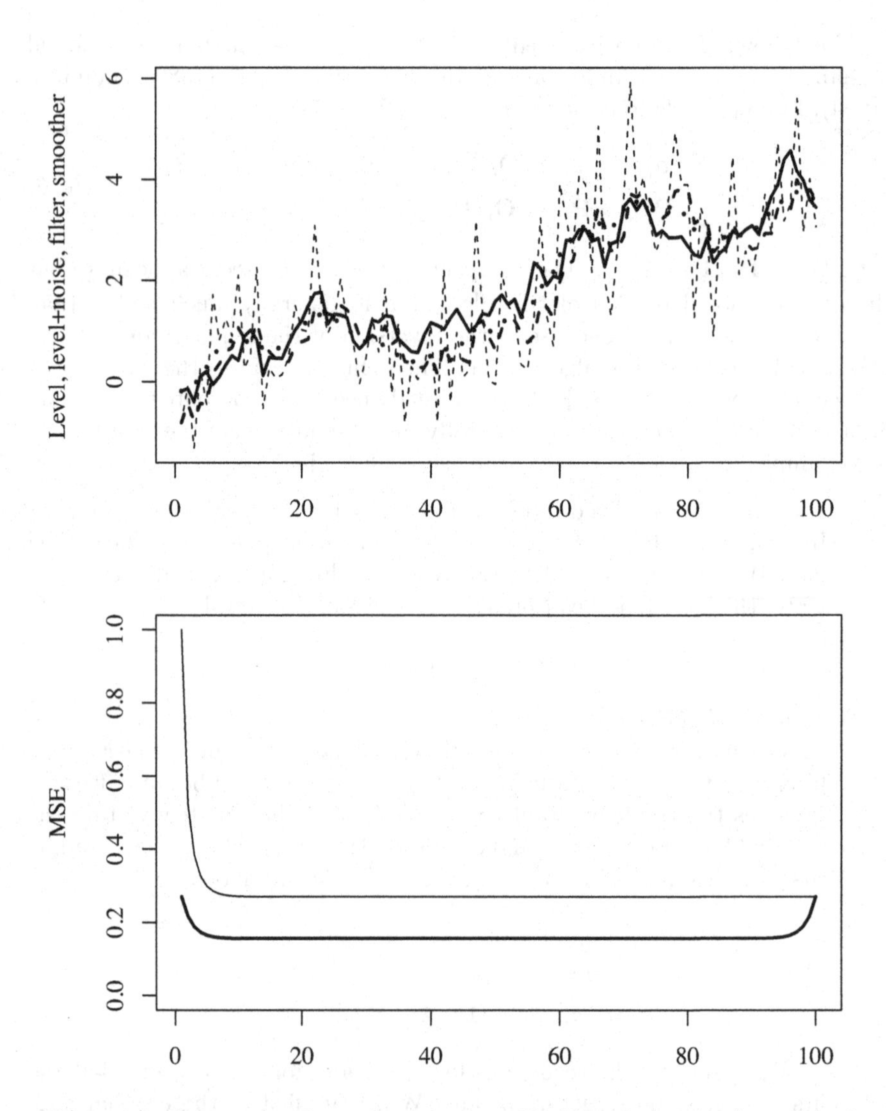

Figure 5.2 *Top graph: observed time series Y_t (thin dashed line), real unobservable level μ_t (line), its filter $a_{t|t}$ (thick dashed line) and smoother $a_{t|n}$ (thick dotted line). Bottom graph: MSE of the filter, $P_{t|t}$ (thinner line), and of the smoother, $P_{t|n}$ (thicker line). The data were simulated from the local level model with $\sigma_\varepsilon^2 = 1$ and $\sigma_\nu^2 = 0.1$. The initial variance for the state variable was $\tau = 10^{10}$.*

that of the smoother: the MSE of the smoother is smaller, being $a_{t|t}$ based on a subset of the observations on which $a_{t|n}$ is projected. The smoother's MSE reaches its maxima at the extremes (i.e., $t = 1$ and $t = n$) of the time series. These features are common to all models that translate into a time-homogeneous state space form without missing observations.

We conclude this section by providing recursions for *disturbance smoothing*, that is, for computing the prediction of the white noise sequences $\{\epsilon_t\}$ and $\{\boldsymbol{\nu}_t\}$ based on the whole time series $\mathcal{Y}_n$. These quantities are extremely useful for identifying outliers of various types, as we will discuss in Chapter 6.

Theorem 5.5 (Disturbance smoother). *Assume the time series $\{\boldsymbol{Y}_1, \ldots, \boldsymbol{Y}_n\}$ is generated by a model in state space form as in Definition 5.1 and that the Kalman filter presented in Theorem 5.3 has run; then, for $t = n, n-1, \ldots, 1$,*

$$
\begin{aligned}
\hat{\boldsymbol{\nu}}_{t|n} &= \mathbb{P}[\boldsymbol{\nu}_t|\mathcal{Y}_n] = \mathbf{Q}_t\boldsymbol{r}_t,\\
\mathbf{P}^{(\nu)}_{t|n} &= \mathbb{E}[\boldsymbol{\nu}_t - \hat{\boldsymbol{\nu}}_{t|n}][\boldsymbol{\nu}_t - \hat{\boldsymbol{\nu}}_{t|n}]^\top = \mathbf{Q}_t - \mathbf{Q}_t\mathbf{N}_t\mathbf{Q}_t,\\
\boldsymbol{u}_t &= \mathbf{F}_t^{-1}\boldsymbol{\iota} - \mathbf{K}_t^\top\boldsymbol{r}_t,\\
\mathbf{D}_t &= \mathbf{F}_t^{-1} + \mathbf{K}_t^\top\mathbf{N}_t\mathbf{K}_t,\\
\hat{\epsilon}_{t|n} &= \mathbb{P}[\epsilon_t|\mathcal{Y}_n] = \mathbf{H}_t\boldsymbol{u}_t,\\
\mathbf{P}^{(\epsilon)}_{t|n} &= \mathbb{E}[\epsilon_t - \hat{\epsilon}_{t|n}][\epsilon_t - \hat{\epsilon}_{t|n}]^\top = \mathbf{H}_t - \mathbf{H}_t\mathbf{D}_t\mathbf{H}_t,
\end{aligned}
\tag{5.17}
$$

where $\boldsymbol{r}_t$ and $\mathbf{N}_t$ were defined in equation (5.15).

The proof for these formulae are based on the same techniques and arguments used for proving Theorem 5.4 and can be found in the book by Durbin and Koopman (2001, Section 4.4) or in the original papers by de Jong (1988), Kohn and Ansley (1989) and Koopman (1993).

As already remarked, these quantities, or better their standardised versions, will be very useful to identify additive outliers, level shifts or slope changes, but the analyst should bear in mind that, differently from the innovation sequence, the sequences of smoothed disturbances are serially correlated.

5.3.3 Forecasting

The problem of forecasting[3] consists in predicting values of the state vector $\boldsymbol{\alpha}_t$ based on the past observations $\mathcal{Y}_s$ with $s < t$. It is true, that generally the aim of forecasting is guessing the future of the observable time series $\boldsymbol{Y}_t$ based on its past $\mathcal{Y}_s$, but once the optimal linear predictor is available for the state vector, it is straightforward to apply the linearity property of linear predictors to obtain forecasts of $\boldsymbol{Y}_t$ based on $\mathcal{Y}_s$:

$$\hat{\boldsymbol{Y}}_{t|s} = \mathbb{P}[\boldsymbol{Y}_t|\mathcal{Y}_s] = \mathbb{P}[\mathbf{Z}_t\boldsymbol{\alpha}_t + \boldsymbol{c}_t + \boldsymbol{\epsilon}_t|\mathcal{Y}_s] = \mathbf{Z}_t\boldsymbol{a}_{t|s} + \boldsymbol{c}_t,$$

where the covariance matrix of the forecast error has to take account of the multiplication by the matrix $\mathbf{Z}_t$ and the additional variability brought by the observation error $\boldsymbol{\epsilon}_t$:

$$\begin{aligned}\mathbb{E}[\boldsymbol{Y}_t - \hat{\boldsymbol{Y}}_{t|s}][\boldsymbol{Y}_t - \hat{\boldsymbol{Y}}_{t|s}]^\top &= \mathbb{E}[\mathbf{Z}_t(\boldsymbol{\alpha}_t - \boldsymbol{a}_{t|s}) + \boldsymbol{\epsilon}_t][\mathbf{Z}_t(\boldsymbol{\alpha}_t - \boldsymbol{a}_{t|s}) + \boldsymbol{\epsilon}_t]^\top \\ &= \mathbf{Z}_t\mathbf{P}_{t|s}\mathbf{Z}_t^\top + \mathbf{H}_t.\end{aligned}$$

Thus, we can concentrate on forecasting $\boldsymbol{\alpha}_t$. If we run the Kalman filter from the first observation to observation s we are provided with the quantities

$$\boldsymbol{a}_{s+1|s}, \qquad \mathbf{P}_{s+1|s}.$$

For obtaining the prediction for $s+2$ we can use the definition of the transition equation:

$$\begin{aligned}\boldsymbol{a}_{s+2|s} = \mathbb{P}[\boldsymbol{\alpha}_{s+2}|\mathcal{Y}_s] &= \mathbb{P}[\mathbf{T}_{s+1}\boldsymbol{\alpha}_{s+1} + \boldsymbol{d}_{t+1} + \boldsymbol{\nu}_{t+1}|\mathcal{Y}_s] \\ &= \mathbf{T}_{s+1}\boldsymbol{a}_{s+1|s} + \boldsymbol{d}_{t+1}.\end{aligned}$$

The same reasoning can be used to sequentially compute $\boldsymbol{a}_{s+3|s}$, $\boldsymbol{a}_{s+4|s}$, etc., obtaining the following recursion: for $k = 1, 2, \ldots, t-1$,

$$\boldsymbol{a}_{s+k+1|s} = \mathbf{T}_{s+k}\boldsymbol{a}_{s+k|s} + \boldsymbol{d}_{t+k}.$$

The covariance matrix of the prediction errors can also be computed sequentially, starting from the Kalman filter output $\mathbf{P}_{s+1|s}$: for $k = 1, 2, \ldots, t-1$

$$\begin{aligned}\mathbf{P}_{s+k+1|s} &= \mathbb{E}[\mathbf{T}_{s+k}\boldsymbol{\alpha}_{s+k} + \boldsymbol{d}_{t+k} + \boldsymbol{\nu}_{t+k} - \mathbf{T}_{s+k}\boldsymbol{a}_{s+k|s} - \boldsymbol{d}_{t+k}][\ldots = \ldots]^\top \\ &= \mathbb{E}[\mathbf{T}_{s+k}(\boldsymbol{\alpha}_{s+k} - \boldsymbol{a}_{s+k|s}) + \boldsymbol{\nu}_{t+k}][\ldots = \ldots]^\top \\ &= \mathbf{T}_{s+k}\mathbf{P}_{s+k|s}\mathbf{T}_{s+k}^\top + \mathbf{Q}_{s+k}.\end{aligned}$$

Example 5.14 (Forecasting for the local level model).

[3]Many textbooks use the word *prediction* instead of *forecasting*, but since in Chapter 1 we already used the first term to indicate a more general concept we prefer to use the second one. In the terminology we used, filtering, forecasting and smoothing are all predictions.

In the random walk plus noise model of Example 5.12 we can easily obtain the forecasting equations as follows: for $k > 0$,

$$\begin{aligned} a_{n+1|n} &= a_{n|n} \\ a_{n+2|n} &= a_{n+1|n} = a_{n|n} \\ \dots &= \dots \\ a_{n+k|n} &= a_{n|n}; \end{aligned}$$

$$\begin{aligned} P_{n+1|n} &= P_{n|n} + \sigma_\nu^2 \\ P_{n+2|n} &= P_{n+1|n} + \sigma_\nu^2 = 2\sigma_\nu^2 \\ \dots &= \dots \\ P_{n+k|n} &= k\sigma_\nu^2; \end{aligned}$$

$$\begin{aligned} \hat{Y}_{n+k|} &= a_{t|t} \\ F_{n+k|n} &= k\sigma_\nu^2 + \sigma_\epsilon^2. \end{aligned}$$

5.3.4 *Initialisation of the state space recursion*

A fundamental task in the definition of the state space form of a model is the individuation of the first two moments of the initial state vector $\boldsymbol{\alpha}_1$.

If some information about the initial values of (some of) the state variables is available, then it should be used by specifying the best guess as mean value and the uncertainty by adjusting the variance: recall that if your prior information about a state variable is well represented by a Gaussian distribution you can use its variance to assign probabilities to symmetric intervals about the guessed value: mean ± 1 standard deviations contains 68.3% of the probability, mean ± 2 standard deviations contains 95.4% of the probability, mean ± 3 standard deviations contains 99.7% of the probability.

If no information about the state variable is available, then stationary and nonstationary state variables should be treated differently. Let us name $\boldsymbol{\alpha}_1^*$ the sub-vector containing only jointly stationary state variables and $\boldsymbol{\alpha}_1^\dagger$ the complementary sub-vector with the nonstationary variables.

Since we are assuming throughout the book that second moments exist, then for the stationary subvector the marginal mean vector and covariance matrix exist and should be used as initial values. So, if we set $\boldsymbol{a}^* = \mathbb{E}[\boldsymbol{\alpha}_t^*]$ and $\mathbf{P}^* = \mathbb{E}[\boldsymbol{\alpha}_t^* - \boldsymbol{a}^*][\boldsymbol{\alpha}_t^* - \boldsymbol{a}^*]^\top$, using the state equation recursion of the stationary subvector we can write

$$\begin{aligned} \mathbb{E}[\boldsymbol{\alpha}_{t+1}^*] &= \mathbb{E}[\mathbf{T}^*\boldsymbol{\alpha}_t^* + \boldsymbol{d}^* + \boldsymbol{\nu}_t^*] \\ \boldsymbol{a}^* &= \mathbf{T}^*\boldsymbol{a}^* + \boldsymbol{d}^* \\ \boldsymbol{a}^* &= (\mathbf{I} - \mathbf{T}^*)^{-1}\boldsymbol{d}^*. \end{aligned} \tag{5.18}$$

Using the same recursion for the covariance matrix we can write

$$\mathbb{E}[\boldsymbol{\alpha}^*_{t+1} - \boldsymbol{a}^*][\boldsymbol{\alpha}^*_{t+1} - \boldsymbol{a}^*]^\top = \mathbb{E}[\mathbf{T}^*\boldsymbol{\alpha}^*_t + \boldsymbol{d}^* + \boldsymbol{\nu}^*_t - \boldsymbol{a}^*][\mathbf{T}^*\boldsymbol{\alpha}^*_t + \boldsymbol{d}^* + \boldsymbol{\nu}^*_t - \boldsymbol{a}^*]^\top$$
$$\mathbf{P}^* = \mathbf{T}^*\mathbf{P}^*\mathbf{T}^{*\top} + \mathbf{Q}^*, \tag{5.19}$$

where we used the identity $\boldsymbol{d}^* = (\mathbf{I} - \mathbf{T}^*)\boldsymbol{a}^*$ of equation(5.18) to substitute $\boldsymbol{d}^*$. Now, to solve for $\mathbf{P}^*$ we need to vectorise the matrices using the *vec operator*, which turns a matrix into a vector by stacking the columns into one long vector and enjoys the property

$$\text{vec}(\mathbf{AXB}) = \left(\mathbf{B}^\top \otimes \mathbf{A}\right)\text{vec}(\mathbf{X}),$$

where $\otimes$ denotes the Kronecker product[4]. Now, by applying this property to our equation we can write

$$\text{vec}(\mathbf{P}^*) = \left(\mathbf{T}^{*\top} \otimes \mathbf{T}^*\right)\text{vec}(\mathbf{P}^*) + \text{vec}(\mathbf{Q}^*)$$
$$\text{vec}(\mathbf{P}^*) = \left(\mathbf{I} - \mathbf{T}^{*\top} \otimes \mathbf{T}^*\right)^{-1}\text{vec}(\mathbf{Q}^*). \tag{5.20}$$

All software languages for numerical analysis and statistics offer functions to *vec* and *unvec* matrices and to carry out Kronecker products. The inversion of the matrix in parenthesis may become computationally demanding if the stationary state vector is long, but generally the number of stationary components in an UCM is not very large.

Alternatively, an approximate solution can be obtained by iterating equation (5.19) a sufficient number of times, starting with an arbitrary matrix as, for example, $\mathbf{P}^*_1 = \mathbf{Q}^*$:

$$\mathbf{P}^*_{i+1} = \mathbf{T}^*\mathbf{P}^*_i\mathbf{T}^{*\top} + \mathbf{Q}^*,$$

for $i = 1, 2, \ldots, m$ and m large enough.

For the nonstationary state variables, a diffuse distribution should be used, and so

$$\boldsymbol{a}^\dagger = \mathbf{0}, \qquad \mathbf{P}^\dagger = \tau\mathbf{I},$$

where τ is a very large number. Some software packages implement exact algorithms such as those by de Jong (1991) and Koopman (1997) for filtering and smoothing that allow $\tau = \infty$,.

---

[4]The Kronecker product of the matrices $\mathbf{A}$ and $\mathbf{B}$ of dimensions, respectively, $m \times n$ and $p \times q$, is defined by

$$\mathbf{A} \otimes \mathbf{B} = \begin{bmatrix} a_{1,1}\mathbf{B} & \ldots & a_{1,n}\mathbf{B} \\ \vdots & \vdots & \vdots \\ a_{m,1}\mathbf{B} & \ldots & a_{m,n}\mathbf{B} \end{bmatrix},$$

where $a_{i,j}$ is the generic element of $\mathbf{A}$. The resulting matrix has dimensions $mp \times nq$.

5.3.5 Adding the Gaussianity assumption

In Chapter 1 we saw that under the assumption of joint Gaussianity of the random variables, the best predictor and the best linear predictor coincide:

$$\mathbb{E}[\boldsymbol{\alpha}_t|\mathcal{Y}_s] = \mathbb{P}[\boldsymbol{\alpha}_t|\mathcal{Y}_s].$$

Moreover, the variance of the variable to be predicted, conditionally on the variables on which the prediction is based, coincides with the covariance matrix of the linear prediction errors:

$$\mathbb{V}\text{ar}(\boldsymbol{\alpha}_t|\mathcal{Y}_s) = \mathbb{E}[\boldsymbol{\alpha}_t - \boldsymbol{a}_{t|s}][\boldsymbol{\alpha}_t - \boldsymbol{a}_{t|s}]^\top.$$

In terms of our results about filtering, smoothing and forecasting, this means that if $\boldsymbol{\alpha}_1 \sim \mathcal{N}(\boldsymbol{a}_{1|0}, \mathbf{P}_{1|0})$, and $\{\boldsymbol{\epsilon}_t\}$ and $\{\boldsymbol{\nu}_t\}$ are jointly Gaussian processes, then also the observable time series $\{\boldsymbol{Y}_t\}$ is Gaussian (because it is a linear combination of jointly Gaussian variables) and

$$\boldsymbol{\alpha}_t|\mathcal{Y}_s \sim \mathcal{N}(\boldsymbol{a}_{t|s}, \mathbf{P}_{t|s}).$$

This allows us to build confidence intervals (CI) for $\boldsymbol{\alpha}_t$ as

$$\Pr\left\{a_{i,t|s} - z_{1-\alpha/2}\sqrt{P_{ii,t|s}} < \alpha_{i,t} < a_{i,t|s} + z_{1-\alpha/2}\sqrt{P_{ii,t|s}}\right\} = 1 - \alpha,$$

where $1-\alpha$ is the level of probability of the CI, z_q is the q-th quantile of the standard normal distribution, $a_{i,t|s}$ and $\alpha_{i,t}$ denote the i-th element of their homonymous vectors, and $P_{ii,t|s}$ is the i-th element on the main diagonal of $\mathbf{P}_{t|s}$. Typical values used for $z_{1-\alpha/2}$ are

$1-\alpha$	0.683	0.800	0.900	0.950	0.954
$z_{1-\alpha/2}$	1.000	1.282	1.645	1.956	2.000

From a frequentist point of view these CI are to be interpreted as: "if the observations are generated according to our model, then $(1-\alpha)\%$ of the times the CI will contain the true realised value $\alpha_{i,t}$". This statement does not necessarily hold for the single sample path, but under all possible samples. However, if the sample path is sufficiently long, then the above statement will hold approximately also for a single sample path.

Notice that generally in UCM the variance $P_{ii,t|n}$ will not converge to zero as n increases, and so $a_{i,t|n}$ will never converge to $\alpha_{i,t}$ in any possible way. Thus, the realisation of $\alpha_{i,t}$ will never be known with certainty.

5.4 Inference for the unknown parameters

Up to now we assumed that all elements in the nonrandom vectors and matrices of the state space form were known. But this is in general not true in real applications. In fact, all UCM we presented in the book entail unknown parameters to be estimated.

5.4.1 *Building and maximising the likelihood function*

If the observations are (at least approximately) jointly normally distributed, then the Kalman filter allows the construction of the likelihood function which can be numerically maximised using algorithms commonly implemented in software packages.

First of all, using the definition of conditional density[5], we can factorise the joint density of the data as

$$f_{\theta}(\boldsymbol{y}_1, \boldsymbol{y}_2, \boldsymbol{y}_3, \ldots, \boldsymbol{y}_n) = f_{\theta}(\boldsymbol{y}_1) f_{\theta}(\boldsymbol{y}_2|\boldsymbol{y}_1) f_{\theta}(\boldsymbol{y}_3|\boldsymbol{y}_2, \boldsymbol{y}_1) \cdots f_{\theta}(\boldsymbol{y}_n|\boldsymbol{y}_{n-1}, \ldots, \boldsymbol{y}_1),$$

where $f_{\theta}(\boldsymbol{y}_t|\boldsymbol{y}_{t-1}, \ldots, \boldsymbol{y}_1)$ represents the density of the random variable $(\boldsymbol{Y}_t|\boldsymbol{Y}_1 = \boldsymbol{y}_1, \boldsymbol{Y}_2 = \boldsymbol{y}_2, \ldots, \boldsymbol{Y}_{t-1} = \boldsymbol{y}_{t-1})$ and $\boldsymbol{\theta}$ is the vector of parameters needed to compute the density.

Now, under the assumption of joint normality of the data, the conditional distribution of $(\boldsymbol{Y}_t|\boldsymbol{Y}_1 = \boldsymbol{y}_1, \boldsymbol{Y}_2 = \boldsymbol{y}_2, \ldots, \boldsymbol{Y}_{t-1} = \boldsymbol{y}_{t-1})$ is also normal and the Kalman filter provides the conditional mean and variance,

$$\begin{aligned}
\mathbb{E}(\boldsymbol{Y}_t|\boldsymbol{Y}_1 = \boldsymbol{y}_1, \boldsymbol{Y}_2 = \boldsymbol{y}_2, \ldots, \boldsymbol{Y}_{t-1} = \boldsymbol{y}_{t-1}) &= \hat{\boldsymbol{y}}_{t|t-1},\\
\mathbb{V}\mathrm{ar}(\boldsymbol{Y}_t|\boldsymbol{Y}_1 = \boldsymbol{y}_1, \boldsymbol{Y}_2 = \boldsymbol{y}_2, \ldots, \boldsymbol{Y}_{t-1} = \boldsymbol{y}_{t-1}) &= \mathbf{F}_t.
\end{aligned}$$

Thus, for $t = 1, 2, \ldots, n$,

$$(\boldsymbol{Y}_t|\boldsymbol{Y}_1 = \boldsymbol{y}_1, \boldsymbol{Y}_2 = \boldsymbol{y}_2, \ldots, \boldsymbol{Y}_{t-1} = \boldsymbol{y}_{t-1}) \sim \mathcal{N}(\hat{\boldsymbol{y}}_{t|t-1}, \mathbf{F}_t),$$

that is,

$$f_{\theta}(\boldsymbol{y}_t|\boldsymbol{y}_{t-1}, \ldots, \boldsymbol{y}_1) = \frac{1}{(2\pi)^{\frac{p}{2}} |\mathbf{F}_t|^{\frac{1}{2}}} \exp\left[-\frac{1}{2}(\boldsymbol{y}_t - \hat{\boldsymbol{y}}_{t|t-1})\mathbf{F}_t^{-1}(\boldsymbol{y}_t - \hat{\boldsymbol{y}}_{t|t-1})^{\top}\right],$$

where p is the number of elements in the vector $\boldsymbol{Y}_t$. Notice that when the initial condition $\boldsymbol{\alpha}_1 \sim \mathcal{N}(\boldsymbol{a}_{1|0}, \boldsymbol{P}_{1|0})$ is specified, then also the $f_{\theta}(\boldsymbol{y}_1)$ is well defined and so we can compute the probability of observing any sample path given the parameters stacked in the vector $\boldsymbol{\theta}$.

The likelihood function is just the joint density of the sample path seen as a function of the unknown parameters $\boldsymbol{\theta}$:

$$L(\boldsymbol{\theta}) = \prod_{t=1}^{n} f_{\theta}(\boldsymbol{y}_t|\boldsymbol{y}_{t-1}, \ldots, \boldsymbol{y}_1);$$

but for theoretical and computational reasons it is generally better to work on the log-likelihood function

$$\begin{aligned}
l(\boldsymbol{\theta}) &= \sum_{t=1}^{n} \log f_{\theta}(\boldsymbol{y}_t|\boldsymbol{y}_{t-1}, \ldots, \boldsymbol{y}_1),\\
&= -\frac{np}{2}\log(2\pi) - \frac{1}{2}\sum_{t=1}^{n}\left(\log|\mathbf{F}_t| + (\boldsymbol{y}_t - \hat{\boldsymbol{y}}_{t|t-1})^{\top}\mathbf{F}_t^{-1}(\boldsymbol{y}_t - \hat{\boldsymbol{y}}_{t|t-1})\right),
\end{aligned} \tag{5.21}$$

[5] i.e., $f(y|x) = f(x, y)/f(x)$.

or on its mean, $\bar{l}(\boldsymbol{\theta}) = l(\boldsymbol{\theta})/n$.

The *maximum likelihood estimator* (MLE) of the vector of unknown parameters $\boldsymbol{\theta}$ is a value $\hat{\boldsymbol{\theta}}$ which maximises the likelihood, or equivalently, the log-likelihood. To show that the maxima of $L(\boldsymbol{\theta})$, $l(\boldsymbol{\theta})$ and $\bar{l}(\boldsymbol{\theta})$ are all reached at the same value $\hat{\boldsymbol{\theta}}$, recall that if these functions are differentiable (and ours are), then $\nabla[L(\boldsymbol{\theta})]_{\hat{\boldsymbol{\theta}}} = \mathbf{0}$ and this implies that

$$[\nabla l(\boldsymbol{\theta})]_{\hat{\boldsymbol{\theta}}} = \frac{[\nabla L(\boldsymbol{\theta})]_{\hat{\boldsymbol{\theta}}}}{L(\hat{\boldsymbol{\theta}})} = \mathbf{0}, \qquad [\nabla \bar{l}(\boldsymbol{\theta})]_{\hat{\boldsymbol{\theta}}} = \frac{[\nabla L(\boldsymbol{\theta})]_{\hat{\boldsymbol{\theta}}}}{nL(\hat{\boldsymbol{\theta}})} = \mathbf{0},$$

where $[\nabla f(\boldsymbol{x})]_{\boldsymbol{x}_0}$ denotes the gradient of the function $f(\boldsymbol{x})$ at the point $\boldsymbol{x} = \boldsymbol{x}_0$.

Most software packages implement Newton or quasi-Newton optimisation algorithms such as the BFGS, that allow the numerical maximisation of $l(\boldsymbol{\theta})$. The numerical optimiser has to be fed with a function that receives a numerical vector with values for $\boldsymbol{\theta}$ and returns the log-likelihood value at that point. Thus, this objective function has to be structured as follows:

1. The function is fed with the vector of values for $\boldsymbol{\theta}$, and the time series observations $\{\boldsymbol{y}_1, \ldots, \boldsymbol{y}_n\}$,
2. The values of the parameters in $\boldsymbol{\theta}$ are assigned to their respective positions in the system matrices of the state space form,
3. The Kalman filter is run to obtain the quantities $\boldsymbol{y}_{t|t-1}$ (or equivalently $\boldsymbol{\iota}_t$) and $\mathbf{F}_t$ for $t = 1, \ldots, n$,
4. The value of the log-likelihood (5.21), or of its mean $\bar{l}(\boldsymbol{\theta})$ is returned.

5.4.2 Large sample properties of the maximum likelihood estimator

As in many non-trivial statistical models, also in the state space framework the exact distribution of the MLE is not known for finite samples. The properties of the MLE are then approximated by its asymptotic (or large sample) properties (i.e., the properties as the time series length n goes to infinity).

Unfortunately, the results we are going to present have been proved only under assumptions that are somewhat too restrictive if we consider the range of models we want to put in state space form. For instance, Caines (1988), who proves the most general result, assumes the stationarity of the state vector, Pagan (1980) covers the case of regression with time-varying coefficients and Ghosh (1989) treats the class of error-shock models. However, the general experience in using UCM is that the classical distributional result in maximum likelihood estimation is a reasonable approximation of the true finite sample distribution also for UCM (compare Figure 5.3).

The classical result on MLE states: if $\boldsymbol{\theta}_0$ is the true (population) parameter vector, the model is identified and no element of $\boldsymbol{\theta}_0$ lies on the boundary of the parameter space, then $\hat{\boldsymbol{\theta}}$ is consistent for $\boldsymbol{\theta}_0$ and

$$\sqrt{n}(\hat{\boldsymbol{\theta}} - \boldsymbol{\theta}_0) \xrightarrow{d} \mathcal{N}(\mathbf{0}, \boldsymbol{\mathcal{I}}^{-1}), \qquad (5.22)$$

where $\mathcal{I}$ is known as the Fisher information matrix and is defined by

$$\mathcal{I} = -\lim_{n\to\infty} \frac{1}{n} \mathbb{E}\left[\sum_{t=1}^{n} \frac{\partial^2 \log f_{\theta}(\mathbf{Y}_t|\mathcal{Y}_{t-1})}{\partial \boldsymbol{\theta} \partial \boldsymbol{\theta}^\top}\bigg|_{\theta=\theta_0} \right], \tag{5.23}$$

provided that the limit exists (this happens for all those models for which the Kalman filter reaches a steady state). This matrix is clearly unknown as it depends on the true parameter vector $\boldsymbol{\theta}_0$. Furthermore, if we exclude trivial models, the expectation is usually impossible to compute analytically. However, if a law of large numbers applies and the gradient of the log likelihood is continuous with respect to $\boldsymbol{\theta}$ (as in our case), we can estimate $\mathcal{I}$ with the mean

$$\hat{\mathcal{I}}_n = -\frac{1}{n} \sum_{t=1}^{n} \frac{\partial^2 \log f_{\theta}(\mathbf{Y}_t|\mathcal{Y}_{t-1})}{\partial \boldsymbol{\theta} \partial \boldsymbol{\theta}^\top}\bigg|_{\theta=\hat{\theta}}, \tag{5.24}$$

where the true parameter value has been substituted with the MLE.

Obtaining an estimate of (5.24) in practice is very simple if the software package we are using can compute the Hessian[6] numerically. In this case, since differentiation is a linear operation

$$\frac{\partial^2}{\partial \boldsymbol{\theta} \partial \boldsymbol{\theta}^\top} \sum_{t=1}^{n} \log f_{\theta}(\mathbf{Y}_t|\mathcal{Y}_{t-1}) = \sum_{t=1}^{n} \frac{\partial^2 \log f_{\theta}(\mathbf{Y}_t|\mathcal{Y}_{t-1})}{\partial \boldsymbol{\theta} \partial \boldsymbol{\theta}^\top},$$

the sample realisation of $\hat{\mathcal{I}}_n$ is just given either by minus the (numerical) Hessian of the log-likelihood $l(\boldsymbol{\theta})$ computed at its maximum $\hat{\boldsymbol{\theta}}$ and divided by n or by minus the (numerical) Hessian of the mean log-likelihood $\bar{l}(\boldsymbol{\theta})$ computed at its maximum $\hat{\boldsymbol{\theta}}$:

$$\hat{\mathcal{I}}_n = -\frac{\partial^2}{\partial \boldsymbol{\theta} \partial \boldsymbol{\theta}^\top} \frac{l(\boldsymbol{\theta})}{n}\bigg|_{\boldsymbol{\theta}=\hat{\boldsymbol{\theta}}} = -\frac{\partial^2}{\partial \boldsymbol{\theta} \partial \boldsymbol{\theta}^\top} \bar{l}(\boldsymbol{\theta})\bigg|_{\boldsymbol{\theta}=\hat{\boldsymbol{\theta}}}.$$

Thus, in practice the distribution of the MLE is approximated with

$$\hat{\boldsymbol{\theta}} \approx \mathcal{N}(\boldsymbol{\theta}_0, (n\hat{\mathcal{I}}_n)^{-1}),$$

and confidence intervals or t-tests can be built in the usual way:

$$\Pr\left\{ \hat{\theta}_i - z_{1-\alpha/2}\sqrt{v_{ii}} < \theta_i < \hat{\theta}_i + z_{1-\alpha/2}\sqrt{v_{ii}} \right\} = 1 - \alpha,$$

where v_{ii} is the i-th element on the main diagonal of $(n\hat{\mathcal{I}}_n)^{-1})$, while for testing the null hypothesis $\theta_i = \theta_{0,i}$, we have that under the null

$$\frac{\hat{\theta}_i - \theta_{0,i}}{\sqrt{v_{ii}}} \approx \mathcal{N}(0,1).$$

[6]The $p \times p$ matrix of second partial derivatives.

Figure 5.3 compares the simulated distribution of the MLE of the three variances of a local linear trend plus noise model with their asymptotic approximation. The approximation seems reasonable most of the time, but for $n = 100$, the distribution of the estimator of the slope disturbance dispersion, the one we set the closest to zero, is rather left-skewed.

One very important thing to stress is that the above result holds only if in the null hypothesis the parameter $\theta_{0,i}$ does not lie on the boundary of the parameter space. Therefore, if the parameter we want to test is a variance, we cannot test the null hypothesis $\theta_{0,i} = 0$ using the above approximation, because zero is on the boundary of the parameter space. Fortunately, there are also results for MLE under these circumstances. If the true value of a single variance parameter is zero and all other conditions hold, then

$$\Pr\left\{\frac{\hat{\theta}_i}{\sqrt{v_{ii}}} = 0\right\} = 0.5, \qquad \Pr\left\{0 < \frac{\hat{\theta}_i}{\sqrt{v_{ii}}} \leq x\right\} = \Phi(x),$$

where $\Phi(x)$ is the standard normal cumulative distribution function. So, if the true parameter of a single variance is zero, then there is 50% chance that the MLE is zero, and if the MLE is greater than zero then it is distributed as a half normal distribution. This means that for testing the hypothesis $\theta_i = 0$, the classical t-ratio statistic rejects at level $\alpha = 5\%$ if it exceeds 1.645 and with level $\alpha = 1\%$ if it is greater than 2.326. If the software package computes the p-value automatically without taking into account the boundary effect on MLE, then the p-value must be doubled. If more than one parameter is on the boundary under the null, the distribution of each of the t-statistics is more involved.

Figure 5.4 reports results based on 10,000 simulations of the MLE of the parameter σ_η^2 in a local level model plus noise with variances $\sigma_\eta^2 = 0$, $\sigma_\zeta^2 = 0.1$, $\sigma_\varepsilon^2 = 1$. The figure depicts the empirical cumulative distribution function of the simulated estimates (i.e., for each real x the relative number of estimates $\hat{\theta}_i$ smaller than or equal to x) and the one of a standard normal distribution (i.e., $\Phi(x) = \Pr\{X \leq x\}$ with X standard normal random variable). The approximation of the asymptotic distribution seems quite accurate: the empirical distribution has a positive jump of probability 0.5 at $x = 0$ and then follows the right side of the normal distribution very closely.

If the assumption of joint normality is not a good approximation to the real distribution of the data, then the asymptotic distribution of the Gaussian maximum likelihood estimator is different from (5.22). The MLE based on the Gaussian distribution when the data are not normally distributed is named Gaussian *quasi-maximum likelihood estimator* (QMLE) and if all other conditions for the MLE hold we have the following result

$$\sqrt{n}(\hat{\boldsymbol{\theta}} - \boldsymbol{\theta}_0) \xrightarrow{d} \mathcal{N}(\mathbf{0}, \boldsymbol{\mathcal{I}}^{-1}\boldsymbol{\mathcal{J}}\boldsymbol{\mathcal{I}}^{-1}), \tag{5.25}$$

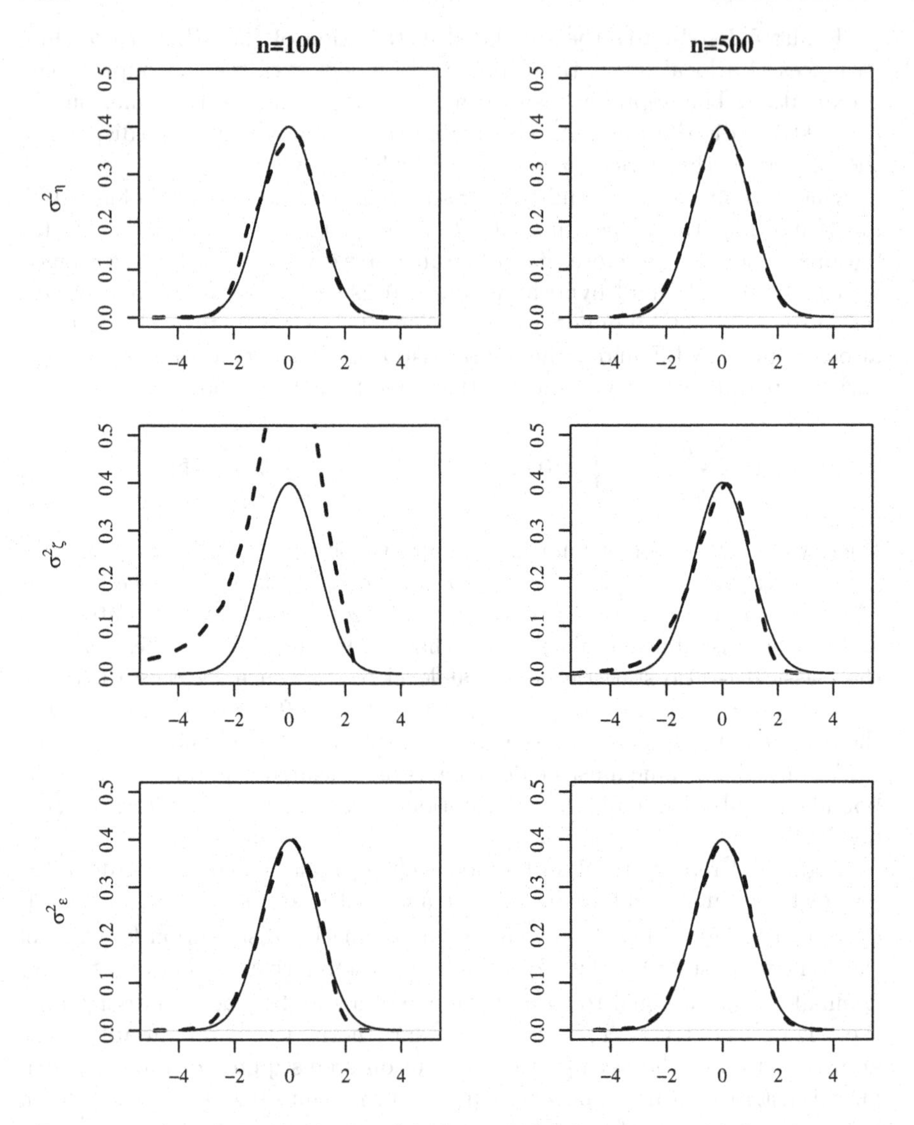

Figure 5.3 *Actual vs. asymptotic distribution of the estimators of the variances in a LLT + noise model with parameters:* $\sigma_\eta^2 = 1$, $\sigma_\zeta^2 = 0.1$, $\sigma_\varepsilon^2 = 1$. *The left column refers to time series with* $n = 100$ *observations, in the right column* $n = 500$. *The simulated density (dashed line) of* $(\hat{\theta}_i - \theta_i)/\sqrt{v_{ii}}$ *is compared with the standard normal density (thin line).*

where

$$\mathcal{J} = \lim_{n\to\infty} \frac{1}{n} \sum_{t=1}^{n} \mathbb{E}\left\{ \left[\left. \frac{\delta \log f(\mathbf{Y}_t|\mathcal{Y}_{t-1})}{\delta \boldsymbol{\theta}} \right|_{\theta=\theta_0} \right] \left[\left. \frac{\delta \log f(\mathbf{Y}_t|\mathcal{Y}_{t-1})}{\delta \boldsymbol{\theta}} \right|_{\theta=\theta_0} \right]^\top \right\}.$$

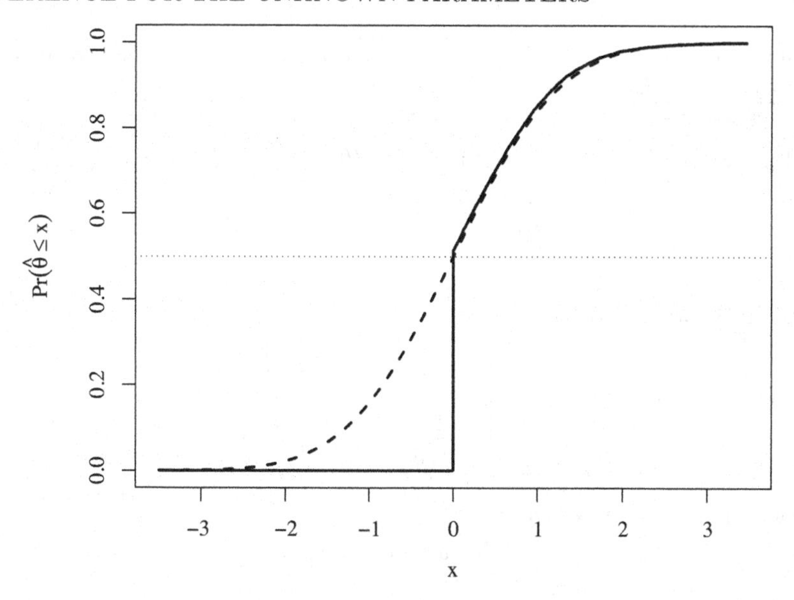

Figure 5.4 *Simulation based empirical cumulative distribution function (line) of the MLE of the variance σ^2_η, under LLT + noise model with variances $\sigma^2_\eta = 0$, $\sigma^2_\zeta = 0.1$, $\sigma^2_\varepsilon = 1$, and standard normal cumulative distribution function (dashed line). 10,000 estimates computed on sample paths of $n = 100$ observations.*

Again, the unknown matrix $\mathcal{J}$ is estimated by the sample mean

$$\hat{\mathcal{J}}_n = \frac{1}{n}\sum_{t=1}^{n}\left[\left.\frac{\delta \log f(\mathbf{Y}_t|\mathcal{Y}_{t-1})}{\delta\boldsymbol{\theta}}\right|_{\theta=\hat{\theta}}\right]\left[\left.\frac{\delta \log f(\mathbf{Y}_t|\mathcal{Y}_{t-1})}{\delta\boldsymbol{\theta}}\right|_{\theta=\hat{\theta}}\right]^\top,$$

where the gradients in square brackets can be computed numerically. If the likelihood is correctly specified, then $\mathcal{I} = \mathcal{J}$ (information matrix identity), and (5.25) reduces to (5.25).

Many parameters to be estimated in UCM live in a bounded or half-bounded parameter space. For example, the parameter space of the variances is $[0, \infty)$, that of the ρ coefficient in the cycle is $[0, 1)$, etc. In this case, one seeks the maximum of the log-likelihood with respect to a transformation of the original parameter that guarantees that the bounds are respected (see Table 5.1).

In order to obtain the distribution of the original parameters when the optimisation is carried out with respect to the transformed parameters $\boldsymbol{\psi}$, with $\boldsymbol{\theta} = \boldsymbol{g}(\boldsymbol{\psi})$, we can use the *delta method*. Assume that $\boldsymbol{g}(\cdot)$ is differentiable and collect the partial derivatives of each element of the vector-valued function

Table 5.1 *Parameter space, transform that maps it to* $[-\infty, \infty]$, *relative anti-transform and Jacobian of the transformation.*

Parameter space	$[0, \infty)$	$[0, \infty)$	(a, b)
Transform	$\psi = \log(\theta)$	$\psi = \sqrt{\theta}$	$\psi = \log\left[\frac{(\theta-a)/(b-a)}{1-(\theta-a)/(b-a)}\right]$
Anti-transform	$\theta = \exp(\psi)$	$\theta = \psi^2$	$a + b[1 + \exp(-\psi)]^{-1}$
Jacobian	$\exp(\psi)$	2ψ	$-b\exp(-\psi) * [1 - \exp(-\psi)]^{-2}$

$\boldsymbol{g}(\cdot)$ in the Jacobian matrix

$$\mathbf{G} = \begin{bmatrix} \frac{\partial g_1}{\partial \psi_1}(\boldsymbol{\psi}) & \cdots & \frac{\partial g_1}{\partial \psi_k}(\boldsymbol{\psi}) \\ \vdots & \vdots & \vdots \\ \frac{\partial g_m}{\partial \psi_1}(\boldsymbol{\psi}) & \cdots & \frac{\partial g_m}{\partial \psi_k}(\boldsymbol{\psi}) \end{bmatrix},$$

where $\boldsymbol{\psi}$ is $k \times 1$ and $\boldsymbol{\theta}$ $m \times 1$.

If $\sqrt{n}(\hat{\boldsymbol{\psi}} - \boldsymbol{\psi}_0) \xrightarrow{d} \mathcal{N}(\mathbf{0}, \mathbf{V})$ as n diverges, then

$$\sqrt{n}(\hat{\boldsymbol{\theta}} - \boldsymbol{\theta}_0) \xrightarrow{d} \mathcal{N}(\mathbf{0}, \mathbf{G}^\top \mathbf{V} \mathbf{G}).$$

Very often each parameter θ_i is obtained through the transformation of just one parameter ψ_i. In this case the Jacobian matrix is diagonal and the variance of each original parameter can be derived directly from the variance of the transformed parameter. Table 5.1 contains common transforms with the relative anti-transforms and the Jacobian of the transformation.

Table 5.2 reports the estimates of the same variances of a LLT + noise model carried out using different transforms of the parameters and the delta method for $n = 500$ (the estimates without transform were obtained using an optimiser that allows the user to restrict the space of each parameter to an interval). The estimates are really very similar whatever the transform. The

Table 5.2 *Estimates and relative standard errors for the three variances in a LLT + noise model using different transforms. The true parameters are* $\sigma_\eta^2 = 1$, $\sigma_\zeta^2 = 0.1$, $\sigma_\epsilon^2 = 1$ *and the sample size is* $n = 500$. *The estimates have been multiplied by 100.*

Transform	Estimate	$100 \cdot \sigma_\eta^2$	$100 \cdot \sigma_\zeta^2$	$100 \cdot \sigma_\epsilon^2$
$\hat{\theta} = \hat{\psi}$	coeff.	118.456	6.317	77.393
	std.err.	26.433	1.886	14.370
$\hat{\theta} = \exp(\hat{\psi})$	coeff.	118.452	6.308	77.350
	std.err.	26.454	1.885	14.378
$\hat{\theta} = \hat{\psi}^2$	coeff.	118.467	6.316	77.389
	std.err.	26.435	1.887	14.370

same can be said for the standard errors. We carried out the same experiment on shorter sample paths ($n = 100$) and the differences in the coefficients and standard errors remain very moderate.

5.4.3 *Coding the maximum likelihood estimator in a model in state space form*

Let us try to put together the results seen in the previous sections by sketching the steps for the implementation of the MLE in a typical numerical or statistical programming language.

First of all, the user should identify a function that numerically maximises or minimises a given objective function and a procedure to compute the numerical Hessian of a given function.

Through its formal parameters or global variables, the objective function should receive the following pieces of information:

- The vector with the parameters with respect to which the objective function is to be optimised,
- The time series observations,
- The system matrices of the state space form,
- An array of indexes that contains information on how to map each parameter in the right element of the right system matrix.

In order to write efficient code, the system matrices should be passed as pointers (if the language allows passing variables through their address) or as global variables. In fact, rebuilding the system matrices every time the objective function is called may be rather time consuming. The array with the indexes should enable the modification of only the values of the system matrices that have to change when the parameters change. The objective function has to run the Kalman filter and return the log-likelihood function or, better, the mean log-likelihood function at the values of the parameters passed as arguments.

Numerical optimisers need initial values of the parameters, and a good choice of these may save time and give more stable solutions. Moreover, the log-likelihood function of a model in state space form may have multiple local maxima, and so different initial values may correspond to different MLE. Among multiple local maxima, the user should select those estimates corresponding to the highest local maximum. Chapter 6 provides some suggestions on how to select initial values.

Once the numerical optimiser has found the (or a) maximum, the function for computing the numerical Hessian should be called by passing the objective function and the vector of maximum likelihood estimates provided by the maximiser. If the objective function returns the log-likelihood, then the sample information matrix is obtained by inverting the Hessian divided by n and multiplied by -1. If the objective function returns the mean log-likelihood, then the Hessian has only to be inverted and multiplied by -1. The square roots of the elements on the diagonal of the sample information matrix contain the standard error of the respective parameters. If some parameters have undergone a transformation, then a delta method step has to be carried out as explained in the previous section and in particular in Table 5.1.

Chapter 6

Modelling

Modelling a time series by UCM entails a number of steps:

- Deciding if the UCM has to be applied to the raw time series or to a transformation;
- Identifying the main components that move the data;
- Setting the model in state space form;
- Finding a sensible set of initial values for the parameters to be estimated by numerical maximum likelihood and estimating them;
- Carrying out diagnostics checks to see if the model is adequate for the time series (are there outliers or structural brakes? are one-step-ahead forecast errors uncorrelated? are they normally distributed?);
- If more than one model seems to fit the data well, finding the model that seems to work the best for our goals.

Modelling is more an artisan type of work rather than an exact science. Thus, the more you do it, the better you get. In this chapter I try to share the reasoning and the techniques I use when building UCM and selecting the best one for a given aim. You may find better ways to do it while working on real time series, but I hope this chapter can save you time, especially if you are new to applied time series modelling.

While theoretical results are very easy to find in the scientific literature, practical suggestions and rules of thumb for modelling time series are rarely encountered in academic papers or books. Indeed, they do not possess the rigour and the novelty necessary to be published in such outlets. It would be very useful for the applied statistician and econometrician to have access to some *Journal of Statistical Practices* or, better, a *Wikipedia of Statistical Practices*, but I am afraid that such outlets have not been invented yet.

6.1 Transforms

The UCM we treat in this book, as well as ARIMA models, are linear and this fact somewhat limits the range of time series we can model. However many time series that show some forms of non-linearity can be transformed so that they look linear or, at least, more linear.

In particular, it happens very often that the trend grows at a geometric rate and the seasonal variations are proportional to the trend level. For instance, if the trend of a time series shows a constant rate of growth, then the resulting time series will increase at geometric rate. In a similar fashion, if in a monthly time series the value of August is only 50% of the level of the series, then the difference with respect to the level will be of 50 units when the level is 100 and of 500 units when the level is 1000. These are examples of time series with multiplicative components,

$$\dot{Y}_t = \dot{\mu}_t \cdot \dot{\psi}_t \cdot \dot{\gamma}_t \cdot \dot{\epsilon}_t,$$

which can be made linear by taking the natural logarithm of both sides,

$$\begin{aligned} \log \dot{Y}_t &= \log \dot{\mu}_t + \log \dot{\psi}_t + \log \dot{\gamma}_t + \log \dot{\epsilon}_t, \\ Y_t &= \mu_t + \psi_t + \gamma_t + \epsilon_t, \end{aligned}$$

with $Y_t = \log \dot{Y}_t$, $\mu_t = \log \dot{\mu}_t$, etc.

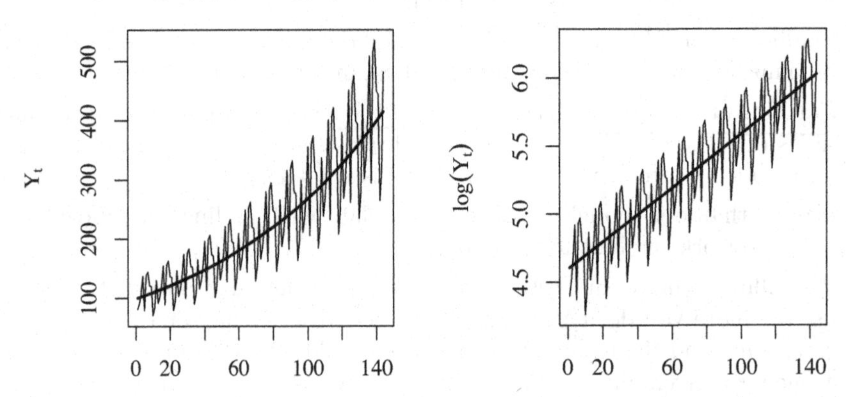

Figure 6.1 *Left: a multiplicative time series with exponential trend and deterministic seasonality. Right: the logarithm of the series on the right.*

Now, suppose that after taking the log the time series is well approximated by a Gaussian model in state space form. In this case, as seen in Chapter 5, for each unobserved component we are able to compute the quantities

$$\dot{a}_{i,t|s} = \mathbb{E}[\dot{\alpha}_{i,t|s}|\mathcal{Y}_s], \qquad \dot{P}_{ii,t|s} = \text{MSE}(\dot{a}_{i,t|s}) = \mathbb{V}\text{ar}(\dot{\alpha}_{i,t}|\mathcal{Y}_s).$$

In order to get estimates of the components before the log, we have to take the exponential of the Gaussian processes and so the original multiplicative components turn out to be log-normal processes. Therefore the inference for the original components is obtained as

$$a_{i,t|s} = \mathbb{E}[\alpha_{i,t|s}|\mathcal{Y}_s] = \exp\left(\dot{a}_{i,t|s} + \frac{1}{2}\dot{P}_{ii,t|s}\right).$$

If the second added in parentheses is omitted, then we are using the conditional median instead of the conditional mean to predict the unobservable components. In fact, since $\alpha_{i,t}$ is normally, and thus symmetrically, distributed then the mean and the median coincide and, the median is preserved by monotonic transforms. Notice that the conditional median is an optimal predictor if we use an absolute-value loss rather than a quadratic loss.

The conditional variances could also be derived using the formula for the second moment of the log-normal, but since these variances are mostly used to build confidence intervals (CI), these can be obtained by applying the exponential transform to the CI computed for the components of $\dot{\alpha}_t$. Indeed, quantiles are preserved by monotonic transforms: a 95% level CI for $\alpha_{i,t|s}$ is generally obtained as

$$\exp\left(\dot{a}_{i,t|s} \pm 1.96\sqrt{\dot{P}_{ii,t|s}}\right).$$

A useful by-product of the log transform is that all the model quantities that measure a change, for example the slope, the seasonal variation, the coefficient of a dummy variable can be interpreted as relative changes[1].

Example 6.1 (Airline passengers from and to Australia).
In Example 4.6 we modelled the log of the number of airline passengers from and to Australia. We estimated a basic structural model (local linear trend plus stochastic seasonal dummies plus noise) with two interventions. Since the estimated value of σ^2_ζ (i.e., the disturbance of the slope component, β_t) is zero, the slope is constant and the smoothed value is $\beta_t = 0.00534$ for all t. This value has to be interpreted as (an approximation of) the monthly average rate of growth of the trend component, which over one year becomes $1.00534^{12} - 1 = 0.066$, that is, an annual growth of 6.6%.

The smoothed estimates of the 9/11 effect on the next three months are -0.099, -0.025 and -0.006, which represent a lost of respectively 9.9%, 2.5% and 0.6% of passengers in those months.

As for the seasonal variation, the month in which the number of airline passengers is largest is January, which at the end of the series has a smoothed seasonal component equal to 0.151, or 15.1%, while

[1]This can be easily proved by recalling that the log is approximately linear around 1: using a first-order Taylor expansion we have that for small x, $\log(1+x) \approx x$. Thus,

$$\log(X_t) - \log(X_{t-1}) = \log\left(\frac{X_t}{X_{t-1}}\right) = \log\left(1 + \frac{X_t - X_{t-1}}{X_{t-1}}\right) \approx \frac{X_t - X_{t-1}}{X_{t-1}}.$$

Analogously, if we set $\log Y_t = \log \mu_t + \beta D_t + \log \varepsilon_t$, with D_t dummy variable for some event, the relative effect of the dummy intervention on Y_t is

$$\frac{\mu_t \exp(\beta D_t)\varepsilon_t - \mu_t\varepsilon_t}{\mu_t\varepsilon_t} \approx \log\left(1 + \frac{\mu_t \exp(\beta D_t)\varepsilon_t - \mu_t\varepsilon_t}{\mu_t\varepsilon_t}\right) = \beta D_t.$$

the month with least passengers is May with a smoothed component of −0.139, that corresponds to a fall of 13.9% with respect to the level.

A popular family of transformations that includes the natural log as a special case is the Box–Cox transform

$$Y_t^{(\lambda)} = \begin{cases} \log Y_t & \text{for } \lambda = 0, \\ \frac{Y_t^\lambda - 1}{\lambda} & \text{otherwise,} \end{cases}$$

where λ is a parameter taking values in $\mathbb{R}$ and the random variable Y_t must be positive with probability 1. Since we want to fit a linear model to $Y_t^{(\lambda)}$, subtracting 1 from Y_t and dividing by λ is really not substantial, but it guarantees the continuity of $Y_t^{(\lambda)}$ with respect to λ, which is very useful if λ is estimated by (possibly numerical) maximum likelihood[2]. So, the Box–Cox family of transformations consists in all the powers and the natural logarithm.

In practice, in the great majority of time series applications, the only transformation taken into consideration is the log, but there may be cases in which it is worth looking at other transforms. If your main interest in UCM is extracting the components and interpreting them and the coefficients, then you should work either on raw data or on the log of them, but if your main interest is in forecasting, then looking for the "right" Box–Cox transform may increase the precision of your predictions. Indeed, one should keep in mind that the predictions of trends like the ones used in UCM are linear. Thus, if the time series tends to grow faster or lower than a straight line, a transform should be taken into consideration. A rough check of the linearity of the trend on a grid of λ values can provide good results. If the time series shows some form of seasonality, then working on yearly averages or sums, or just concentrating on the same period of the year (e.g., the same month or a quarter) is a good idea (see Figure 6.2).

Another reason for transforming a time series is stabilising the variance if this varies over time. If there is a regular relation between the level of the series and the standard deviation, then it is possible to find a transform that makes a heteroskedastic time series homoskedastic. A practical way to find the Box–Cox transform that makes the variance constant over time is by plotting an estimate of the standard deviation of the series (y-axis) against an estimate of the level (x-axis). If the time series is monthly or quarterly this can be easily accomplished by taking sample means and standard deviation of every year of the time series for grid λ-values. If there is a λ value that makes the standard deviation constant, or at least not related to the level, then that transform is the one that best stabilises the variance (see Figure 6.3).

The goals of stabilising the variance and linearising the trend may imply contradictory choices of the λ parameter. If this happens you can pick a value

[2]The continuity of $Y_t^{(\lambda)}$ in $\lambda = 0$ is due to the fact that $\lim_{\lambda\to 0^-} Y^{(\lambda)} = \log Y_t = \lim_{\lambda\to 0^+} Y^{(\lambda)}$.

that is a decent compromise between the two goals, but if your aim is short-term forecasting then stabilising the variance (and the seasonal variation) should prevail, while if you need to forecast the trend for a mid- to long-term horizon, then the λ that makes the trend grow linearly should be chosen.

Example 6.2 (Box–Cox transform for the U.S. airline time series).
In the famous Airline model fitted by Box and Jenkins (1976) on this time series (see Figure 2.1), the data had been previously transformed using the log. However, the exponential growth implied by a (locally) linear trend on the logarithm seems to be somewhat too optimistic.

In order to see which Box–Cox transform makes the time series grow approximately linearly we compute the total annual passengers, Box–Cox transform the time series for a grid of λ values $(0.0, 0.1, \ldots, 1.0)$, and fit a straight line to each transformed series. The highest coefficient of determination R^2 (which in this case is just the square of the correlation coefficient between the series and a linear trend) is obtained for $\lambda = 0.5$ ($R^2 = 99.56\%$). Figure 6.2 depicts the series and the linear trend for $\lambda = 0, 0.5, 1$ (i.e., log, square root, raw data). The line is clearly closest when $\lambda = 0.5$.

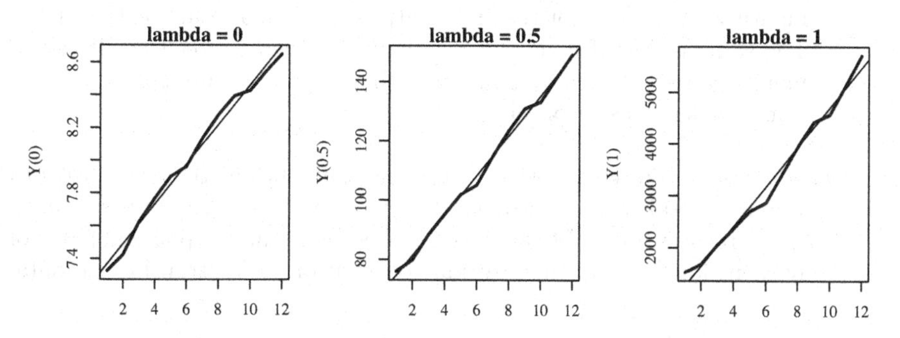

Figure 6.2 *Annual U.S. airline passengers after Box–Cox transforms* ($\lambda = 0.0, 0.5, 1.0$) *with fitted linear trend.*

Let us now assess what value of λ makes the series homoskedastic. Using a grid search strategy we found that the value of λ that makes the series look more homoskedastic is $\lambda = -0.25$. Figure 6.3 compares the mean-variance scatter plots of various transformations of the series: a log transform (i.e., $\lambda = 0$) makes the variance somewhat more stable than without any transform (i.e., $\lambda = 1$), but for $\lambda = -0.25$ the result is much better.

Which transformation should we then apply? The answer depends on our primary goal. If we want to give an interpretation to the components the UCM provides us with, then the log transform is the best compromise because it is better than working on raw data, and it allows a multiplicative interpretation of the extracted components. If we need

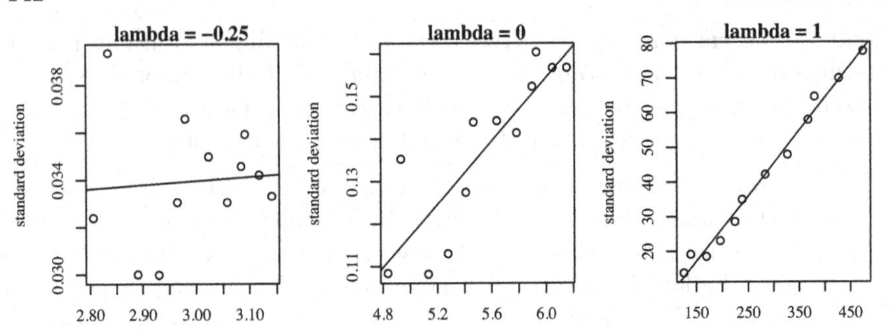

Figure 6.3 *Scatter plots of the annual means and standard deviations of the U.S. airline passengers series after Box–Cox transforms ($\lambda = -0.25, 0.50, 1.00$) with fitted linear trend.*

to forecast in the short run (up to one year) and we need good prediction of the monthly figures, then $\lambda = -0.25$ would guarantee accurate forecasts for the evolution of the seasonal component. If we are more interested in the prediction of the trend and accurate forecasts of the monthly figures are not that relevant, then $\lambda = 0.5$ would be the choice. The same is true if mid- and long-term forecasts are needed. If two or more of these goals are competing, then a compromise such as $\lambda = 0$ would be the reasonable choice.

In some particular cases, other transformations such as the *logit*, *probit* or similar could be taken into account. If Y_t takes values in the interval $(0, 1)$, such as if it records a market share, the consensus towards a political party or the sentiment for a brand, then working on logit or probit transformed data

$$\text{logit}(Y_t) = \log\left(\frac{Y_t}{1 - Y_t}\right), \qquad \text{probit}(Y_t) = F^{-1}(Y_t),$$

with $F^{-1}(\cdot)$ quantile function of a probability distribution (generally the standard normal), guarantees that the forecasts will be in the unitary interval.

6.2 Choosing the components

Building an UCM consists in choosing the components that should be present in the process that has generated the time series we need to analyse and forecast. In this book the reader is introduced to the most frequently used components, but he/she can modify them or think of alternative ones in order to match the features of the time series.

The first components one should consider are those that "move" the time series the most, or stated in a more statistical way, the ones that account for the largest share of the variance of the time series. Thus, the trend and the seasonal components are the first ones to consider.

If the time series is monthly or quarterly a seasonal component has to be present in the model. In Section 3.4 we introduced two alternatives: the stochastic dummy and the trigonometric seasonal components. The main difference between these two components is in how they evolve over time. The trigonometric seasonal component evolves more smoothly than its competitor. Furthermore, if the stochastic sinusoids of the trigonometric seasonal component are not all necessary, some of them can be eliminated from the model, making the state vector smaller and the computational burden lighter. For example, for smooth seasonal movements as those related to temperatures one or two sinusoids can be sufficient.

Example 6.3 (Monthly mean temperature in New York City).
We estimated an UCM with local linear trend, stochastic trigonometric seasonal component and noise on the monthly time series of the mean temperatures in New York City from Jan-1980 to Dec-2013. The six stochastic sinusoids are depicted in Figure 6.4 with their 95% confidence intervals (grey shades).

It is evident how the greatest part of the variance of the seasonal component is accounted for by the first sinusoid. The 95% confidence intervals for the sinusoids 4, 5, and 6 always include the zero line and, thus, are not significantly different than zero. As a result, the state space form of the model can be restated without these last three sinusoids, reducing the dimension of the state vector by five units.

As for the choice of the trend component, since the local linear trend (LLT) embeds I(0) and I(1) processes with or without a deterministic linear trend, and an I(2) process, some econometricians would suggest testing for a unit root on the original series and on its first difference using tests such as the Augmented Dickey–Fuller (Dickey and Fuller, 1979; Said and Dickey, 1984) or variations of it (Phillips and Perron, 1988; Elliott et al., 1996) or, alternatively, testing for the stationarity of the same time series and assuming a unit root when the KPSS test (Kwiatkowski et al., 1992) or its variations reject (de Jong et al., 2007; Pelagatti and Sen, 2013).

Although unit root and stationarity testing is an extremely fascinating research area, I am not convinced that it is that useful in modelling with UCM. Indeed, unit root tests usually provide a clear sentence when the visual inspection of the line plot already reveals that the time series is integrated, while it gives weak answers (e.g., p-values close to the typical test levels, answers that change with some options you have to fix in the unit root tests, etc.) when the graph does not give a clear picture. Furthermore, by estimating the complete LLT, which is an I(2) process, there is a positive probability that if some variance is zero, and thus the order of integration of the trend may be smaller than 2, this parameter is estimated to be zero, or very small. Finally, if the variance of the slope disturbance σ_η^2 of the LLT is small compared to the other variances (σ_η^2, σ_ϵ^2 in particular), as usually happens in real time series, then the

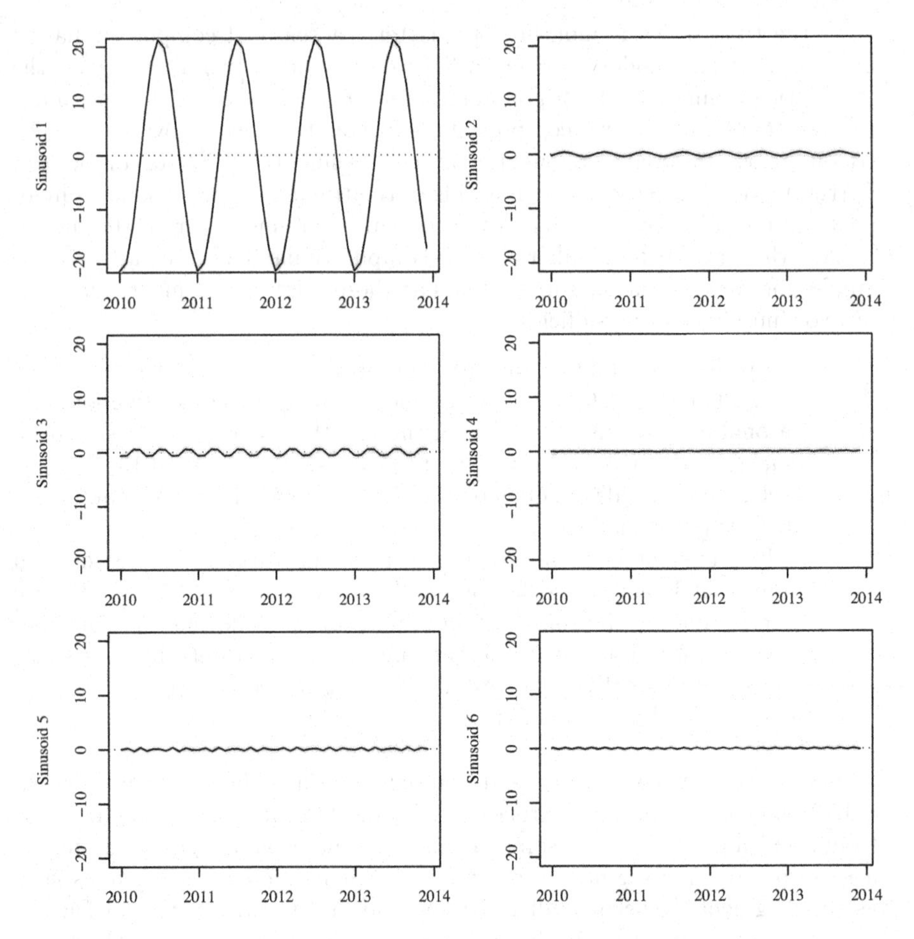

Figure 6.4 *Smoothed sinusoids forming the seasonal component of the monthly mean temperature in New York City. The grey shades represent the 95% CI.*

reduced form of this model has a moving average part with one characteristic root very close to -1, and in this condition all unit root tests tend to reject the unit root hypothesis with high probability, or in statistical words, they are severely over-sized. In order to prove this statement empirically, 10,000 time series of length $n = 100$ were generated from a LLT plus noise model with parameters $\sigma_\epsilon = 2$, $\sigma_\eta = 1$, $\sigma_\zeta = 0.2$. The Augmented Dickey–Fuller (ADF) test (with automatic BIC-based lag selection and a linear trend) was applied to the raw time series and on their first difference. Since the simulated time series are I(2) processes, the null hypothesis that (at least) one unit root is present in the data is true both on raw data and first differences, but at a 5% nominal level the test rejected 5.7% of the times on the levels and 99.2% of the times on the differences. The conclusion is that the ADF is not able to find the second unit root and selects a random walk plus noise model, when the real model is LLT plus noise.

Thus, unless it is evident that a time series is I(0) or I(1), the LLT with unconstrained variances should be fitted to the data. If some estimated variance is zero or very close to zero, and/or the relative component does not seem to evolve significantly over time, then the model can be refitted imposing the constraint to that variance.

On the contrary, if we need to fit an UCM with stochastic cycle, then it could become necessary to set the variance of the level disturbance σ_ζ^2 equal to zero. Indeed, the integrated random walk (IRW) produces a smoother trend that is simpler to discriminate from the business cycle component. So, while it is very hard to estimate the parameters of a LLT plus cycle model, the task is much simpler for an IRW plus cycle. The technical reason for this to happen is that in a LLT plus cycle model the log-likelihood function can have ridges on which different values of the parameters provide an almost identical log-likelihood. We will discuss the stochastic cycle at large in Chapter 8 and so we will not spend more time on the stochastic cycle here.

Before introducing real regressors in the UCM, it is important to reflect on a few issues.

First of all, if you want to use such a model for forecasting then you will need future values of the regressors, and this is usually not the case unless you use lagged regressors, or regressors that are available enough time before the dependent variable is. If you think you can have access to reliable predictions of future values of the regressors, as for example when they are meteorological variables, then you should keep in mind that substituting known past values with future predicted values adds uncertainty to your forecasts that are hard to assess. The best approach would be using forecasts also in the past. For example, suppose that every day you have to forecast the daily demand of electricity of the next day using the average daily temperature as a regressor. The best way to attack this problem is collecting the time series of one-day-ahead temperature forecasts and use it as a regressor instead of the observed temperature, which anyway will not be available in the future. This approach is not always feasible as not many past forecasts are available, and in this case the second best is using observed data in fitting the model and forecasts in predicting future values of the time series and, if possible, adding an estimate of the MSE of the regressor forecasts to the variance of prediction errors (i.e., the diagonal elements of the matrix $\mathbf{F}_{n+k|n}$ in Section 5.3.3) provided by the forecasting equations (but this correction is very rarely applied in practice).

Secondly, the long-run relation between the dependent variable and the regressor(s) should be considered. If in the UCM model an I(1) (i.e., random walk possibly with drift) or I(2) (i.e., LLT or IRW) trend is present and the regressors are stationary, then using the regressors in the observation equation of the UCM does not affect the long-run behaviour of the trend, but if the regressors are not zero-mean, it affects the level of the estimated trend component. For this reason, many software packages have the option of plotting or saving the component *trend plus regressor effects.* If, instead, you want to meaningfully isolate the trend from the regression effects, then you should subtract an estimate of the mean of the regressors from themselves before using them in the UCM. Somewhat more tricky is the case in which the regressors are trending. If the dependent variable is stationary, then the regressors have to be differenced until they become stationary otherwise the regression is nonsense: indeed, how can the conditional mean of a mean-reverting variable linearly depend on values that never revert to any constant value? If the dependent variable and the regressors are trending, then there is the possibility that they share the same trend(s), in which case the time series are said to be *cointegrated* (i.e., we are assuming that the trends are well approximated by a LLT or some particular cases of it). In this case, since the trend of Y_t is modelled though the regressors, then there is no need to put a LLT (or its special cases) in the UCM which should be substituted with a constant. Otherwise, if the dependent variable and the regressors do not share a common trend, then a LLT (or its special cases) should be put in the UCM. In order to assess if the dependent variable and the regressors share a common trend, one way to proceed is though a formal *cointegration test*. The simplest choice is the Engle–Granger test, also known as Phillips–Ouliaris test (Engle and Granger, 1987; Phillips and Ouliaris, 1990), which consists in the application of a unit root test (with modified critical values) to the residual of an ordinary least square regression of the variable Y_t on the regressors, but tests with better properties such as Johansen (1991) or Nyblom and Harvey (2000) may be available in statistical software packages.

Again, testing before modelling is not necessarily the best approach to build a UCM. In fact, estimating a UCM with a LLT and trending regressors may give some hints on the presence of cointegration. If the dependent variable is cointegrated with the regressors it can happen that the estimated variances of the LLT are either zero or very small and the movement of the smoothed trend component may appear approximately constant or stationary. These are all signs of cointegration and when they appear the UCM can be estimated again without the stochastic trend component. In case it is not clear if the regressors are cointegrated with the dependent variable, a good strategy is estimating two models, one with trend and one without trend. Selecting the best one is discussed in Section 6.5.

Artificial regressors as dummy variables, structural changes, sinusoids (see Chapter 4) are much less problematic. They can be built to model facts known to the analyst or they can be inferred in the process of modelling as explained below in Section 6.4. It is important to remember that components that have a known value at time $t = 1$, as for the case of intervention variables, need this information specified in the distribution of the initial state vector. For instance, in the model with dynamic intervention,

$$\begin{aligned}
Y_t &= \mu_t + \lambda_t + \epsilon_t, \quad \epsilon_t \sim \text{WN}, \\
\mu_{t+1} &= \mu_t + \eta_t, \quad \eta_t \sim \text{WN}, \\
\lambda_{t+1} &= \delta\lambda_t + \omega\mathbb{I}(t = \tau),
\end{aligned}$$

with $\mathbb{I}$ indicator for the event $t = \tau$ (i.e., it takes the value 1 when $t = \tau$ and zero otherwise), the dynamic response variable λ_t is equal to 0 up to the event time $t = \tau$, and this information must be included in the state space specification by assigning to λ_1 the degenerate distribution with both mean and variance equal to zero. Notice that if the transition equation is in *future form* as in this book, the first effect of the event at $t = \tau$ will be on $Y_{\tau+1}$.

When we choose the components of a UCM we have to take care that the complete model is *identifiable*, that is, different values of the unobserved components and unknown parameters must generate different joint probability distributions of the observations. Some excellent texts such as Harvey (1989) and Anderson and Moore (1979) spend many pages on the concept of *identifiability* and the related conditions. However, in practice it is very rare that, while modelling, one carries out a formal verification of these conditions. Usually common sense is enough. When choosing the components of a UCM you should take care that the same feature is not modelled by two or more components. For example, the model

$$Y_t = \mu_t + \mu_t^* + \epsilon_t,$$

Where the two μ's are random walks with unknown variances, is not identified because the sum of two independent random walks is a random walk. Indeed, the reader can easily verify that $\nu_t = \mu_t + \mu_t^*$ is a random walk and the variance of its increment $\Delta\nu_t$ is given by the sum of the variances of the increments of μ_t and μ_t^*. Therefore, any combination of these two variances that sums up to the same value generates observationally identical processes. The problem is analogous to the solution of one linear equations in two unknowns, say $x + y = 3$: there are infinite x and y from which the identity holds. Probably the most common mistake, as far as identifiability is concerned, is letting two different variables or parameters take care of the initial level of a time series. For instance, the regression plus random walk plus noise model,

$$\begin{aligned}
Y_t &= \mu_t + \beta_{0,t} + \beta_{1,t}X_t + \epsilon, \quad \eta_t \sim \text{WN}, \\
\mu_{t+1} &= \mu_t + \eta_t, \quad \eta_t \sim \text{WN}, \\
\beta_{0,t+1} &= \beta_{0,t}, \\
\beta_{1,t+1} &= \beta_{1,t}
\end{aligned}$$

is not identifiable when the state vector at time $t = 1$ is assigned a diffuse distribution. In this case the level of the series at time t is given by

$$\beta_{0,1} + \mu_1 + \sum_{s=2}^{t} \eta_{s-1} + \beta_{1,1}X_t$$

and the initial level of the series is given by $\beta_{0,1} + \mu_1$, which takes the same value for infinite choices of the two addends. The same problem arises when you specify an ARMA component with non-zero mean over a model that already includes a level component, as for example

$$\begin{aligned} Y_t &= \mu_t + \psi_t + \epsilon, \quad \eta_t \sim \text{WN}, \\ \mu_{t+1} &= \mu_t + \eta_t, \quad \eta_t \sim \text{WN}, \\ \psi_{t+1} &= d + \phi\psi_t + \zeta, \quad \zeta_t \sim \text{WN}. \end{aligned}$$

Again, we have that at time $t = 1$ the level of Y_1 is given by $\mu_1 + d/(1 - \phi)$, where ϕ, if the AR is stationary is identified through the short-term dynamics of Y_t, while infinite choices of μ_1 and d yield the same level. If $\phi = 1$ the identification problems become even worse and it may be a good idea to bound the parameter to $(-1, 1)$ in the numerical estimation.

The use of multiple stochastic cycle components does not cause identification problems if the frequency parameters are different.

6.3 State space form and estimation

Stating UCM in state space form has been treated in detail in Section 5.2. Here we just want to share some practical suggestions for choosing the initial distribution of the state variables and the starting values for parameters to be estimated numerically.

If you have some information about single state variables at time $t = 1$ you should try to formalise it in the mean-variance pair. For example, if you are sure that the trend component μ_t at time $t = 1$ is between 90 and 110 and that the normal distribution is a decent approximation of your uncertainty, then you can exploit the fact that 99.7% of the probability is contained in the interval $\mu \pm 3\sigma$, where μ is the mean and σ the standard deviation of the distribution. So, setting $\mu = 100$ and $\sigma = 10/3$ is a simple way to state the initial moments of the trend component exploiting that piece of information. If the shape of Gaussian distribution is not a good approximation of your uncertainty about the initial value, you may want to increase the value of the standard deviation obtained in that formula, for example setting σ equal to the distance between the maximum value (110 in the example) and the mid-point (100 in the example).

If no information is available on the starting value of the state variables, then the stationary components should be assigned their marginal distribution, using the formulae discussed in Section 5.3.4. For the nonstationary components, if the software package offers exact algorithm for diffuse initial conditions, then you can just set the mean of the state variables equal to zero and their variance equal to infinity. If, instead, only the standard Kalman filter is available, then we can use the observations to derive sensible initial conditions for some common components.

For example, for the level component we can set

$$\mathbb{E}[\mu_1] = Y_1, \qquad \mathbb{V}\mathrm{ar}(\mu_1) = \frac{1}{n}\sum_{t=1}^{n}(Y_t - Y_1)^2,$$

or

$$\mathbb{E}[\mu_1] = \frac{1}{s}\sum_{t=1}^{s} Y_t, \qquad \mathbb{V}\mathrm{ar}(\mu_1) = \frac{1}{n}\sum_{t=1}^{n}\left(Y_t - \mathbb{E}[\mu_1]\right)^2,$$

if we want to account for a seasonal component of period s in estimating the mean value of μ_1. Similarly, for the slope component β_t we can set

$$\mathbb{E}[\beta_1] = \Delta Y_2, \qquad \mathbb{V}\mathrm{ar}(\mu_1) = \frac{1}{n-1}\sum_{t=2}^{n}(\Delta Y_t - \mathbb{E}[\beta_1])^2,$$

or

$$\mathbb{E}[\beta_1] = \frac{1}{s}\Delta_s Y_{s+1}, \qquad \mathbb{V}\mathrm{ar}(\mu_1) = \frac{1}{n-s}\sum_{t=s+1}^{n}\left(\frac{1}{s}\Delta_s Y_t - \mathbb{E}[\beta_1]\right)^2,$$

if we want to account for a seasonal component of period s. The variances computed in this way are probably larger than necessary, but smaller than infinity as in the diffuse case. Another drawback is that they depend on the sample size, and generally they grow with n. This can be easily solved by using only the initial part of the sample, for instance the first four or five years, if the time series is monthly or quarterly.

As for the seasonal component, for the stochastic dummy specification we can regress the first year of data on a linear trend and use the residuals to obtain the expected value of each "season". For the trigonometric approach, the same residuals can be regressed on the six sinusoids and their value at $t = 1$ used as initial value. The variance of each component can be set equal to the sample variance among the seasonal state variables.

Notice that these are just suggestions based on common sense and they can be criticised as the data are used twice: once for determining the initial values and once while running the Kalman filter and building the likelihood. But this critique, sound from the theoretical point of view, is very weak from the practical one. In fact, when we build a model for time series, we use the data many times. First we watch a line plot to decide if we have to take transformations and determine which components to include in the UCM. Then, we try different specifications and choose which one fits the data the best and look for outliers observing the model residuals (see next section). Finally, when we have a working model we estimate it every time new observations are available and carry out some maintenance.

Example 6.4 (Effect of initial conditions on a basic structural model fitted to the Airline data).

We applied a basic structural model to the U.S. airline time series using

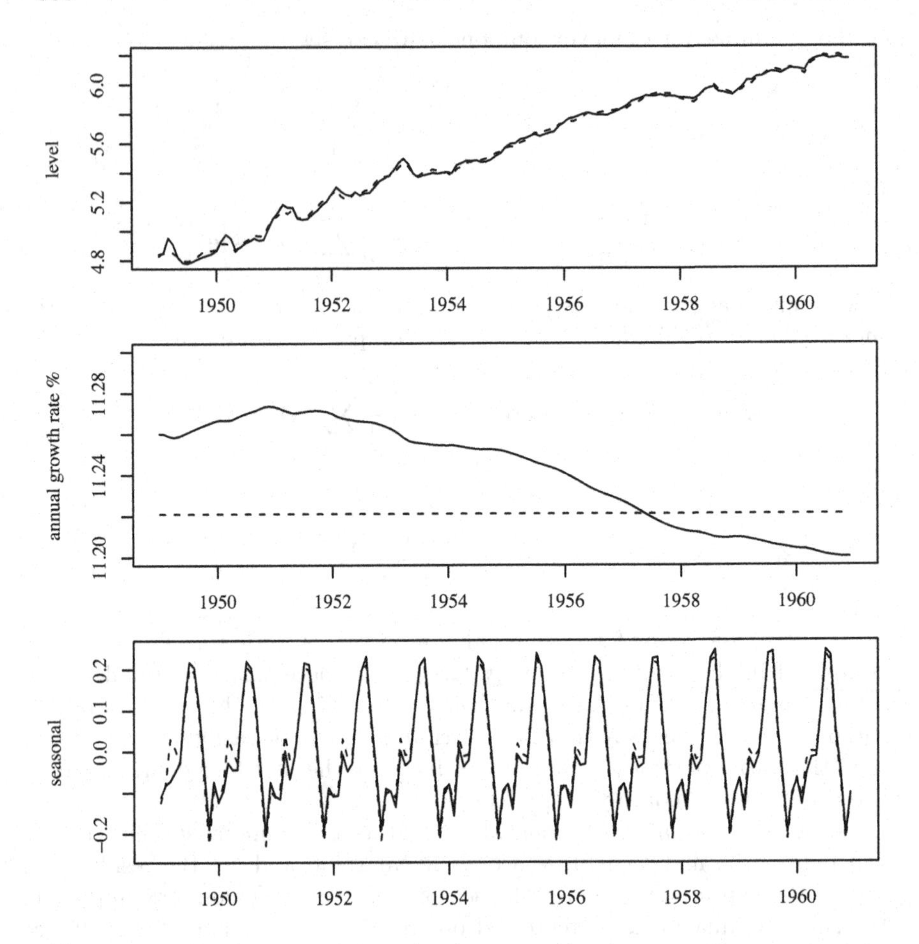

Figure 6.5 *Comparison of the smoothed components of a BSM fitted to the Airline time series using diffuse (dashed line) and informative (line) initial conditions obtained as discussed in this section.*

diffuse and informative initial conditions determined as explained in this section. The components are slightly different for two reasons: the initial conditions and their influence on the estimates of the unknown variances.

As one can see from Figure 6.5 the slope components are very close to each other, but one is constant (the one estimated with diffuse initial conditions), while the other is slowly evolving. My experience with modelling this time series makes me express my preference for the time-varying slope solution.

Another important issue is the choice of starting values for the unknown parameters to be estimated by maximum likelihood (ML) or quasi-maximum likelihood (QML). In fact, for many UCM the log-likelihood function may have multiple local maxima and the algorithm for numerical optimisation may get stacked into one of these instead of reaching the global maximum. Furthermore, if starting values are badly chosen, the computer may fail to compute the log-likelihood or its numerical derivatives, that are necessary to all Newton or quasi-Newton type optimisers. In UCM many parameters are variances, and a general suggestion about starting values for variances is that larger is better than smaller. The reason is easily explained with the plot in Figure

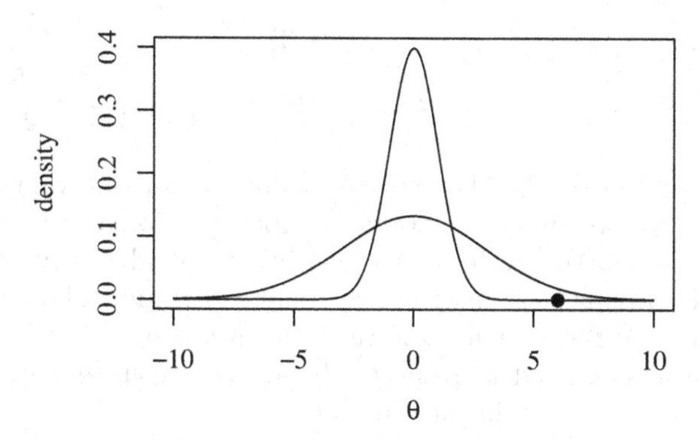

Figure 6.6 *Density of one point with respect to different variances.*

6.6: if we consider the distribution with smaller variance, the black point has a density value extremely close to zero, and given the limited precision of a number in the representation used by a computer, this number may not be distinguishable from zero. Furthermore, since we work on log-likelihoods, the zero is mapped into $-\infty$ and this makes the numerical optimisation fail at the very beginning. On the contrary if we consider the same point with respect to the larger variance distribution, the density is no longer so close to zero.

Now, let us see how to fix sensible starting values. Again, this is just a suggestion in case you don't find better ways. If the observations are generated by a basic structural model (with stochastic dummy seasonal component), then the process $\{\Delta\Delta_s Y_t\}$ has the following autocovariance function

$$\begin{aligned}
\gamma(0) &= 2\sigma_\eta^2 + s\sigma_\zeta^2 + 6\sigma_\omega^2 + 4\sigma_\epsilon^2 \\
\gamma(1) &= (s-1)\sigma_\zeta^2 - 4\sigma_\omega^2 - 2\sigma_\epsilon^2 \\
\gamma(2) &= (s-2)\sigma_\zeta^2 + \sigma_\omega^2 \\
\gamma(k) &= (s-k)\sigma_\zeta^2, \qquad \text{for } k = 3, 4, \dots, s-2 \\
\gamma(s-1) &= \sigma_\zeta^2 + \sigma_\epsilon^2 \\
\gamma(s) &= -\sigma_\eta^2 - 2\sigma_\epsilon^2 \\
\gamma(s+1) &= \sigma_\epsilon^2 \\
\gamma(k) &= 0, \qquad \text{for } k \geq s+2.
\end{aligned}$$

Now, solving the first four autocovariance equations for the unknown variances, with the sample autocovariances substituting the population ones, we obtain

$$\begin{aligned}
\sigma_\zeta^2 &= \frac{1}{s-3}\gamma(3)\\
\sigma_\omega^2 &= \gamma(2) - (s-2)\sigma_\zeta^2 = \gamma(2) - \frac{s-2}{s-3}\gamma(3)\\
\sigma_\epsilon^2 &= \frac{s-1}{2}\sigma_\zeta^2 - 2\sigma_\omega^2 - \frac{1}{2}\gamma(1) = \frac{5s-9}{2(s-3)}\gamma(3) - 2\gamma(2) - \frac{1}{2}\gamma(1)\\
\sigma_\eta^2 &= \frac{1}{2}\gamma(0) - \frac{s}{2}\sigma_\zeta^2 - 3\sigma_\omega^2 - 2\sigma_\epsilon^2 = \frac{1}{2}\gamma(0) + \gamma(1) + \gamma(2) - \frac{5s-6}{2(s-3)}\gamma(3).
\end{aligned}$$

One problem of using the above equations to get starting values for the variances is that no positivity constraint is imposed, and so you could have nonsense negative starting values. An easy workaround is using the recursive scheme (the formula between the two equal signs) and taking the maximum between zero and the formula. Alternatively, if the signs of $\gamma(s-1)$, $\gamma(s)$ and $\gamma(s+1)$ are correct, that is, respectively $+$, $-$, $+$, you can solve with respect to the variances and obtain the alternative formulae

$$\begin{aligned}
\sigma_\epsilon^2 &= \gamma(s+1)\\
\sigma_\eta^2 &= -\gamma(s) - 2\sigma^2\epsilon = -\gamma(s) - 2\gamma(s+1)\\
\sigma_\zeta^2 &= \gamma(s-1) - \sigma_\epsilon^2 = \gamma(s-1) - \gamma(s+1).
\end{aligned}$$

The variance of the seasonal component can be obtained from the formula for $\gamma(0)$,

$$\sigma_\omega^2 = \frac{1}{6}\left(\gamma(0) - 2\sigma_\eta^2 - s\sigma_\zeta^2 - 4\sigma_\epsilon^2\right),$$

if positive, or set equal to zero (or close to zero) if negative. Notice that if in your model there is no seasonal component, or the trend is either a random walk or an integrated random walk the autocovariance equations above can still be used by setting the respective variances to zero.

Table 6.1 compares the starting values obtained by applying the two alternative set of formulae (with the $\max(x, 0)$ trick) with estimates of a BSM applied to 100 times the log-airline time series, carried out using different software packages. The starting values are not distant from the final estimates, so these strategies seem to be useful. However, coherently with the first advice (i.e., for variances larger is better than smaller), taking the maximum between the pairs of starting values generated by the two sets of formulae should be a better strategy. Finally, my experience with the numerical estimation of UCM suggests that a zero starting value for a variance is not always a good idea and a small but positive number can avoid that the optimiser leaves those zero variances unchanged. Summing a small proportion of $\gamma(0)$, like some 1%, to all zero variances (or to all the variances) is a good idea.

There is a whole chapter (Chapter 10) dedicated to comparing different

	σ^2_η	σ^2_ζ	σ^2_ω	σ^2_ϵ
Formulae 1	3.85	0.00	2.19	0.00
Formulae 2	1.74	0.00	0.79	3.16
EViews	4.54	0.00	0.74	1.89
Gretl	6.97	0.00	0.64	1.32
Ox/SsfPack	6.99	0.00	0.64	1.29
R/FKF	6.99	0.00	0.64	1.30
R/StructTS	7.71	0.00	13.97	0.00
SAS/ETS	6.99	0.00	0.64	1.30
STAMP*	2.98	0.00	0.04	2.34
Stata	6.99	0.00	0.64	1.30

*STAMP implements a trigonometric seasonal.

Table 6.1 *Starting values computed with the two methods proposed in the text compared with some estimates for a BSM applied to the log-airline series.*

software packages, but Table 6.1 calls for a short comment about the concordance of the results: Gretl, Ox/SsfPack, R/FKF, SAS/ETS and Stata provide identical estimates, while EViews and R/StructTS deviate from those. In the table we included also the estimates of STAMP, even though it uses a trigonometric seasonal component, because we wanted to show that the initial values are reasonable also for that kind of specification.

If in the UCM a stochastic cycle is present, as already mentioned, the trend should be specified as an integrated random walk in order to facilitate the separation of the cycle from the trend. If the stochastic cycle is used to model a business cycle, then the typical period is some 4–5 years, which can be converted in frequency as $\lambda = 2\pi/p$, where p is the period (e.g., if the cycle repeats itself every 5 years and the observations are monthly then $p = 12 \cdot 5 = 60$). As for the *damping factor* ρ, which determines the persistence of the cycle, 0.9 seems to be a good choice in many situations. Less trivial is the choice of the variance of the disturbance or of the cycle. Recall that these variances are related *via* the relation $\mathbb{V}\mathrm{ar}(\psi_t) = \sigma^2_\kappa/(1-\rho^2)$, and take care to check which implementation uses your software package (for example SAS and SsfPack use the variance of the cycle). A reasonable starting value for $\mathbb{V}\mathrm{ar}(\psi_t)$ can be one half of the variance of the series after taking the two differences $\Delta\Delta_s$, that is $\gamma(0)/2$, or if you need the variance of the disturbance: $\sigma^2_\kappa = (1-\rho^2)\gamma(0)/2$.

Other components, regressors or intervention variables may be present, but once the parameters of the components responsible for the greatest part of the total variance have been given starting values, common sense should suffice.

To check the robustness of your estimations with respect to the choice of starting values, you can try to carry out the optimisation with different sets of starting values. If the estimates change with the starting values, then in principle you should pick the estimates that yield the highest value of the log-likelihood. However, if a model with a slightly smaller likelihood gives components that are more natural and simpler to interpret, I would consider keeping it in the list of candidate models.

Some software packages offer routines for global optimisation (e.g., simulated annealing) that, at the cost of an additional computational effort, are generally able to find the global maximum of the likelihood regardless of the starting values.

6.4 Diagnostics checks, outliers and structural breaks

Most diagnostic checks are based on the standardised innovations (or residuals)

$$\iota_{i,t}^{(s)} = \iota_{i,t}/\sqrt{F_{ii,t}},$$

which can be computed from the Kalman filter output. In fact, if the model is well specified they should be (approximately) serially uncorrelated, and if the assumption of Gaussianity of the data is reasonable, they should be (approximately) standard normal. Both these hypotheses can be assessed by visual inspection of some plots and formally tested. Thus, it is generally a good idea to plot the innovations against time and to plot their sample autocorrelation functions with 95% confidence bands for the zero (cf. Section 2.2). Portmanteau statistics such as those in Corollary 2.6 are very useful to test for the serial uncorrelation of the residuals. These Q-statistics, when computed on the residuals of a UCM, under the null of zero correlation are chi-square distributed as with $h-p+1$ degrees of freedom, where h is the maximum order of the autocorrelation on which $Q(h)$ is computed and p the number of estimated parameters in the UCM.

As for the normality of the innovations, the histogram, possibly with the density of the standard normal superimposed, is a very useful tool. The third and fourth moments of the standardised innovations, respectively sample skewness and kurtosis,

$$S = \frac{1}{n}\sum_{t=1}^{n}\left(\frac{\iota_{i,t}^{(s)} - \bar{\iota}_{i,t}^{(s)}}{\bar{\sigma}^2}\right)^3, \qquad K = \frac{1}{n}\sum_{t=1}^{n}\left(\frac{\iota_{i,t}^{(s)} - \bar{\iota}_{i,t}^{(s)}}{\bar{\sigma}^2}\right)^4,$$

with $\bar{\iota}_{i,t}^{(s)}$ and $\bar{\sigma}^2$ sample mean and variance of the standardised innovations[3], can be extremely useful, as under the null of normality,

$$\sqrt{\frac{n}{6}}S \xrightarrow{d} \mathcal{N}(0,1), \qquad \sqrt{\frac{n}{24}}(K-3) \xrightarrow{d} \mathcal{N}(0,1).$$

These two statistics are often used jointly in the Jarque–Bera (sometimes also

[3]Even though the innovations are standardised they do not need to have zero mean and unit variance in finite sample. That's the reason for this second standardisation.

Bowman–Shenton) test statistic

$$JB = \frac{n}{6}\left(S^2 + \frac{1}{4}(K-3)^2\right),$$

which under the null of normality is asymptotically chi-square distributed with 2 degrees of freedom. This normality test can be carried out in many software packages, but be aware that if the sample size is small (less than 100 observations) it is oversized (i.e., it rejects the null of normality more often than it should). Substituting in the definition of JB the factor n with $n-p+1$, can improve the finite sample approximation.

If the hypothesis of normality is rejected, this can be due to the non Gaussianity of the observations or to the presence of some outliers. The difference between these two concepts is very labile, but we can think of outliers as events to which we can give an explanation and, if repeated in the future, they are expected to have similar effects on the time series. Even if the too-extreme-to-be-Gaussian observations cannot be given an explanation in terms of exogenous effects, it may be useful to model them though intervention variables because least squares methods (as Kalman filtering) are rather sensitive to this kind of departure from normality. In this case the confidence intervals associated with predictions are associated with smaller probabilities than the nominal (e.g., typically software packages produce 95% CI, but if the real distribution of the data has heavy tails, the real coverage probability will be less than 95%).

The best tool for identifying outliers of various natures are the quantities computed by the *disturbance smoother* of Theorem 5.5. In particular, the standardised statistic

$$r_{i,t}^{(s)} = r_{i,t}/\sqrt{N_{ii,t}},$$

with $r_{i,t}$ and $N_{ii,t}$ respectively representing the i-th element of $\boldsymbol{r}_t$ and on the diagonal of $\mathbf{N}_t$, both defined in Theorem 5.4, can be used to test the null hypothesis

$$H_0 : [\boldsymbol{\alpha}_{t+1} - \boldsymbol{d}_t - \mathbf{T}_t\boldsymbol{\alpha}_t - \boldsymbol{\nu}_t]_i = 0,$$

where $[\cdot]_i$ indicates the i-th element of the contained vector. Stated in words, the null hypothesis holds that no outlier is present in the i-th state variable at time $t+1$. Similarly, the statistic

$$u_{i,t}^{(s)} = u_{i,t}/\sqrt{D_{ii,t}}$$

where the quantities $\boldsymbol{u}_t$ and $\mathbf{D}_t$ have been defined in Theorem 5.5, can be used to test the hypothesis

$$H_0 : [\boldsymbol{Y}_t - \boldsymbol{c}_t - \mathbf{Z}_t\boldsymbol{\alpha}_t - \boldsymbol{\epsilon}_t]_i = 0.$$

Under the assumption of Gaussianity of the state space form, and so, of the data, when the respective null hypotheses hold, the statistics $u_{i,t}^{(s)}$ and $r_{i,t}^{(s)}$ are distributed as standard normal, and so it is customary to plot them against time together with the 95% two-tail critical values for the null, ± 1.96 (or more often ± 2). These peculiar t-statistics are generally referred to as *auxiliary residuals*. Two things the analyst should always bear in mind when observing this graph is that i) the statistics $u_{i,t}^{(s)}$ and $r_{i,t}^{(s)}$ are serially correlated, ii) the 95% critical values interval will not contain the statistic, in the average, 5% of the times and so you shouldn't consider all the values outside that interval as outliers unless they are rather extreme (in some packages the value is considered outlying if in absolute value it is greater the 2.5).

If the statistic $u_{i,t}^{(s)}$ is large in absolute value, then a probable additive outlier (AO) affected the time series. The interpretation of the rejection of the null by the statistic $r_{i,t}^{(s)}$ depends on the component. This is how you should interpret the outlier according to the component:

level μ_t : the outlier is a level shift (LS);

slope β_t : the outlier is a slope shift (SS);

seasonal γ_t : the outlier is a change in the seasonal pattern and it may affect the time series for more than one period of time;

cycle ψ_t : the outlier can be interpreted as an exceptional shock affecting the (business) cycle (also the statistic for orthogonal component ψ_t^* should be checked).

In all these cases, the simplest way to adapt the UCM to the finding is adding a dummy variable multiplied by a coefficient to be estimated in the equation where the outlier has revealed itself. Refer to Chapter 4 and in particular to Section 4.2 for more details.

If in the software package you use there is no easy way to obtain the state disturbances or the auxiliary residuals, then a simple workaround that provides satisfactory results in finding outliers and structural breaks is using the smoothed state variables as if they were the real quantities and solving the transition equation for the disturbance:

Measurement error $\hat{\epsilon}_t = \mathbf{Y}_t - \mathbf{c}_t - \mathbf{Z}_t\boldsymbol{\alpha}_t$;

Level $\hat{\eta}_t = \mu_{t+1|t+1} - \mu_{t|t} - \beta_{t|t}$;

Slope $\hat{\zeta}_t = \beta_{t+1|t+1} - \beta_{t|t}$.

For building t-statistics like in auxiliary residuals, one can drop few values of these time series at the very beginning and end of the sample (because their MSE is higher) and divide $\hat{\epsilon}_t$, $\hat{\eta}_t$ and $\hat{\zeta}_t$ by their respective sample standard deviations.

Example 6.5 (Break in Nile flow).
A LLT plus noise model was fitted to the Nile flow data and the estimates of the three variances are $\sigma_\eta^2 = 1753$, $\sigma_\zeta^2 = 0$ and $\sigma_\epsilon^2 = 14678$, which reduce the UCM to a random walk with drift plus noise.

Figure 6.7 depicts the standardised innovations (top-left), their histogram with a kernel estimate of their density and the standard normal

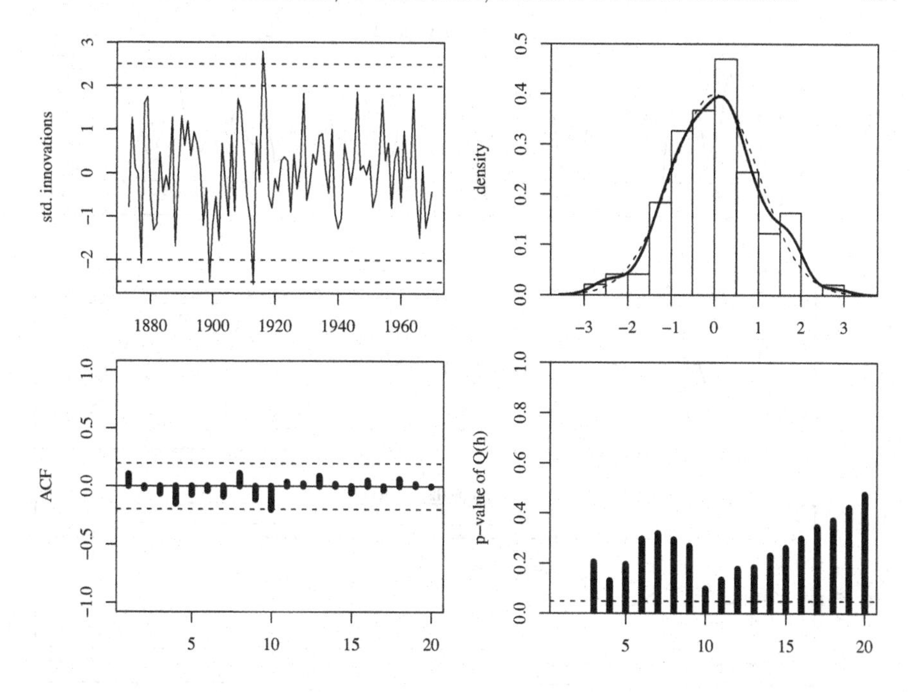

Figure 6.7 *Innovation plots of the LLT plus noise model fitted to the Nile flow time series.*

density (top-right), the sample autocorrelation function (bottom-left) and the p-values of the Ljung–Box statistics for a range of values of the maximum lag h (bottom-right). The analysis of these plots does not reveal particular problems with the fitted model: i) there are four innovations outside the ± 2 bounds, but this is not unexpected since the observations are 98, and circa 5 (i.e., 5% of 98) of them are expected to fall outside those limits; ii) the histogram reveals a good approximation to normality, confirmed by a Jarque–Bera test statistic of 0.21 (p-value $= 0.90$); iii) the correlogram does not reveal significant correlations (again, one falls outside the bounds, but being the 5% of 20 values equal to 1, this event was expected); iv) the absence of linear memory in the innovations is also confirmed by the Ljung–Box statistics that, whatever maximum lag h we pick, cannot reject the null of white-noise residuals at the 5% level.

In order to further investigate for possible outliers and breaks Figure 6.8 depicts the auxiliary residuals of each component with bounds at ± 2 and ± 2.5. The absolute value of the t-stat relative to the measurement error exceeds 2.5 in years 1877 and 1913 indicating the presence of additive outliers. The absolute value of the auxiliary residuals for the level disturbance is greater than 2.5 only in 1899 indicating a (negative) jump in the level of the flow. As discussed by Cobb (1978), there are meteoro-

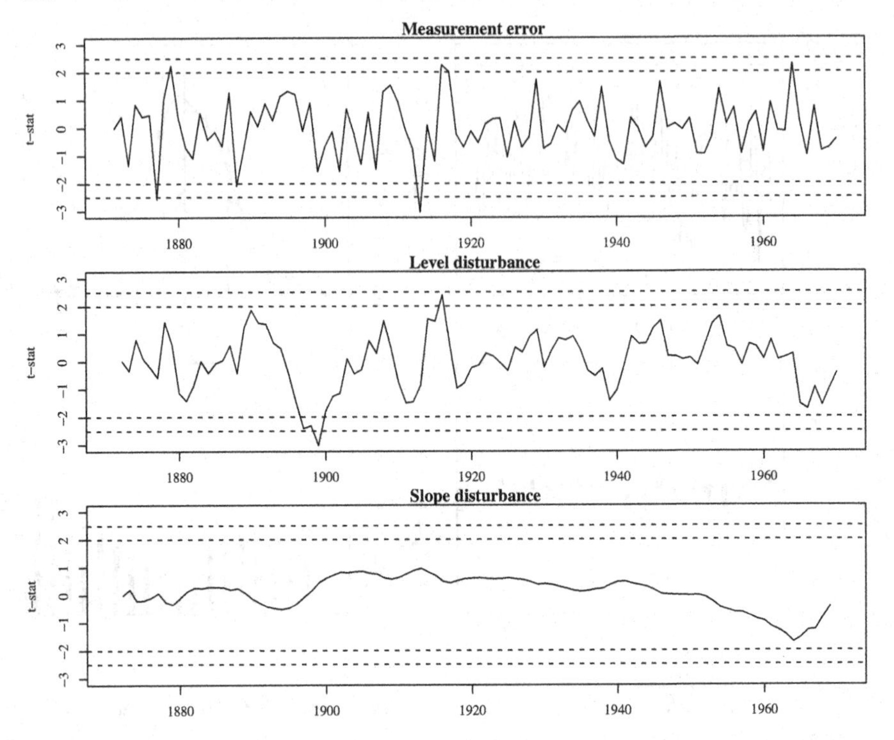

Figure 6.8 *Auxiliary residuals of the LLT plus noise model fitted to the Nile flow time series.*

logical reasons that suggest a drastic drop in the river flow starting from about that year. For the model refitted with the level change dummy added to the UCM see Example 4.2. In the model, the slope coefficient, which is constant and slightly negative (−3.41, with standard error 4.32 and, thus, non significant), does not seem to be affected by any changes.

As explained in the main text, if the software package does not provide algorithms for computing auxiliary residuals we can use the workaround described above. Figure 6.9 depicts the approximated t-statistics for the measurement error and for the level. Since the variance of the slope is zero, the smoothed estimates of it are constant and therefore it is not possible to use them to test for a break. The graphs in Figure 6.9 are almost identical to the corresponding ones in Figure 6.8.

6.5 Model selection

In time series analysis, especially when the main goal is producing forecasts, it is common to collect more models that pass all the diagnostic checks to compare their fitting and predictive capabilities.

The selection of one model can be based on *in-sample* or *out-of-sample*

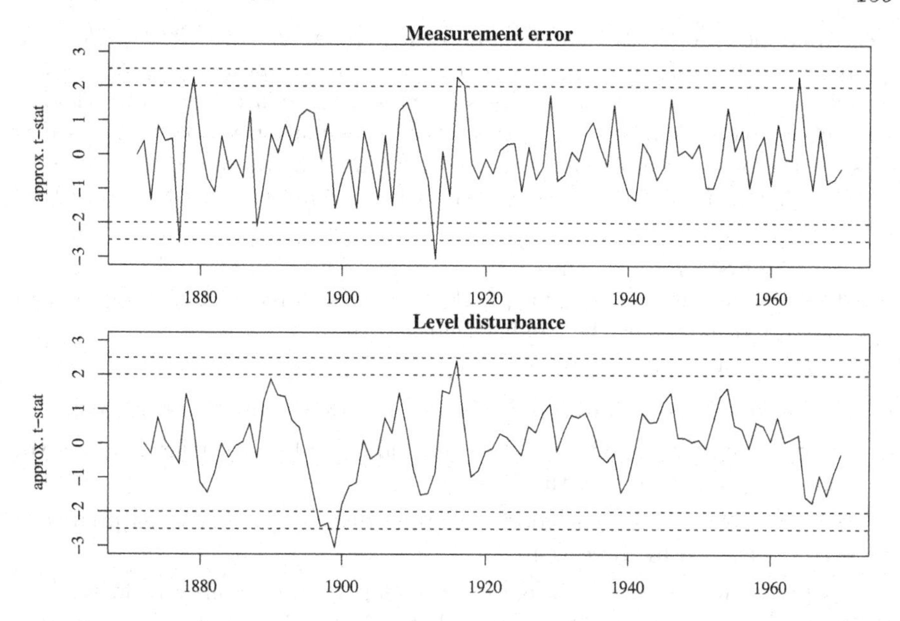

Figure 6.9 *Approximated auxiliary residuals of the LLT plus noise model fitted to the Nile flow time series.*

criteria. In the first case the same observations are used both for estimating the model and for evaluating their fit and forecasts, while in the second case the model is evaluated on a sample of observations not used for estimation purposes.

The main problem in selecting a model using an in-sample strategy is *overfitting*. In fact, if the sample is finite, it is always possible to build a model with many parameters that achieves a *perfect fit*, where in time series a perfect fit is a sequence of one-step-ahead forecasts that matches the predicted observations without error. Thus, in principle, the more the model is complicated and parametrised, the better the in-sample fit. But if the data are generated by a stochastic mechanism, achieving a perfect fit means that the model is fitting features of the data that will never appear again in the future, and thus the forecasts will generally be terrible. We want our model to fit only the predictable features of the data, and not those that cannot be predicted. Therefore, in-sample selection criteria are based on measures of *goodness of fit* (i.e., closeness of the predicted values to the actual observations) penalised by the number of estimated parameters.

If the full time series sample is large enough, an alternative selection strategy can be based on splitting the sample into a fit window and an evaluation window: the model is fit using data only from the first window and the goodness of fit statistics are computed using the second window. This procedure can be complicated further by choosing many different set of windows. In this case one speaks of *cross-validation*, and one popular version of it is the *leave-h-out*, in which the full sample is split into n/h sub-samples of (approximately) equal length h. Then the estimation is carried out n/h times, and each time a sub-sample of length h is kept out of the estimation window and used to assess the out-of-sample performance of the model. Another cross-validation strategy, which I find particularly useful when the principal goal is forecasting, consists in using the models as if they were in production:

1. Fix the first n_0 (for example, $n_0 = \lfloor n/2 \rfloor$ or $n_0 = \lfloor 2n/3 \rfloor$) observations of the time series as fit window, and the longest forecast horizon h you are interested in (h must be smaller than $n - n_0$);
2. Estimate the models using the current fit window;
3. Produce forecasts up to h-step-ahead and save the prediction errors;
4. Add the next observation to the fit window and go to point 2 unless you have reached the end of the series;
5. Compute mean losses or goodness-of-fit statistics for the saved prediction errors grouped by forecast horizon.

This procedure allows to compare the models by their ability to forecast on different horizons: in fact it could be that one model is good at forecasting in the short term and a different model is better in forecasting in the long run. Of course, you can carry out similar comparisons using the leave-h-out cross validation, but the proposed procedure simulates the work the forecaster will do with the selected model(s).

The principal loss statistics used in literature and implemented by software packages are:

Mean Square Error (MSE)

$$\frac{1}{m} \sum_{t=t_0+1}^{t_0+m} (y_t - \hat{y}_t)^2;$$

Root Mean Square Error (RMSE) the square root of the MSE

$$\sqrt{\frac{1}{m} \sum_{t=t_0+1}^{t_0+m} (y_t - \hat{y}_t)^2};$$

Mean Absolute Error (MAE)

$$\frac{1}{m} \sum_{t=t_0+1}^{t_0+m} |y_t - \hat{y}_t|;$$

Mean Absolute Percentage Error (MAPE)

$$\frac{100}{m} \sum_{t=t_0+1}^{t_0+m} \frac{|y_t - \hat{y}_t|}{|y_t|}.$$

The formulae have to be adapted to the context: if we apply them to in-sample one-step-ahead forecast errors (innovations), then $t_0 = 0$, $m = n$ and $\hat{y}_t = \hat{y}_{t|t-1}$, while if we apply them to out of sample forecasts up to h-step-ahead, then $m = h$, $\hat{y}_t = \hat{y}_{t|t_0}$ and t_0 is the last observation used to fit the model before the forecast period. In my view, the natural choice is the MSE or the RMSE as our predictors are built to minimise them, but the MAE has an even easier interpretation, and this is a good aspect when you have to communicate the results to non technical people. The MAPE is also a number many people understand with no problem, but in many situations it can be misleading when used to select a model. In fact the denominator can have strong effect in inflating or deflating the errors. For example, the time series of the Italian industrial production index drops every August some 40% with respect to the level; using the MAPE in this situation leads to choosing the model that forecasts August the best with high probability because its absolute error is inflated 2.5 times with respect to the other errors: are we sure this is the model we want to select? Furthermore, suppose you have a time series that shows a strong positive growth, in this case the denominator inflates the errors in the remote past and deflates the errors in the recent past and present, so the MAPE tends to select the models that predict the past better than the future, and this does not sound like a good criterion for forecasters. Finally, the MAPE weights the errors asymmetrically: for example if $y_t = 150$ and $\hat{y}_t = 100$, then the ratio in the sum is equal to $50/150 = 33\%$, while if the figures are swapped, $y_t = 100$ and $\hat{y}_t = 150$, the same ratio becomes $50/100 = 50\%$. Symmetric versions of the MAPE have been proposed but they are not popular in software for estimating UCM[4].

As a final remark on these loss statistics, notice that if you select the model on the log-transformed data rather than on raw data, then measures such as the MSE and MAE become relative measures, closer to the spirit of MAPE. Indeed,

$$|\log y_t - \log \hat{y}_t| = \left|\log\left(\frac{y_t}{\hat{y}_t}\right)\right| = \left|\log\left(1 + \frac{y_t - \hat{y}_t}{\hat{y}_t}\right)\right| \approx \left|\frac{y_t - \hat{y}_t}{\hat{y}_t}\right|,$$

but because of the absolute value also

$$|\log y_t - \log \hat{y}_t| = |\log \hat{y}_t - \log y_t| \approx \left|\frac{y_t - \hat{y}_t}{y_t}\right|.$$

Thus, the MAE computed on log-transformed observations produces something that we could call *symmetric MAPE* with respect to the raw data. It appears clear that the MAPE applied to log-transformed data is really nonsense (i.e., it is the percentage of a percentage).

The principal goodness-of-fit statistics used in the literature and implemented by software packages are:

[4]For an interesting discussion on symmetric MAPE read the post "Errors on percentage errors" on Professor Hyndman's blog: http://robjhyndman.com/hyndsight/smape/

R-square is the proportion of total variance of the time series explained by the fitted model,

$$R^2 = 1 - \frac{\hat{\sigma}^2}{\frac{1}{n}\sum_{t=1}^{n}(y_t - \bar{y})^2}, \qquad \text{with } \bar{y} = \frac{1}{n}\sum_{t=1}^{n} y_t;$$

Random Walk R-square is the complement to 1 of the ratio between the residual variance explained by the model and the residual variance of a random walk with drift,

$$R_D^2 = 1 - \frac{\hat{\sigma}^2}{\frac{1}{n}\sum_{t=1}^{n}(\Delta y_t - \bar{\Delta y})^2}, \qquad \text{with } \bar{\Delta y} = \frac{1}{n}\sum_{t=1}^{n} \Delta y_t;$$

Random Walk + Seasonal R-square is the complement to 1 of the ratio between the residual variance explained by the model and the residual variance of a random walk with drift and deterministic seasonal component,

$$R_S^2 = 1 - \frac{\hat{\sigma}^2}{\frac{1}{n}\sum_{t=1}^{n}(\Delta y_t - \bar{\Delta y}_t)^2},$$

with $\bar{\Delta y}_t$ mean of the time series values in the "season" to which t belongs. The quantity named $\hat{\sigma}^2$ in the above equations is generally the sample variance of the innovations of the model,

$$\hat{\sigma}^2 = \frac{1}{n - p + 1}\sum_{t=1}^{n} \iota_t,$$

but in some software packages (e.g., STAMP), if the state space is time invariant, $\hat{\sigma}^2$ is set equal to the steady-state variance of the innovation (i.e., the diagonal elements of F_t, once F_t has become constant). This choice is motivated by the fact that one-step-ahead forecasts of future values will have MSE which are equal to the steady-state innovation errors, while the first in-sample-predictions have very high MSE due to the uncertainty associated with the first realisations of the state variables, specially when diffuse distributions are used to initialise the state variables. Notice that if d state variables are assigned diffuse initial distributions, then the sums in the R-squares begin from $d + 1$ and the divisor n becomes $n - d$.

The following quantities penalise the goodness of fit statistics for the number of parameters estimated in the models. Let $l(\hat{\theta})$ be the log-likelihood of the model at the estimated parameters:

Adjusted R-square

$$\bar{R}^2 = 1 - (1 - R^2)\frac{n-1}{n-p} = R^2 - (1 - R^2)\frac{p}{n-p};$$

Akaike Information Criterion (AIC)

$$\text{AIC} = -2l(\hat{\theta}) + 2p$$

Akaike Information Criterion corrected (AICc)

$$\text{AICc} = -2l(\hat{\theta}) + \frac{2pn}{n-p-1}$$

Bayesian Information Criterion (BIC)

$$\text{BIC} = -2l(\hat{\theta}) + p \log n$$

Hannan–Quinn Information Criterion (HQ)

$$\text{HQ} = -2l(\hat{\theta}) + 2p \log \log n.$$

Again, if d state variables are assigned diffuse initial conditions, then n has to be substituted with $n - d$. The values of all the information criteria (IC) are not interesting *per se*, but only when compared in different models: once a criterion is picked, the model with lowest IC (i.e., closer to $-\infty$ not to 0) is to be selected. There is never-ending debate about which criterion should be used to select the best model: some criteria are consistent, some are efficient and some are less biased than others. Brockwell and Davis (1991) in their authoritative and widely cited volume recommend the use of the AICc. If you are interested in understanding the pros and cons of these and other IC, then you should consult the monograph by McQuarrie and Tsai (1998). In practice, the AIC tends to select models with more parameters than the ones selected by the BIC, while the HQ and AICc select models of intermediate size. Notice that some packages implement rescaled versions of the information criteria (divided by n) and sometimes the negative log-likelihood is substituted by the steady-state innovation variance, but this does not change the way they are used. This makes information criteria difficult to compare across software packages: you should always check which definition is used.

Clearly, loss and goodness-of-fit statistics are to be used out-of-sample, while the adjusted R-square and the IC are to be used to select models if all observations belong to the fit window (in-sample).

Example 6.6 (Selection of a model for the airline time series).
We fitted three models to the logarithm of the U.S. air passengers time series: i) BSM with diffuse initial conditions, ii) BSM with initial conditions inferred from the first observations, iii) the Airline model of Box and Jenkins (1976) (i.e., ARIMA(0,1,1)(0,1,1)).

In Table 6.2 we report the information criteria and the results of applying the out-of-sample strategy described in Section 6.5 to the forecasts produced by the two models. The evaluation is carried out by comparing the actual data with forecasts that are anti-transformed with the formulae $\exp(\hat{y}_{t+h|t})$, which we refer to as L1, and $\exp(\hat{y}_{t+h|t} + 0.5\hat{\sigma}^2_{t+h})$, which is named L2.

According to all information criteria the best model in the set is the Airline, which turns out to be the most accurate also in one-step-ahead

Table 6.2 *RMSE of h-step-ahead forecasts and information criteria for three models fitted to the log-airline time series. Smallest values in bold and second smallest in italics.*

	BSM diffuse		BSM inform.		Airline	
Step	L1	L2	L1	L2	L1	L2
1	12.84	12.83	12.58	12.56	*10.43*	**10.42**
2	34.54	34.57	**33.50**	*33.52*	38.03	38.06
3	56.36	56.42	**54.54**	*54.60*	61.96	62.03
4	71.63	71.75	**69.28**	*69.39*	77.80	77.91
5	81.24	81.42	**78.45**	*78.63*	87.45	87.61
6	84.06	84.33	**80.88**	*81.16*	90.51	90.73
7	86.74	87.13	**83.12**	*83.53*	93.29	93.59
8	89.31	89.84	**85.01**	*85.60*	96.30	96.70
9	92.07	92.76	**87.11**	*87.90*	98.45	98.96
10	88.97	89.83	**83.64**	*84.65*	93.01	93.65
11	80.20	81.29	**74.85**	*76.15*	80.97	81.80
12	63.57	65.02	**57.80**	*59.59*	61.82	62.97
AIC	*756*		858		**721**	
AICC	*756*		858		**721**	
BIC	*767*		870		**727**	
HQ	*760*		862		**724**	

forecasts. For all other prediction horizons the best model is the BSM with informative initial conditions. If we exclude one-step-ahead predictions, L1 forecasts seem to give better results than L2. These conclusions are not contradictory as the likelihood, and thus the information criteria, is built using one-step-ahead predictions, which in both cases select the Airline model. We computed also the MAPE and the conclusions are the same: the Airline is best for one-step with a MAPE of 0.8%, the BSM with informative initial conditions (L1) is the best choice for all other horizons: the range of the MAPE is 0.9% – 7.0%. The RMSE increases with the forecast horizon, but unexpectedly after the 9th step it decreases again to a level close to the 3-step-ahead.

Chapter 7

Multivariate Models

UCM can be easily extended to model a vector $\boldsymbol{Y}_t$ of N time series:

$$\boldsymbol{Y}_t = \boldsymbol{\mu}_t + \boldsymbol{\psi}_t + \boldsymbol{\gamma}_t + \boldsymbol{\epsilon}_t, \tag{7.1}$$

where the components on the rhs of the equal sign are all $N \times 1$ vectors. Regressors and interventions can be added as in the univariate case.

Unlike the vector autoregressions seen in Section 2.5, multivariate UCM are not able to model (Granger) causalities among time series because, unless otherwise specified, they do not include lagged endogenous variables on the right side of the equal sign. However, they are particularly fit to extract common features, such as common trends, cycles or seasonal components from a panel of time series.

As for many other models in multivariate statistics, the curse of dimensionality affects multivariate UCM as well. Indeed, unless strong common factor restrictions are imposed on the components, the number of parameters grows with the square of the number of time series.

7.1 Trends

The multivariate extension of the local linear trend (LLT) component is

$$\begin{aligned} \boldsymbol{\mu}_{t+1} &= \boldsymbol{\mu}_t + \boldsymbol{\beta}_t + \boldsymbol{\eta}_t, & \boldsymbol{\eta}_t &\sim \mathrm{WN}(\mathbf{0}, \boldsymbol{\Sigma}_\eta) \\ \boldsymbol{\beta}_{t+1} &= \boldsymbol{\beta}_t + \boldsymbol{\zeta}_t, & \boldsymbol{\zeta}_t &\sim \mathrm{WN}(\mathbf{0}, \boldsymbol{\Sigma}_\zeta), \end{aligned} \tag{7.2}$$

where the two covariance matrices can have full or reduced rank. When the rank of the covariance matrices is less than N, then some common movements in the trends are observed.

In order to investigate the common features that can be obtained as special cases of reduced-rank local linear trends, let us analyse the nested trends.

When $\boldsymbol{\Sigma}_\zeta = \mathbf{0}$ and $\boldsymbol{\beta}_1 = \mathbf{0}$ the trend reduced to the multivariate random walk,

$$\boldsymbol{\mu}_{t+1} = \boldsymbol{\mu}_t + \boldsymbol{\eta}_t, \quad \boldsymbol{\eta}_t \sim \mathrm{WN}(\mathbf{0}, \boldsymbol{\Sigma}_\eta),$$

and if $\boldsymbol{\Sigma}_\eta$ has rank $K_\mu < N$, then N random walks are driven by only r shocks. In this case $N - K_\mu$ trend components can be obtained as linear transformations of the other K_μ random walks, and one speaks of *common levels*.

It is easier to see the common features of a multivariate random walk with shocks having reduced-rank covariance matrices, when it is reparametrised as

$$\begin{aligned}\boldsymbol{\mu}_t &= \boldsymbol{\Theta}_\mu \boldsymbol{\mu}_t^\dagger + \boldsymbol{\theta}_\mu \\ \boldsymbol{\mu}_{t+1}^\dagger &= \boldsymbol{\mu}_t^\dagger + \boldsymbol{\eta}_t^\dagger, \qquad\qquad \boldsymbol{\eta}_t^\dagger \sim \mathrm{WN}(\mathbf{0}, \boldsymbol{\Sigma}_\eta^\dagger),\end{aligned}$$

where $\boldsymbol{\mu}_t^\dagger$ and $\boldsymbol{\eta}_t^\dagger$ are $K_\mu \times 1$ vectors, $\boldsymbol{\Theta}_\mu$ is a $N \times K_\mu$ matrix with structure

$$\boldsymbol{\Theta}_\mu = \begin{bmatrix}\mathbf{I}_{K_\mu} \\ \boldsymbol{\Theta}\end{bmatrix},$$

$\boldsymbol{\theta}_\mu$ is a $N \times 1$ vector with the first K_μ elements equal to zero and the covariance matrix $\boldsymbol{\Sigma}_\eta^\dagger$ has full rank. The original parametrisation can be obtained by setting

$$\boldsymbol{\Sigma}_\eta = \boldsymbol{\Theta}_\mu \boldsymbol{\Sigma}_\eta^\dagger \boldsymbol{\Theta}_\mu^\top \quad \text{and} \quad \mathbb{E}[\boldsymbol{\mu}_1] = \boldsymbol{\Theta}_\mu \mathbb{E}[\boldsymbol{\mu}_1^\dagger] + \boldsymbol{\theta}_\mu$$

but in this form we clearly see that K_μ random walks are distributed via the matrix of *loadings* $\boldsymbol{\Theta}_\mu$ to create the original trends $\boldsymbol{\mu}_t$. The first K_μ random walks drive the first K_μ levels, while the remaining $N - K_\mu$ levels are obtained as linear functions of the first random walks. Since the random walks $\boldsymbol{\mu}_t^\dagger$ define only r initial values $\boldsymbol{\mu}_1^\dagger$, while $\boldsymbol{\mu}_t$ may have N independent initial states, the vector $\boldsymbol{\theta}_\mu$ supplies the $N - K_\mu$ missing initial state values.

If we suppose the vectorial time series $\mathbf{Y}_t$ is generated by this reduced-rank level plus a noise then the model can be rewritten as

$$\begin{aligned}\mathbf{Y}_t &= \boldsymbol{\Theta}_\mu \boldsymbol{\mu}_t^\dagger + \boldsymbol{\theta}_\mu + \boldsymbol{\epsilon}_t, \quad \boldsymbol{\epsilon}_t \sim \mathrm{WN}(\mathbf{0}, \boldsymbol{\Sigma}_\epsilon), \\ \boldsymbol{\mu}_{t+1}^\dagger &= \boldsymbol{\mu}_t^\dagger + \boldsymbol{\eta}_t^\dagger, \qquad\qquad \boldsymbol{\eta}_t^\dagger \sim \mathrm{WN}(\mathbf{0}, \boldsymbol{\Sigma}_\eta^\dagger).\end{aligned} \tag{7.3}$$

Similarly, if we consider the multivariate integrated random walk,

$$\begin{aligned}\boldsymbol{\mu}_{t+1} &= \boldsymbol{\mu}_t + \boldsymbol{\beta}_t \\ \boldsymbol{\beta}_{t+1} &= \boldsymbol{\beta}_t + \boldsymbol{\zeta}_t, \quad \boldsymbol{\zeta}_t \sim \mathrm{WN}(\mathbf{0}, \boldsymbol{\Sigma}_\zeta),\end{aligned}$$

with $\boldsymbol{\Sigma}_\zeta$ of rank $K_\beta < N$, we can reparametrise it as

$$\begin{aligned}\boldsymbol{\mu}_{t+1} &= \boldsymbol{\mu}_t + \boldsymbol{\Theta}_\beta \boldsymbol{\beta}_t^\dagger + \boldsymbol{\theta}_\beta \\ \boldsymbol{\beta}_{t+1}^\dagger &= \boldsymbol{\beta}_t^\dagger + \boldsymbol{\zeta}_t^\dagger, \qquad\qquad \boldsymbol{\zeta}_t^\dagger \sim \mathrm{WN}(\mathbf{0}, \boldsymbol{\Sigma}_\beta^\dagger),\end{aligned}$$

where we have set $\boldsymbol{\beta}_t = \boldsymbol{\Theta}_\beta \boldsymbol{\beta}_t^\dagger + \boldsymbol{\theta}_\beta$. Again, $\boldsymbol{\beta}_t^\dagger$ is a $N \times 1$ vector, $\boldsymbol{\Theta}_\beta$ is a $N \times K_\beta$ matrix whose top $K_\beta \times K_\beta$ sub-matrix is the identity matrix, $\boldsymbol{\theta}_\beta$ is a $N \times 1$ vector with the first K_β elements equal to zero and the covariance matrix $\boldsymbol{\Sigma}_\beta^\dagger$ has full rank. From the latter formula, it is clear that the N trends in $\boldsymbol{\mu}_t$ are driven by K_β random slopes that are assigned to each trend through the matrix of loadings $\boldsymbol{\Theta}_\beta$, while the vector $\boldsymbol{\theta}_\beta$ supplies $N - K_\beta$ additional initial values for the slopes, which are only K_β in $\boldsymbol{\beta}_1^\dagger$, but can be different in each of the N trends. If the observed time series are given by this reduced-rank integrated random walk model plus noise, we can write

$$\begin{aligned}
\mathbf{Y}_t &= \boldsymbol{\Theta}_\beta \boldsymbol{\mu}_t^{\dagger\dagger} + \boldsymbol{\theta}_t + \boldsymbol{\epsilon}_t, & \boldsymbol{\epsilon}_t &\sim \mathrm{WN}(\mathbf{0}, \boldsymbol{\Sigma}_\epsilon),\\
\boldsymbol{\mu}_{t+1}^{\dagger\dagger} &= \boldsymbol{\mu}_t^{\dagger\dagger} + \boldsymbol{\beta}_t^{\dagger\dagger}, & & \qquad (7.4)\\
\boldsymbol{\beta}_{t+1}^{\dagger\dagger} &= \boldsymbol{\beta}_t^{\dagger\dagger} + \boldsymbol{\zeta}_t^{\dagger\dagger}, & \boldsymbol{\zeta}_t^\dagger &\sim \mathrm{WN}(\mathbf{0}, \boldsymbol{\Sigma}_\beta^\dagger),
\end{aligned}$$

with $\boldsymbol{\theta}_t = \boldsymbol{\theta}_\mu + \boldsymbol{\theta}_\beta t$, where $\boldsymbol{\theta}_\mu$ and $\boldsymbol{\theta}_\beta$ are $N \times 1$ vectors each with K_β zeros as first elements.

If we consider the multivariate local linear trend, things complicate a bit as we have two sets of correlated shocks with possibly different ranks: $\boldsymbol{\eta}_t$ and $\boldsymbol{\zeta}_t$. For the common trend representation of this case we can exploit the fact that a local linear trend can also be written as the sum of a random walk plus an integrated random walk, and so we can write

$$\begin{aligned}
\boldsymbol{\mu}_t &= \boldsymbol{\Theta}_\mu \boldsymbol{\mu}_t^\dagger + \boldsymbol{\Theta}_\beta \boldsymbol{\mu}_t^{\dagger\dagger} + \boldsymbol{\theta}_\mu + \boldsymbol{\theta}_\beta t, & &\\
\boldsymbol{\mu}_{t+1}^\dagger &= \boldsymbol{\mu}_t^\dagger + \boldsymbol{\eta}_t^\dagger, & \boldsymbol{\eta}_t^\dagger &\sim \mathcal{N}(\mathbf{0}, \boldsymbol{\Sigma}_\eta^\dagger),\\
\boldsymbol{\mu}_{t+1}^{\dagger\dagger} &= \boldsymbol{\mu}_t^{\dagger\dagger} + \boldsymbol{\beta}_t^{\dagger\dagger}, & & \qquad (7.5)\\
\boldsymbol{\beta}_{t+1}^{\dagger\dagger} &= \boldsymbol{\beta}_t^{\dagger\dagger} + \boldsymbol{\zeta}_t^{\dagger\dagger}, & \boldsymbol{\zeta}_t^{\dagger\dagger} &\sim \mathcal{N}(\mathbf{0}, \boldsymbol{\Sigma}_\zeta^{\dagger\dagger}),
\end{aligned}$$

where all the symbols have already been defined above. In particular, $\boldsymbol{\theta}_\mu$ is N with the first K_μ elements equal to zero, and $\boldsymbol{\theta}_\beta$ is also $N \times 1$, but with the first K_β elements equal to zero. If the observable vectorial time series $\mathbf{Y}_t$ is given by this reduced-rank local linear trend plus noise then we can write

$$\mathbf{Y}_t = \boldsymbol{\Theta}_\mu \boldsymbol{\mu}_t^\dagger + \boldsymbol{\Theta}_\beta \boldsymbol{\mu}_t^{\dagger\dagger} + \boldsymbol{\theta}_\mu + \boldsymbol{\theta}_\beta t + \boldsymbol{\epsilon}_t.$$

Particularly interesting is when $K_\mu = K_\beta$ and $\boldsymbol{\Theta}_\mu = \boldsymbol{\Theta}_\beta$. In this case there are K_μ local linear trends that drive N trends. We can make this more evident by writing the model also as

$$\begin{aligned}
\mathbf{Y}_t &= \boldsymbol{\Theta}_\mu \boldsymbol{\mu}_t^\ddagger + \boldsymbol{\theta}_\mu + \boldsymbol{\theta}_\beta t + \boldsymbol{\epsilon}_t, & \boldsymbol{\epsilon}_t &\sim \mathrm{WN}(\mathbf{0}, \boldsymbol{\Sigma}_\epsilon),\\
\boldsymbol{\mu}_{t+1}^\ddagger &= \boldsymbol{\mu}_t^\ddagger + \boldsymbol{\beta}_t^{\dagger\dagger} + \boldsymbol{\eta}_t^\dagger, & \boldsymbol{\eta}_t^\dagger &\sim \mathrm{WN}(\mathbf{0}, \boldsymbol{\Sigma}_\eta^\dagger), \qquad (7.6)\\
\boldsymbol{\beta}_{t+1}^{\dagger\dagger} &= \boldsymbol{\beta}_t^{\dagger\dagger} + \boldsymbol{\zeta}_t^{\dagger\dagger}, & \boldsymbol{\zeta}_t^{\dagger\dagger} &\sim \mathrm{WN}(\mathbf{0}, \boldsymbol{\Sigma}_\zeta^{\dagger\dagger}).
\end{aligned}$$

All these different common trend configurations entail various forms of *cointegration* (Engle and Granger, 1987). A set of time series which are all order-d integrated, or shortly $\mathrm{I}(d)$, are said to be cointegrated if there are linear combinations of them which are integrated of order $d-c$, where $0 < c \leq d$. This property is synthetically expressed with the notation $\mathrm{CI}(d, c)$. Let us analyse the various variants of the local linear trend with reduced rank covariance matrices within the cointegration framework.

The time series in the reduced rank random walk plus noise model (7.3) are clearly I(1), but since there are N_μ random walks driving N time series there are $N - N_\mu$ independent linear combinations of them that cancel out the trends. Indeed, for any $N \times K_\mu$ matrix $\mathbf{\Theta}_\mu$ there exists a $(N - K_\mu) \times N$ matrix $\mathbf{B}$ such that $\mathbf{B\Theta}_\mu = \mathbf{0}$.[1] So, if we premultiply the time series $\mathbf{Y}_t$ in equation (7.3) by such a $\mathbf{B}$ we get

$$\mathbf{B}\mathbf{Y}_t = \mathbf{B}\boldsymbol{\theta}_\mu + \mathbf{B}\boldsymbol{\epsilon}_t,$$

which is a stationary process. Thus, the vector of time series generated by the reduced-rank local level model (7.3) is CI(1, 1).

If we consider the reduced-rank integrated random walk plus noise of equation (7.4), we see that all time series are I(2) processes but, again, we can find a $(N - K_\beta) \times N$ matrix $\mathbf{B}$ that annihilates $\mathbf{\Theta}_\beta$, obtaining

$$\mathbf{B}\mathbf{Y}_t = \mathbf{B}\boldsymbol{\theta}_t + \mathbf{B}\boldsymbol{\epsilon}_t,$$

which is trend-stationary. In case also $\mathbf{B}\boldsymbol{\theta}_t = \mathbf{0}$, then the time series share both the stochastic and the deterministic components of the trends and $\mathbf{BY}_t$ is stationary. In symbols, we can write $\mathbf{Y}_t \sim$ CI(2, 2), because there are linear combinations of the I(2) time series that are I(0).

The local linear trend plus noise (7.5) generates $I(2)$ time series, however, in general we cannot find a linear combination of them which is $I(0)$ because we can select a matrix $\mathbf{B}$, which either solves $\mathbf{B\Theta}_\mu = \mathbf{0}$ or $\mathbf{B\Theta}_\beta = \mathbf{0}$. If we choose a $\mathbf{B}$ which solves the former equation, then we can annihilate the component $\boldsymbol{\mu}_t^\dagger$ and $\mathbf{BY}_t$ follows an integrated random walk plus noise model, which is still I(2). On the contrary, if we choose $\mathbf{B}$ that solves $\mathbf{B\Theta}_\beta = \mathbf{0}$, we can annihilate $\boldsymbol{\mu}_t^{\dagger\dagger}$ and $\mathbf{Y}_t$ becomes I(1). So, in symbols we can write $\mathbf{Y}_t \sim$ CI(2, 1). In the special case in which $\mathbf{\Theta}_\mu = \mathbf{\Theta}_\beta$, as in the model in equation (7.6), there is one $\mathbf{B}$ that solves both equations and so we obtain

$$\mathbf{B}\mathbf{Y}_t = \mathbf{B}\boldsymbol{\theta}_\mu + \mathbf{B}\boldsymbol{\theta}_\beta t + \mathbf{B}\boldsymbol{\epsilon}_t,$$

which is trend-stationary, and so $\mathbf{Y}_t \sim$ CI(2,2). Again, if $\mathbf{B}$ solves also $\mathbf{B}\boldsymbol{\theta} = \mathbf{0}$, then also the deterministic component of the trend is common and $\mathbf{Y}_t$ becomes stationary.

Example 7.1 (Brent, futures and interest rates).
Let us consider the spot price of the Brent crude oil and a future contract on it. From financial theory, if there are no transaction costs, we know that the future price F_t should be related to the spot price P_t and the risk-free (continuously compounding per annum) interest rate r_t by the equation

$$F_t = P_t \exp(\delta r_t),$$

[1]The $(N - K) \times N$ matrix $\mathbf{B}$ that solves $\mathbf{BM} = \mathbf{0}$, where $\mathbf{M}$ is a $N \times K$ matrix, provides through its rows a basis for the *null space* of $\mathbf{M}$. Thus, if you are looking for such a function in the software package you are using, search for the keywords "null space".

where δ represents the time to delivery expressed in years. Taking logs, we can write

$$f_t = p_t + \delta r_t,$$

with $f_t = \log F_t$ and $p_t = \log P_t$. Now, for many reasons that include transaction costs, noisy trading, measured quantities not corresponding to the theoretical ones (in particular risk-free interest rates), nobody expects this identity to hold exactly. However, this equation suggests that in the long run only two trends should drive all three time series: in fact the value of one variable can be approximately computed from the values of the other two variables.

Since the dynamics of each of the three time series is well approximated by that of a random walk, we fit a multivariate local level model, where we expect the random walk covariance matrix to be rank-deficient. Furthermore, to allow short-term deviations from the equilibrium implied by the financial model, we add an AR(1) noise with possibly correlated shocks:

$$\begin{aligned} \mathbf{Y}_t &= \boldsymbol{\mu}_t + \boldsymbol{\varphi}_t, \\ \boldsymbol{\mu}_{t+1} &= \boldsymbol{\mu}_t + \boldsymbol{\eta}_t, & \boldsymbol{\eta}_t &\sim \mathrm{WN}(\mathbf{0}, \boldsymbol{\Sigma}_\eta), \\ \boldsymbol{\varphi}_{t+1} &= \mathbf{D}\boldsymbol{\varphi}_t + \boldsymbol{\zeta}_t, & \boldsymbol{\zeta}_t &\sim \mathrm{WN}(\mathbf{0}, \boldsymbol{\Sigma}_\zeta), \end{aligned}$$

with $\mathbf{D}$ diagonal matrix with autoregressive coefficients. The estimates based on the time sample 1st Jan 2009 – 2nd Dec 2014 give the following correlation matrices of the two disturbance vectors

$$\hat{\mathbf{C}}_\eta = \begin{bmatrix} 1.00 & 1.00 & 0.08 \\ 1.00 & 1.00 & 0.08 \\ 0.08 & 0.08 & 1.00 \end{bmatrix}, \quad \hat{\mathbf{C}}_\zeta = \begin{bmatrix} 1.00 & -1.00 & -0.05 \\ -1.00 & 1.00 & 0.05 \\ -0.05 & 0.05 & 1.00 \end{bmatrix},$$

where the order of the variables is p_t, f_t, r_t. The three AR(1) coefficients on the diagonal of $\mathbf{D}$ are 0.19 (for p_t), 0.82 (for f_t), 0.72 (for r_t). It is clear from the values of the correlation coefficients that both covariance matrices have rank 2. We estimate the model again imposing the reduced rank restriction and letting the trend of the series f_t be a linear combination of the two random walks driving the other two time series. The estimated vector of loadings for f_t is $[1.06 \;\; 12.79]$, where the first element refers to the trend of p_t and the second to the trend of r_t. The cointegration vector implied by this loading is the vector $\boldsymbol{b}$ that solves

$$\boldsymbol{b}^\top \begin{bmatrix} 1.00 & 0.00 \\ 1.06 & 12.79 \\ 0.00 & 1.00 \end{bmatrix} = [0 \;\; 0].$$

The estimated cointegration vector is $[1.00 \;\; -0.94 \;\; 12.03]$, which implies that $p_t - 0.94 f_t + 12.03 r_t$ is stationary.

Putting the local liner trend in state space form is straightforward: the transition equation is given by

$$\begin{bmatrix}\mu_{t+1}\\ \beta_{t+1}\end{bmatrix} = \begin{bmatrix}\mathbf{I}_N & \mathbf{I}_N\\ \mathbf{0} & \mathbf{I}_N\end{bmatrix}\begin{bmatrix}\mu_t\\ \beta_t\end{bmatrix} + \begin{bmatrix}\eta_t\\ \zeta_t\end{bmatrix},$$

with

$$\mathbb{E}\left(\begin{bmatrix}\eta_t\\ \zeta_t\end{bmatrix}\begin{bmatrix}\eta_t & \zeta_t\end{bmatrix}\right) = \begin{bmatrix}\Sigma_\eta & \mathbf{0}\\ \mathbf{0} & \Sigma_\zeta\end{bmatrix};$$

and the portion of the matrix $\mathbf{Z}$ in the observation equation that selects the components from the multivariate local linear trend is the $N \times 2N$ matrix

$$\mathbf{Z}_\mu = \begin{bmatrix}\mathbf{I}_N & \mathbf{0}\end{bmatrix}.$$

If rank restrictions are imposed as in model (7.5), then the transition equation takes the form

$$\begin{bmatrix}\mu^\dagger_{t+1}\\ \mu^{\dagger\dagger}_{t+1}\\ \beta^{\dagger\dagger}_{t+1}\end{bmatrix} = \begin{bmatrix}\mathbf{I}_N & \mathbf{0} & \mathbf{0}\\ \mathbf{0} & \mathbf{I}_N & \mathbf{I}_N\\ \mathbf{0} & \mathbf{0} & \mathbf{I}_N\end{bmatrix}\begin{bmatrix}\mu^\dagger_t\\ \mu^{\dagger\dagger}_t\\ \beta^{\dagger\dagger}_t\end{bmatrix} + \begin{bmatrix}\eta^\dagger_t\\ \mathbf{0}\\ \zeta^{\dagger\dagger}_t\end{bmatrix},$$

with disturbance covariance matrix

$$\begin{bmatrix}\Sigma^\dagger_\eta & \mathbf{0} & \mathbf{0}\\ \mathbf{0} & \mathbf{0} & \mathbf{0}\\ \mathbf{0} & \mathbf{0} & \Sigma^{\dagger\dagger}_\zeta\end{bmatrix}.$$

The selection matrix in the observation equation becomes

$$\mathbf{Z}^\dagger_\mu = \begin{bmatrix}\Theta_\mu & \Theta_\beta & \mathbf{0}\end{bmatrix},$$

where the zero matrix has the same dimensions as Θ_β, and the vector c_t is now needed to contain the additional trends:

$$c_t = \theta_\mu + \theta_\beta t.$$

7.2 Cycles

In economics, the most cited definition of business cycles is the one given by Burns and Mitchell (1946):

> Business cycles are a type of fluctuation found in the aggregate economic activity of nations that organize their work mainly in business enterprises: a cycle consists of expansions occurring at about the same time in many economic activities, followed by similarly general recessions, contractions, and revivals which merge into the expansion phase of the next cycle; in duration, business cycles vary from more than one year to ten or twelve years; they are not divisible into shorter cycles of similar characteristics with amplitudes approximating their own.

According to this definition, business cycles are fluctuations that occur similarly, but not identically, in many time series. The periodicity and the persistence of the cycles should be approximately the same for all time series, while the amplitude can be different because the business cycle affects various aspects of the economic activity with different strength. Thus, the natural generalisation of the univariate cycle seen in Section 3.3 to the multivariate case is represented by the *similar cycles*: each series in the model has its own cycle component, but all these stochastic cycles share the same frequency λ and the same damping factor ρ and the shocks driving each cycle component can be correlated. In formulae, the cycle of the i-th time series is given by

$$\begin{bmatrix}\psi_{i,t+1}\\ \psi^*_{i,t+1}\end{bmatrix} = \begin{bmatrix}\rho\cos\lambda & \rho\sin\lambda\\ -\rho\sin\lambda & \rho\cos\lambda\end{bmatrix}\begin{bmatrix}\psi_{i,t}\\ \psi^*_{i,t}\end{bmatrix} + \begin{bmatrix}\kappa_{i,t}\\ \kappa^*_{i,t}\end{bmatrix}, \qquad i = 1,\dots,N,$$

where, if we put all the $\kappa_{i,t}$ disturbances in the vector $\boldsymbol{\kappa}_t$ and the $\kappa^*_{i,t}$ disturbances in the vector $\boldsymbol{\kappa}^*_t$, the covariance structure of the disturbances is given by

$$\mathbb{E}\left[\boldsymbol{\kappa}_t\boldsymbol{\kappa}_t^\top\right] = \mathbb{E}\left[\boldsymbol{\kappa}^*_t\boldsymbol{\kappa}^{*\top}_t\right] = \boldsymbol{\Sigma}_\kappa, \qquad \mathbb{E}\left[\boldsymbol{\kappa}_t\boldsymbol{\kappa}^{*\top}_t\right] = \mathbf{0}.$$

The cross-covariance function of the cycle vector is given by

$$\boldsymbol{\Gamma}(h) = \mathbb{E}\left[\boldsymbol{\psi}_t\boldsymbol{\psi}_t^\top\right] = \frac{\rho^h\cos(h\lambda)}{1-\rho^2}\boldsymbol{\Sigma}_\kappa,$$

from which we can derive the correlation function between any cycle component pairs:

$$\rho_{ij}(h) = \rho^h\cos(h\lambda)\rho_{ij}(0), \qquad i,j = 1,\dots,N.$$

We can write the transition equation of the similar cycles as

$$\begin{bmatrix}\boldsymbol{\psi}_{t+1}\\ \boldsymbol{\psi}^*_{t+1}\end{bmatrix} = \left(\begin{bmatrix}\rho\cos\lambda & \rho\sin\lambda\\ -\rho\sin\lambda & \rho\cos\lambda\end{bmatrix}\otimes\mathbf{I}_N\right)\begin{bmatrix}\boldsymbol{\psi}_t\\ \boldsymbol{\psi}^*_t\end{bmatrix} + \begin{bmatrix}\boldsymbol{\kappa}_t\\ \boldsymbol{\kappa}^*_t\end{bmatrix}, \tag{7.7}$$

with

$$\mathbb{E}\left(\begin{bmatrix}\boldsymbol{\kappa}_t\\ \boldsymbol{\kappa}^*_t\end{bmatrix}\begin{bmatrix}\boldsymbol{\kappa}_t & \boldsymbol{\kappa}^*_t\end{bmatrix}\right) = \begin{bmatrix}\boldsymbol{\Sigma}_\kappa & \mathbf{0}\\ \mathbf{0} & \boldsymbol{\Sigma}_\kappa\end{bmatrix}.$$

The portion of the matrix $\mathbf{Z}$ in the observation equation that selects the cycles from the above representation is the $N\times 2N$ matrix

$$\mathbf{Z}_\psi = \begin{bmatrix}\mathbf{I}_N & \mathbf{0}\end{bmatrix}.$$

As we have seen for the trend component, also the covariance matrix $\boldsymbol{\Sigma}_\kappa$ can have rank K_ψ smaller than N, and in this case one speaks of *common cycles*. Moreover, there are situations in which it is opportune to assume the presence of just one common cycle, and thus $K_\psi = 1$. In presence of K_ψ common cycles the model can be reparametrised as

$$\boldsymbol{\psi}_t = \boldsymbol{\Theta}_\psi\boldsymbol{\psi}^\dagger_t + \boldsymbol{\theta}_\psi$$

$$\begin{bmatrix}\boldsymbol{\psi}^\dagger_{t+1}\\ \boldsymbol{\psi}^{*\dagger}_{t+1}\end{bmatrix} = \left(\begin{bmatrix}\rho\cos\lambda & \rho\sin\lambda\\ -\rho\sin\lambda & \rho\cos\lambda\end{bmatrix}\otimes\mathbf{I}_{K_\beta}\right)\begin{bmatrix}\boldsymbol{\psi}^\dagger_t\\ \boldsymbol{\psi}^{*\dagger}_t\end{bmatrix} + \begin{bmatrix}\boldsymbol{\kappa}^\dagger_t\\ \boldsymbol{\kappa}^{*\dagger}_t\end{bmatrix},$$

where $\psi_t^\dagger$ and $\psi_t^{*\dagger}$ are $K_\beta \times 1$ matrices, $\mathbb{E}[\kappa_t^\dagger \kappa_t^{\dagger\top}] = \mathbb{E}[\kappa_t^{*\dagger} \kappa_t^{*\dagger\top}] = \Sigma_\kappa^\dagger$, the $N \times K_\psi$ matrix Θ_ψ has the top $K_\psi \times K_\psi$ sub-matrix equal to the identity matrix and, finally, the vector θ_ψ has the first K_ψ elements equal to zero.

As in the univariate case, the cyclical component can be made smoother by using the higher-order cycle covered in Section 3.3.3, but, for various reasons, in the multivariate context the use of higher-order cycles may present more disadvantages than advantages: in fact, the number of state variable increases drastically and, if some reduced-rank restrictions are imposed, the averaging over many time series implied by the common cycle(s) tends to generate smooth cycles also using (7.7).

A very useful extension of the similar cycle model is the one proposed by Rünstler (2004), which *rotates* cycles in order to let leading, coincident and lagging variables enter the same common cycle model. This extension works for any $K_\psi < N$, but the only applications available in the literature (Azevedo et al., 2006; Koopman and Azevedo, 2008) are to the common cycle ($K_\psi = 1$), and so we just limit the presentation to this case. So, there is only one stochastic cycle pair $[\psi_t \;\; \psi_t^*]$, which is attributed to one of the time series, say the first: $Y_{1,t} = \psi_t + \ldots$. All the other time series rescale and shift the common cycle as

$$Y_{i,t} = \delta_i \cos(\xi_i \lambda)\psi_t + \delta_i \sin(\xi_i \lambda) + \ldots,$$

where δ_i adjust the amplitude of the cycle, while $-\pi/2 < \xi_i < \pi/2$ shift the cycle to the left ($\xi_i < 0$) or to the right ($\xi_i > 0$) of $|\xi_i|$ time periods.

We dedicate the whole of Chapter 8 to the analysis of business cycles using UCM, thus we conclude this section with a simple example and refer the reader to that chapter for more information and applications.

Example 7.2 (Business cycle indicator for the United States).
The Conference Board produces a monthly coincident indicator of the U.S. business cycle based on four seasonally adjusted time series: Employees on nonagricultural payrolls (EMP), Personal income less transfer payments (INC), Industrial production index (IPI), Manufacturing and trade sales (SAL). Here we fit a linear trends plus similar cycles model to the monthly (symmetric) growth rate of the four time series[2].

The correlation matrix of the similar cycles is

$$\begin{bmatrix} 1.000 & 0.969 & 0.912 & 0.899 \\ 0.969 & 1.000 & 0.985 & 0.979 \\ 0.912 & 0.985 & 1.000 & 1.000 \\ 0.899 & 0.979 & 1.000 & 1.000 \end{bmatrix},$$

and its eigenvalues, [3.873 0.127 0.000 0.000], reveal the rank of this matrix is 2. Furthermore, by dividing the eigenvectors by their sum we

[2]For an explanation of how symmetric growth rates are defined and the composite index computed refer to the web page www.conference-board.org/data/bci/index.cfm?id=2154

obtain [0.9680.0320.0000.000], which indicates that one common cycle takes account of some 97% of the variance of the four (standardised) cycles, and so we set the rank equal to one (i.e., common cycle). The smoothed common cycle component is depicted in Figure 7.1 together with the official NBER dating of the U.S. recessions. The synchronisation between the official recessions and the common cycle downturns is very clear. The peaks of our cycle component tend to anticipate the official ones because the NBER dating is based on the levels of the time series (business cycle), while we are using growth rate data (growth cycle). My opinion is that when extracting the cycle from a pool of time series, the growth cycle definition is more meaningful because, unless the levels of the time series are driven by one common trend, there is really no unique way to define a common level to which the common cycle has to be summed.

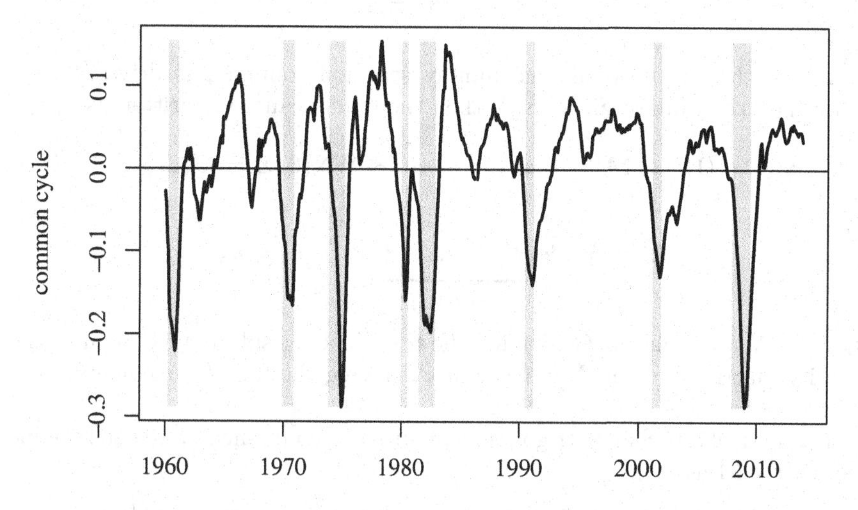

Figure 7.1 *Common cycle extracted from the symmetric growth rate of the four time series used by the Conference Board with NBER recession dating (grey shade).*

7.3 Seasonalities

When working with a vector of seasonal time series, it can be a good idea to seasonally adjust them before building the multivariate UCM. Of course, this operation is statistically suboptimal, but the number of state variables that multiple seasonal components bring about in the state space form can be very relevant and make the numerical estimation of the unknown parameters slow and unstable. But when the time series are not many (e.g., only 2 or 3) and/or the seasonal period, s, not too long (e.g., as in quarterly observations), multiple seasonal components can be put in the UCM.

For the construction of the stochastic dummy seasonal component, let us define the $(s-1) \times (s-1)$ matrix

$$\mathbf{M} = \begin{bmatrix} -\mathbf{1}^\top & -1 \\ \mathbf{I}_{s-2} & \mathbf{0} \end{bmatrix},$$

and let us name $\mathbf{\Sigma}_\omega$ the covariance matrix of the N seasonal disturbances. The transition equation of this component is

$$\boldsymbol{\gamma}_{t+1} = (\mathbf{I}_N \otimes \mathbf{M})\boldsymbol{\gamma}_t + \boldsymbol{\omega}_t, \qquad \boldsymbol{\omega}_t \sim \mathrm{WN}\big(\mathbf{0}, \mathbf{\Sigma}_\omega \otimes \mathbf{J}_{s-1}\big), \tag{7.8}$$

where $\mathbf{J}_N$ is a $N \times N$ matrix of zeros except for the first element which is equal to one. The portion of the matrix $\mathbf{Z}$ in the observation equation relative to the seasonal component is the $N \times (s-1)N$ matrix

$$\mathbf{Z}_\gamma = \mathbf{I}_N \otimes \underbrace{\begin{bmatrix} 1 & 0 & \dots & 0 \end{bmatrix}}_{1 \times (s-1)}.$$

If only $K_\gamma < N$ stochastic dummy seasonal components drive N time series (i.e., the rank of $\mathbf{\Sigma}_\omega$ is K_γ), then the model can be rewritten as

$$\boldsymbol{\gamma}^\dagger_{t+1} = (\mathbf{I}_{K_\gamma} \otimes \mathbf{M})\boldsymbol{\gamma}^\dagger_t + \boldsymbol{\omega}^\dagger_t, \qquad \boldsymbol{\omega}^\dagger_t \sim \mathrm{WN}\big(\mathbf{0}, \mathbf{\Sigma}^\dagger_\omega \otimes \mathbf{J}_{s-1}\big), \tag{7.9}$$

and

$$\mathbf{Z}^\dagger_\gamma = \mathbf{\Theta}_\gamma \otimes \underbrace{\begin{bmatrix} 1 & 0 & \dots & 0 \end{bmatrix}}_{1 \times (s-1)}, \qquad \boldsymbol{c}_\gamma = \boldsymbol{\theta}_\gamma$$

with the $N \times K_\gamma$ matrix $\mathbf{\Theta}_\gamma$ having the top $K_\gamma \times K_\gamma$ sub-matrix equal to the identity matrix, and the $N \times 1$ vector $\boldsymbol{\theta}_\gamma$ having the first K_γ elements equal to zero.

For multiple stochastic trigonometric seasonal components, we can redefine the matrix $\mathbf{M}$ as

$$\mathbf{M} = \begin{bmatrix} \mathbf{R}(1 \cdot 2\pi/s) & \mathbf{0} & \dots & \mathbf{0} \\ \mathbf{0} & \mathbf{R}(2 \cdot 2\pi/s) & \dots & \mathbf{0} \\ \vdots & \vdots & \ddots & \vdots \\ \mathbf{0} & \mathbf{0} & \dots & \mathbf{R}\big(\lfloor s/2 \rfloor \cdot 2\pi/s\big) \end{bmatrix},$$

where $\mathbf{R}(\cdot)$ is the 2×2 rotation matrix defined in equation (3.6). Remember that when s is even, the rotation matrix $\mathbf{R}\big(\lfloor s/2 \rfloor \cdot 2\pi/s\big)$ in our model collapses to the scalar $[-1]$. Thus, the matrix $\mathbf{M}$ is $(s-1) \times (s-1)$ both when s is even or odd. The transition equation becomes

$$\boldsymbol{\gamma}_{t+1} = (\mathbf{I}_N \otimes \mathbf{M})\boldsymbol{\gamma}_t + \boldsymbol{\omega}_t, \qquad \boldsymbol{\omega}_t \sim \mathrm{WN}\big(\mathbf{0}, \mathbf{\Sigma}_\omega \otimes \mathbf{I}_{s-1}\big), \tag{7.10}$$

and the part of the matrix $\mathbf{Z}$ relative to these cycles is

$$\mathbf{Z}_\gamma = \mathbf{I}_N \otimes \boldsymbol{m}^\top,$$

where $\boldsymbol{m}$ is a vector of $s-1$ elements alternating 1 and 0's. If the rank of $\boldsymbol{\Sigma}_\omega$ is $K_\gamma < N$, then we can rewrite the seasonal component in the common components form:

$$\gamma_{t+1}^\dagger = (\mathbf{I}_{K_\gamma} \otimes \mathbf{M})\gamma_t^\dagger + \boldsymbol{\omega}_t^\dagger, \qquad \boldsymbol{\omega}_t^\dagger \sim \mathrm{WN}(\mathbf{0}, \boldsymbol{\Sigma}_\omega^\dagger \otimes \mathbf{I}_{s-1}), \tag{7.11}$$

and

$$\mathbf{Z}_\gamma^\dagger = \boldsymbol{\Theta}_\gamma \otimes \boldsymbol{m}^\top, \qquad \boldsymbol{c}_\gamma = \boldsymbol{\theta}_\gamma,$$

with the theta matrix and vector defined as above.

Notice that in both forms, the common seasonal components do not impose to the seasonal patterns to be similar. They only share the same shocks that make them evolve over time.

7.4 State space form and parametrisation

The assemblage of the multivariate components into an UCM is not very different than in the univariate case. The transition matrices are to be diagonally concatenated, while the $\mathbf{Z}$ matrix of the observation equation is obtained by stacking side by side all the sub-matrices. If there are reduced ranks components, all the deterministic parts contained in the various $\boldsymbol{\theta}$ vectors are all summed up into the $\boldsymbol{c}_t$ vector of the state space form. For example, if we want to set up a model with all the above components having reduced ranks:

$$\begin{bmatrix} \boldsymbol{\mu}_{t+1}^\dagger \\ \boldsymbol{\mu}_{t+1}^{\dagger\dagger} \\ \boldsymbol{\beta}_{t+1}^{\dagger\dagger} \\ \boldsymbol{\psi}_{t+1}^\dagger \\ \boldsymbol{\psi}_{t+1}^{*\dagger} \\ \boldsymbol{\gamma}_{t+1}^\dagger \end{bmatrix} = \begin{bmatrix} \mathbf{I} & \mathbf{0} & \mathbf{0} & \mathbf{0} & \mathbf{0} & \mathbf{0} \\ \mathbf{0} & \mathbf{I} & \mathbf{I} & \mathbf{0} & \mathbf{0} & \mathbf{0} \\ \mathbf{0} & \mathbf{0} & \mathbf{I} & \mathbf{0} & \mathbf{0} & \mathbf{0} \\ \mathbf{0} & \mathbf{0} & \mathbf{0} & \rho\cos\lambda\otimes\mathbf{I} & \rho\sin\lambda\otimes\mathbf{I} & \mathbf{0} \\ \mathbf{0} & \mathbf{0} & \mathbf{0} & -\rho\sin\lambda\otimes\mathbf{I} & \rho\cos\lambda\otimes\mathbf{I} & \mathbf{0} \\ \mathbf{0} & \mathbf{0} & \mathbf{0} & \mathbf{0} & \mathbf{0} & \mathbf{I}\otimes\mathbf{M} \end{bmatrix} \begin{bmatrix} \boldsymbol{\mu}_t^\dagger \\ \boldsymbol{\mu}_t^{\dagger\dagger} \\ \boldsymbol{\beta}_t^{\dagger\dagger} \\ \boldsymbol{\psi}_t^\dagger \\ \boldsymbol{\psi}_t^{*\dagger} \\ \boldsymbol{\gamma}_t^\dagger \end{bmatrix} + \begin{bmatrix} \boldsymbol{\eta}_t^\dagger \\ \mathbf{0} \\ \boldsymbol{\zeta}_t^{\dagger\dagger} \\ \boldsymbol{\kappa}_t^\dagger \\ \boldsymbol{\kappa}_t^{*\dagger} \\ \boldsymbol{\omega}_t^\dagger \end{bmatrix},$$

with covariance matrix of the disturbance vector

$$\mathbf{Q} = \begin{bmatrix} \boldsymbol{\Sigma}_\eta^\dagger & \mathbf{0} & \mathbf{0} & \mathbf{0} & \mathbf{0} & \mathbf{0} \\ \mathbf{0} & \mathbf{0} & \mathbf{0} & \mathbf{0} & \mathbf{0} & \mathbf{0} \\ \mathbf{0} & \mathbf{0} & \boldsymbol{\Sigma}_\zeta^{\dagger\dagger} & \mathbf{0} & \mathbf{0} & \mathbf{0} \\ \mathbf{0} & \mathbf{0} & \mathbf{0} & \boldsymbol{\Sigma}_\kappa^\dagger & \mathbf{0} & \mathbf{0} \\ \mathbf{0} & \mathbf{0} & \mathbf{0} & \mathbf{0} & \boldsymbol{\Sigma}_\kappa^\dagger & \mathbf{0} \\ \mathbf{0} & \mathbf{0} & \mathbf{0} & \mathbf{0} & \mathbf{0} & \boldsymbol{\Sigma}_\omega^\dagger \otimes \mathbf{I}_{s-1}, \end{bmatrix}$$

where we selected the trigonometric seasonal component instead of the dummy based. The $\mathbf{Z}$ matrix in the observation equation for this model is

$$\mathbf{Z} = \begin{bmatrix} \boldsymbol{\Theta}_\mu & \boldsymbol{\Theta}_\beta & \mathbf{0} & \boldsymbol{\Theta}_\psi & \mathbf{0} & \mathbf{I}\otimes\boldsymbol{m}^\top \end{bmatrix},$$

and the vector $\mathbf{c}_t$ in the observation equation is given by

$$\boldsymbol{c}_t = \boldsymbol{\theta}_\mu + \boldsymbol{\theta}_\beta t + \boldsymbol{\theta}_\gamma.$$

For the numerical estimation of the numerous covariance matrices in the model, that must be positive semi-definite, it is useful to put the free parameters in a lower triangular matrix, say $\mathbf{\Pi}$, which multiplied by its own transpose produces by construction a unique positive semi-definite matrix (Cholesky decomposition): $\mathbf{\Pi}\,\mathbf{\Pi}^\top = \mathbf{\Sigma}$.

Alternatively, all the covariance matrices can be taken as identity matrices and the matrices of coefficients $\mathbf{\Theta}$ redefined as

$$\mathbf{\Theta} = \begin{bmatrix} \theta_{1,1} & 0 & 0 & \dots & 0 \\ \theta_{2,1} & \theta_{2,2} & 0 & \dots & 0 \\ \theta_{3,1} & \theta_{3,2} & \theta_{3,3} & \dots & 0 \\ \vdots & \vdots & \vdots & \ddots & \vdots \\ \theta_{K,1} & \theta_{K,2} & \theta_{K,3} & \dots & \theta_{K,K} \\ \vdots & \vdots & \vdots & \vdots & \vdots \\ \theta_{N,1} & \theta_{N,2} & \theta_{N,3} & \dots & \theta_{N,K} \end{bmatrix}.$$

This alternative parametrisation is useful when using a software package, like EViews, where the Cholesky decomposition of the covariance matrices in the state space representation is not simple to implement.

Part III

Applications

Chapter 8

Business Cycle Analysis with UCM

Business cycle analysis is a field of economic time series analysis in which UCM really excel. Very often applied economists use fixed linear filters, such as Hodrick and Prescott (1997), Baxter and King (1999), Christiano and Fitzgerald (2003) to extract the business cycle from a time series. The advantages of UCM with respect to the fixed linear filtering approach are that

- Extracting the business cycle and forecasting is done contemporaneously in a consistent way;
- The extraction of the common business cycle from a pool of time series can be carried out within the same methodology;
- The definition of the business cycle characteristics, such as the mean periodicity/frequency and persistence, can be fixed as in the fixed linear filtering approach or estimated from the data.

For a better comprehension of the properties of the methods for extracting the business cycle signal from a time series, it is necessary to gain some knowledge in *spectral analysis*. We did not include a chapter on it in the first part of this book because we wanted to keep the volume short and because not everybody interested in UCM is also involved in business cycle analysis. Thus, this chapter begins with a section that presents the spectral analysis tools we need in an intuitive manner. For rigorous and extensive treatment of the spectral analysis of time series the reader should refer to the classic volumes by Jenkins and Watts (1968), Koopmans (1995), Bloomfield (2000) (the last one being the most accessible).

8.1 Introduction to the spectral analysis of time series

In this section it is simpler to index a time series of length n as $y_0, y_1, \ldots, y_{n-1}$.

The main ingredient of spectral analysis is the sinusoid

$$a\cos(\lambda t) + b\sin(\lambda t),$$

that we already encountered in Section 3.3.1. In particular, we are interested in the sinusoids at Fourier frequencies $\lambda_j = 2\pi j/n$, for $j = 1, 2, \ldots, \lfloor s/2 \rfloor$, that enjoy the nice zero mean property,

$$\frac{1}{n}\sum_{t=0}^{n-1}\cos(\lambda_j t) = 0, \qquad \frac{1}{n}\sum_{t=0}^{n-1}\sin(\lambda_j t) = 0, \qquad j = 1, 2, \dots, \lfloor n/2 \rfloor,$$

and orthogonality property,

$$\frac{1}{n}\sum_{t=0}^{n-1}\cos(\lambda_k t)\cos(\lambda_j t) = 0, \qquad \forall j \neq k,$$

$$\frac{1}{n}\sum_{t=0}^{n-1}\sin(\lambda_k t)\sin(\lambda_j t) = 0, \qquad \forall j \neq k,$$

$$\frac{1}{n}\sum_{t=0}^{n-1}\sin(\lambda_k t)\cos(\lambda_j t) = 0, \qquad \forall j, k.$$

Finally, the variances of the sinusoids are given by

$$\frac{1}{n}\sum_{t=0}^{n-1}\cos(\lambda_j t)^2 = \frac{1}{n}\sum_{t=0}^{n-1}\sin(\lambda_j t)^2 = \frac{1}{2}, \qquad 0 < j < \frac{1}{2},$$

and if n is even also

$$\frac{1}{n}\sum_{t=0}^{n-1}\cos(\lambda_j t)^2 = 1.$$

Notice that when n is even, then $\sin(\lambda_{n/2} t) = \sin(\pi t) = 0$ for all t, and so the last sine vanishes.

Let us consider the time series $\{y_0, y_1, \dots, y_{n-1}\}$ and the linear regression

$$y_t = a_0 + \sum_{j=1}^{\lfloor n/2 \rfloor} a_j \cos(\lambda_j t) + b_j \sin(\lambda_j t). \tag{8.1}$$

From elementary regression theory we know that, by computing the coefficients using least squares, that is by choosing the values of $a_0, a_1, \dots, a_{\lfloor n/2 \rfloor}$, $b_1, \dots, b_{\lfloor n/2 \rfloor}$ that minimise the sum of squared errors

$$S = \sum_{t=0}^{n-1}\left[y_t - a_0 - \sum_{j=1}^{\lfloor n/2 \rfloor}\left(a_j \cos(\lambda_j t) + b_j \sin(\lambda_j t)\right)\right]^2,$$

the following properties hold:

- if there are so many regressors (including the constant) as observations, then the fit is perfect (i.e., $S = 0$),
- if the regressors are orthogonal, then the estimate of each regressor coefficient is independent from the other ones.

In the regression (8.1) both conditions hold and the regression coefficients that solve the least squares problem are given by

$$\hat{a}_0 = \frac{1}{n}\sum_{t=0}^{n-1} y_t = \bar{y},$$

$$\hat{a}_j = \frac{2}{n}\sum_{t=0}^{n-1} y_t \cos(\lambda_j t), \qquad 0 < j < n/2,$$

$$\hat{b}_j = \frac{2}{n}\sum_{t=0}^{n-1} y_t \sin(\lambda_j t), \qquad 0 < j < n/2,$$

and if n is even, also

$$\hat{a}_{n/2} = \frac{2}{n}\sum_{t=0}^{n-1} y_t(-1)^t.$$

Since the fit is perfect, it is clear that there is no loss of information if one decides to use the sequence of coefficients $\{\hat{a}_j\}$ and $\{\hat{b}_j\}$ instead of the time series $\{y_t\}$. One advantage of the coefficients over the time series is that, while the latter is generally autocorrelated, all the coefficients are orthogonal. Furthermore, if in equation (8.1) we subtract $\hat{a}_0$ from both sides of the equation sign, then by taking the square and averaging over $t = 1, 2, \ldots, n$, and exploiting the orthogonality properties of the sinusoids and their variances, we can write

$$\begin{aligned}\frac{1}{n}\sum_{t=0}^{n-1}(y_t - \bar{y})^2 &= \frac{1}{n}\sum_{t=0}^{n-1}\left[\sum_{j=1}^{\lfloor n/2 \rfloor} \hat{a}_j \cos(\lambda_j t) + \hat{b}_j \sin(\lambda_j t)\right]^2 \\ &= \frac{1}{2}\sum_{0<j<n/2} (\hat{a}_j^2 + \hat{b}_j^2) + \hat{a}_{n/2}^2,\end{aligned}$$

where the term $\hat{a}_{n/2}$ exists only when n is even. Thus, the coefficient pairs $(\hat{a}_j^2 + \hat{b}_j^2)/2$ decompose the total variance of the time series, in the sum of the variances due to the sinusoids at frequencies $2\pi j/n$, for each j in the range $[1, \lfloor n/2 \rfloor]$.

One of the main tools in spectral analysis is the *periodogram*, which is a function that maps the frequency of each sinusoid to its (rescaled) contribution to the total variance:

$$I(\lambda_j) = \frac{n}{8\pi}(\hat{a}_j^2 + \hat{b}_j^2). \tag{8.2}$$

If we define the sample autocovariance function of the time series $\{y_t\}$

$$\hat{\gamma}(k) = \frac{1}{n}\sum_{t=k}^{n-1}(y_t - \bar{y})(y_{t-k} - \bar{y}),$$

for $k = 1, 2, \ldots, n-1$, then the periodogram can also be computed as[1]

$$I(\lambda_j) = \frac{1}{2\pi}\left[\hat{\gamma}(0) + 2\sum_{k=1}^{n-1}\hat{\gamma}(k)\cos(\lambda_j k)\right]. \tag{8.3}$$

Example 8.1.
Figure 8.1 depicts the periodograms of time series we have already seen in previous examples: (i) average monthly temperature in New York from Jan 1895 to Dec 2013; (ii) monthly growth rate of the (seasonally adjusted) number of employees on nonagricultural payrolls in U.S. from Jan 1950 to Dec 2013; (iii) logarithm of the number (in thousands) of total monthly passengers on international airlines from and to U.S. from Jan 1949 to Dec 1960; (iv) monthly growth rates of the airline passengers time series at point (iii).

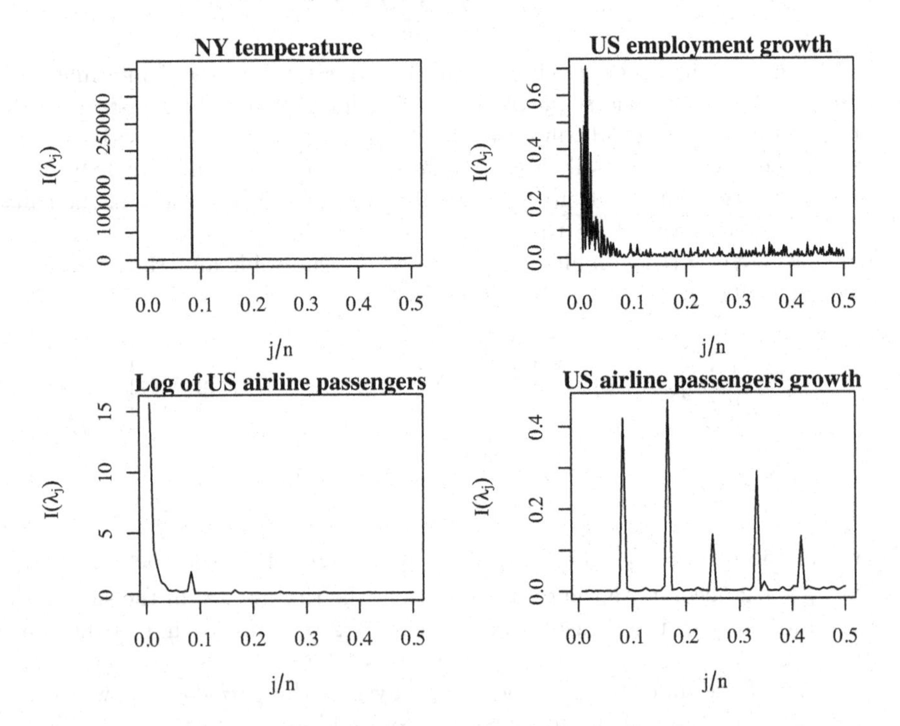

Figure 8.1 *Periodograms of time series already encountered in this book.*

In the plots, on the x-axis the frequency is represented as j/n instead of the full $2\pi j/n$. This choice is common to many software packages (sometimes, the frequency is represented also as $2j/n$), and in the comment to the plots we will refer to this definition of frequency.

[1]The proof is simple but cumbersome.

The top-left periodogram shows a strong peak at frequency $1/12 = 0.08\bar{3}$, which corresponds to sinusoids of period 12 months. In fact, as we saw in Example 6.3, the seasonal component of this temperature is well represented by just one sinusoid of one-year period.

The top-right periodogram shows that a large part of the variance is explained by sinusoids around the peak frequency of $1/81 = 0.0123$, which corresponds to a period of $81/12 = 6.75$ years. The sinusoids around this frequency can be attributed to the business cycle that the economists usually indicate as fluctuations in the range 1.5–8 years.

The bottom graphs are computed on different transformations of the same time series. The lhs plot represents the periodogram of the series in levels (after a log transform): here most of the variance is concentrated on very low frequencies. This fact is due to the strongly tranding behaviour of the time series. Indeed, if a time series grows (or decreases), then its sample variance diverges with the length of the data, and these smooth, long-run movements of the time series are well represented by low frequency sinusoids. On the rhs, the periodogram is computed on the monthly growth rates of the same series, which no not show any strong tendency to grow or decrease. Now the low-frequency sinusoids explain only a little part of the variance, while the sinusoids at seasonal frequencies, 1/12, 2/12, 3/12, 4/12, 5/12, and 6/12 are responsible for most of it. Actually, the sinusoid at frequency $6/12 = 0.5$ does not contribute much, and this suggests that in a UCM, the state variable of the stochastic trigonometric component taking care of this sinusoid could be omitted without significant losses.

Now, it is natural to ask what happens to the periodogram when it is computed on a random sample path $\{Y_1, Y_2, \ldots, Y_n\}$ of a stochastic process $\{Y_t\}$ and not on a sequence of numbers $\{y_1, y_2, \ldots, y_n\}$.

Theorem 8.1 (Spectrum of a stationary process). *Let $\{Y_t\}$ be a weakly stationary process with absolutely summable autocovariance function, (i.e., $\sum_{k=-\infty}^{\infty} |\gamma(k)| < \infty$), then the* spectrum

$$g(\lambda) = \frac{1}{2\pi}\left[\gamma(0) + 2\sum_{k=1}^{\infty} \gamma(k)\cos(\lambda k)\right],$$

is a well-defined function with the following properties:

- *$g(\lambda)$ is real and continuous;*
- *$g(\lambda) \geq 0$;*
- *$g(\lambda) = f(-\lambda)$;*
- *$g(\lambda) = f(\lambda + 2\pi)$.*

Furthermore, the autocovariance function can be recovered from $f(\lambda)$ as

$$\gamma(k) = \int_{-\pi}^{\pi} g(\lambda)\cos(\lambda k)\,\mathrm{d}\lambda.$$

For a proof the reader should refer to the cited bibliography on spectral analysis, even though, with the exception of the inversion formula, all the results are straightforward to obtain.

For the interpretation of the spectrum, or also *spectral density*, of a stationary process, first of all notice that, by the inversion formula,

$$\gamma(0) = \int_{-\pi}^{\pi} g(\lambda)\cos(\lambda 0)\,\mathrm{d}\lambda = \int_{-\pi}^{\pi} g(\lambda)\,\mathrm{d}\lambda,$$

that is, the spectrum integrated over an interval 2π wide is equal to the variance of the process. Twice the integral of the spectrum computed over the interval $[\lambda_1, \lambda_2]$ contained in $[0, \pi]$ represents the variance of the process generated by sinusoids in that frequency range.

Usually, the spectrum and also the periodogram are computed using the complex exponential function instead of the cosine:

$$\mathrm{e}^{\mathrm{i}\lambda} = \cos(\lambda) + \mathrm{i}\sin(\lambda),$$

where i is the imaginary unit defined as $\mathrm{i} = \sqrt{-1}$. The following identities can turn useful in the rest of the section:

$$\cos(\lambda) = \frac{\mathrm{e}^{\mathrm{i}\lambda} + \mathrm{e}^{-\mathrm{i}\lambda}}{2}, \qquad \sin(\lambda) = \frac{\mathrm{e}^{\mathrm{i}\lambda} - \mathrm{e}^{-\mathrm{i}\lambda}}{2\mathrm{i}}.$$

Using the complex exponential, we can define the *discrete Fourier transform* (DFT) of the sequence $\{y_0, y_1, \ldots, y_{n-1}\}$,

$$J(\lambda) = \frac{1}{n}\sum_{t=0}^{n-1} y_t \mathrm{e}^{-\mathrm{i}\lambda t}.$$

We can now define the periodogram either through the squared modulus of the DFT of the data sequence or through the DFT of the sample autocovariance function:

$$\begin{aligned} I(\lambda_j) &= \frac{n}{2\pi}|J(\lambda_j)|^2 \\ &= \frac{1}{2\pi}\sum_{k=-(n-1)}^{n-1} \hat{\gamma}(k)\mathrm{e}^{-\mathrm{i}\lambda_j k}. \end{aligned}$$

Similarly, the spectrum can be defined also as DFT of the population autocovariance function:

$$g(\lambda) = \frac{1}{2\pi}\sum_{k=-\infty}^{\infty} \gamma(k)\mathrm{e}^{-\mathrm{i}\lambda k}.$$

The autocovariance function is recovered from the spectrum using the continuous (inverse) Fourier transform:

$$\gamma(k) = \int_{-\pi}^{\pi} g(\lambda)\mathrm{e}^{i\lambda k}\,\mathrm{d}\lambda.$$

The spectrum of a stationary causal ARMA process is given by

$$g_{\text{ARMA}}(\lambda) = \frac{\sigma^2}{2\pi} \frac{\theta_q(\mathrm{e}^{-\mathrm{i}\lambda})\theta_q(\mathrm{e}^{\mathrm{i}\lambda})}{\phi_p(\mathrm{e}^{-\mathrm{i}\lambda})\phi_p(\mathrm{e}^{\mathrm{i}\lambda})}. \tag{8.4}$$

Thus, the white noise has spectrum

$$g_{\text{WN}}(\lambda) = \frac{\sigma^2}{2\pi},$$

where all the frequencies equally contribute to the variance of the process.

The spectrum of the MA(1) process is

$$\begin{aligned} g_{\text{MA}(1)}(\lambda) &= \frac{\sigma^2}{2\pi}(1 + \theta\mathrm{e}^{-\mathrm{i}\lambda})(1 + \theta\mathrm{e}^{\mathrm{i}\lambda}) \\ &= \frac{\sigma^2}{2\pi}[1 + \theta(\mathrm{e}^{-\mathrm{i}\lambda} + \mathrm{e}^{\mathrm{i}\lambda}) + \theta^2] \\ &= \frac{\sigma^2}{2\pi}[1 + 2\theta\cos(\lambda) + \theta^2]. \end{aligned}$$

Using similar computations, we can compute the spectrum of the AR(1) process:

$$g_{\text{AR}(1)}(\lambda) = \frac{\sigma^2}{2\pi} \frac{1}{[1 - 2\phi\cos(\lambda) + \phi^2]}.$$

The spectrum of the AR(2) is (you can prove it)

$$g_{\text{AR}(2)}(\lambda) = \frac{\sigma^2}{2\pi} \frac{1}{1 + \phi_1^2 + \phi_2^2 - 2\phi_1\cos(\lambda) + 2\phi_1\phi_2\cos(\lambda) - 2\phi_2\cos(2\lambda)},$$

and if the roots of $1 - \phi_1 x - \phi_2 x^2 = 0$ are complex, it reaches its maximum at frequency

$$\lambda_{\max} = \arccos\left(\frac{\phi_1\phi_2 - \phi_1}{4\phi_2}\right),$$

while if the roots are real it peaks at frequency 0 or π.

The spectrum of the stochastic cycle used in UCM is

$$g_{\psi}(\lambda) = \frac{\sigma_\kappa^2}{2\pi} \frac{1 + \rho^2 - 2\rho\cos(\lambda_c)\cos(\lambda)}{1 + \rho^4 + 4\rho^2\cos(\lambda_c)^2 - 4\rho(1 + \rho^2)\cos(\lambda_c)\cos(\lambda) + 2\rho^2\cos(2\lambda)},$$

where we used λ_c to denote the cycle frequency. The spectrum reaches its maximum at $\lambda = \lambda_c$ and the damping factor ρ determines the concentration of the spectral density around this frequency.

Figure 8.2 depicts the spectra of the four processes, and the only two able to produce peaks in the business cycle range (i.e., 2–8 years) are the AR(2) with complex roots and the stochastic cycle introduced in Section 3.3. The parameter λ_c of the stochastic cycle determines the frequency of the peak in the spectrum, while the damping factor ρ is responsible for the concentration

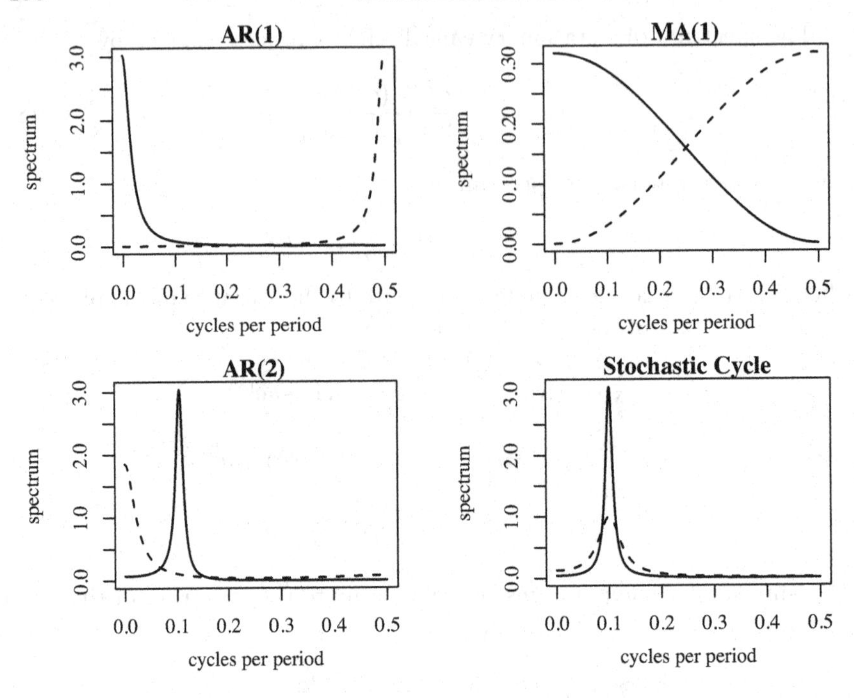

Figure 8.2 *Spectra of various processes: i) AR(1) with coefficient* $\phi = 0.9$ *and* $\phi = -0.9$ *(dashed line), ii) MA(1) with coefficients* $\theta = 0.9$ *and* $\theta = -0.9$ *(dashed line), iii) AR(2) with coefficient pair* $\phi_1 = 1.5, \phi_2 = -0.9$ *(complex roots) and* $\phi_1 = 0.4, \phi_2 = 0.4$ *(real roots, dashed line), iv) stochastic cycle with frequency* $\lambda_c = 2\pi/10$ *and damping factor* $\rho = 0.95$ *and* $\rho = 0.85$ *(dashed line). Here, as in many software packages the frequency is defined as* $\lambda/(2\pi)$.

of the spectral density in a neighbourhood of λ_c: the higher $\rho \in (0,1)$, the more concentrated the spectral density. As ρ approaches 1, the density at λ_c diverges. Indeed, the spectrum is only defined for weakly stationary processes with absolutely summable autocovariance functions. However, when unit roots are present in the AR part of the process, one can define the *pseudo spectrum*, by letting the spectrum admit a finite number of singularities (poles). For example, the pseudo spectrum of a random walk can be obtained from $g_{\text{AR}(1)}(\lambda)$ by setting $\phi = 1$:

$$g_{\text{RW}}(\lambda) = \frac{\sigma^2}{2\pi} \frac{1}{2[1 - \cos(\lambda)]},$$

with a pole at frequency $\lambda = 0$.

The two seasonal components used in UCM have pseudo spectra with poles at seasonal frequencies $\lambda_j = 2\pi j/s$, for $j = 1, 2, \ldots, \lfloor s/2 \rfloor$, while seasonal integrated ARMA processes, that is ARIMA$(p, 0, q)(P, 1, Q)_s$, have poles at the seasonal frequencies and at $\lambda = 0$.

In order to estimate the spectrum of the process based on a finite sample path (time series), the periodogram may seem a natural candidate. In fact, because in equation (8.3) the sample autocovariances substitute the population autocovariances of the spectrum (8.1), $\{I(\lambda_j)\}$ can be considered an estimator based on the method of moments. However, $I(\lambda_j)$ is not consistent for $g(\lambda_j)$, but instead, under regularity conditions (that hold if the observations are generated by a linear process)

$$I(\lambda_j) \xrightarrow{d} g(\lambda_j)\frac{\chi_2^2}{2},$$

where χ_2^2 denotes a chi-square distribution with two degrees of freedom. $\{I(\lambda_j)\}$ is not consistent because the parameters to be estimated, that is the periodogram values, diverge at the same rate as n.

There are many alternative ways to estimate the spectral density of a process, but two of the most popular are:

- smoothing the periodogram (nonparametric way),
- fitting an ARMA model to the time series to use the spectrum $g_{\text{ARMA}}(\lambda)$ with the estimated parameters substituting the population parameters (parametric way).

Most statistical software packages have functions or procedures that estimate the spectrum by smoothing the periodogram.

In this chapter we are interested in linear filters that extract the business cycle from a time series. A very useful property of the DFT reveals the effect of linear filters on time series. A *linear filter* is an absolutely summable sequence of constants $\{h_j\}$ that map a process $\{Y_t\}$ in a new process $\{W_t\}$ by the convolution

$$W_t = \sum_{j=-\infty}^{\infty} h_j Y_{t-j} = h(\mathbb{B})Y_t,$$

with $h(\mathbb{B}) = \sum_{j=-\infty}^{\infty} \mathbb{B}^j$ and $\sum_{j=-\infty}^{\infty}|h_j| < \infty$. Since the Fourier transform turns convolutions into products, the DFT of $\{W_t\}$ is given by the DFT of $\{h_j\}$ times the DFT of $\{Y_t\}$, and, therefore, the spectrum of $\{W_t\}$ is

$$g_W(\lambda) = h(\mathrm{e}^{-\mathrm{i}\lambda})h(\mathrm{e}^{\mathrm{i}\lambda})g_Y(\lambda), \tag{8.5}$$

with $h(\mathrm{e}^{-\mathrm{i}\lambda}) = \sum_{j=-\infty}^{\infty} h_j \mathrm{e}^{-\mathrm{i}\lambda}$ called *frequency response function*. The function

$$G^2(\lambda) = h(\mathrm{e}^{-\mathrm{i}\lambda})h(\mathrm{e}^{\mathrm{i}\lambda}) = |h(\mathrm{e}^{-\mathrm{i}\lambda})|^2,$$

is called the *squared gain function* or *power transfer function* of the filter $h(\mathbb{B})$, and its square root $G(\lambda)$ takes the name of *gain* of the filter.

From equation (8.5) it is clear that the sinusoids at frequencies for which $G^2(\lambda) > 1$ are amplified by the filter, while the sinusoids at frequencies for which $G^2(\lambda) < 1$ are attenuated by the filter.

For example, the difference filter is $h(\mathbb{B}) = 1 - \mathbb{B}$ and its effect on the spectrum is given by

$$G_{\Delta}^2(\lambda) = (1 - \mathrm{e}^{-\mathrm{i}\lambda})(1 - \mathrm{e}^{\mathrm{i}\lambda}) = -(\mathrm{e}^{-\mathrm{i}\lambda} + \mathrm{e}^{\mathrm{i}\lambda}) = 2[1 - \cos(\lambda)];$$

and the seasonal difference filter $h(\mathbb{B}) = 1 - \mathbb{B}^s$ has squared gain

$$G_{\Delta_s}^2(\lambda) = (1 - \mathrm{e}^{-s\mathrm{i}\lambda})(1 - \mathrm{e}^{s\mathrm{i}\lambda}) = -(\mathrm{e}^{-\mathrm{i}\lambda} + \mathrm{e}^{\mathrm{i}\lambda}) = 2[1 - \cos(s\lambda)].$$

The two squared gain functions are depicted in Figure 8.3.

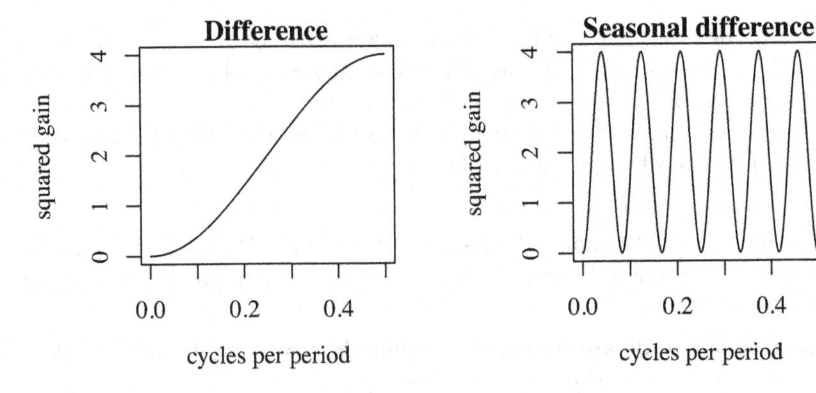

Figure 8.3 *Squared gain functions of the filters difference and seasonal difference with* $s = 12$. *Here, as in many software packages the frequency is defined as* $\lambda/(2\pi)$.

The *ideal filter* to extract the movement of a time series over a band of frequencies $[\lambda_1, \lambda_2]$ has gain function which is equal to one over those frequencies and zero elsewhere. An ideal *low-pass* filter extracts all and only the sinusoids in the band $[0, \lambda_2]$, an ideal *high-pass* filter does the same over the band $[\lambda_1, \pi]$ and when the frequency band does not contain the extremes 0 and π, the filter is called *band-pass.* Unfortunately, ideal filters need infinitely many coefficients $\{h_j\}$, and, thus, infinite-length time series. So ideal filters can only be approximated in finite samples. The properties of some business cycle filters will be discussed in the next section through the gain function.

8.2 Extracting the business cycle from one time series

For a long time, the standard tool for extracting the business cycle from a time series was the Hodrick–Prescott (HP) filter (the 1981 discussion paper has been published as Hodrick and Prescott, 1997), or better, the time series detrended by Hodrick–Prescott filtering. The HP filter can be shown to be equivalent to the smoothed trend in the integrated random walk plus noise model

$$\begin{aligned} Y_t &= \mu_t + \epsilon_t, & \epsilon &\sim \mathrm{WN}(0, \sigma_\epsilon^2), \\ \mu_{t+1} &= \mu_t + \beta_t \\ \beta_{t+1} &= \beta_t + \zeta_t, & \zeta_t &\sim \mathrm{WN}(0, \lambda_{\mathrm{HP}} \cdot \sigma_\epsilon^2), \end{aligned}$$

and the gain function of the smoother (in the middle of the sample path) is

$$\begin{aligned} G_{\mathrm{HP}}(\lambda) &= \frac{1}{1 + 4\lambda_{\mathrm{HP}}^{-1}[1 - \cos(\lambda)]^2}, \\ &= \frac{1}{1 + 16\lambda_{\mathrm{HP}}^{-1} \sin(\lambda/2)^4} \end{aligned}$$

This gain function is monotonically decreasing from 1 to 0 on the domain $[0, \pi]$, and so, if we call *cut-off frequency*, the frequency $\lambda_{1/2}$ for which the gain is equal to 1/2, then the value of the parameter λ_{HP} that yields the desired cut-off is given by

$$\bar{\lambda}_{\mathrm{HP}} = [2\sin(\lambda_{1/2}/2)]^4, \qquad 0 < \lambda_{1/2} < \pi.$$

The HP filter is a low-pass filter and the business cycle is obtained by subtracting the filtered series from the original one. Using the state space terminology, we can say that the business cycle is defined as $Y_t - \hat{\mu}_{t|n}$, where $\hat{\mu}_{t|n}$ denotes the smoothed inference of μ_t (based on all the observations). The gain of this business cycle filter, which is now high-pass, is just $1 - G_{\mathrm{HP}}(\lambda)$. The business cycle extracted using this approach is not smooth because the high-pass filter leaves all the high-frequency sinusoids unaltered. Sample gain functions of the low-pass and high-pass HP filters are depicted in Figure 8.4 (top-left). It is clear that the high-pass HP filter can be used to extract the business cycle only when the spectrum of the time series is almost zero at high frequencies. Alternatively you can pass the low-pass filter to extract frequencies below λ_2 and run the high-pass filter on the filtered series to extract the remaining frequencies above λ_1 (with $\lambda_1 < \lambda_2$).

In the last few years the filter proposed by Baxter and King (1999) has become very popular and can be found in many software packages. The Baxter–King (BK) filter is obtained by approximating the infinite-length ideal band-pass filter with a $2K + 1$-length symmetric filter. The filter is given by $h(\mathbb{B}) = \sum_{j=-K}^{K} h_j$ with

$$\begin{aligned} h_j &= b_j + \theta \\ b_j &= \begin{cases} \frac{\lambda_2 - \lambda_1}{\pi}, & \text{for } j = 0; \\ \frac{\sin(\lambda_2 j) - \sin(\lambda_1 j)}{\pi j}, & \text{for } j = 1, \ldots, K; \end{cases} \\ \theta &= -\frac{1}{2K+1} \sum_{j=-K}^{K} b_j. \end{aligned}$$

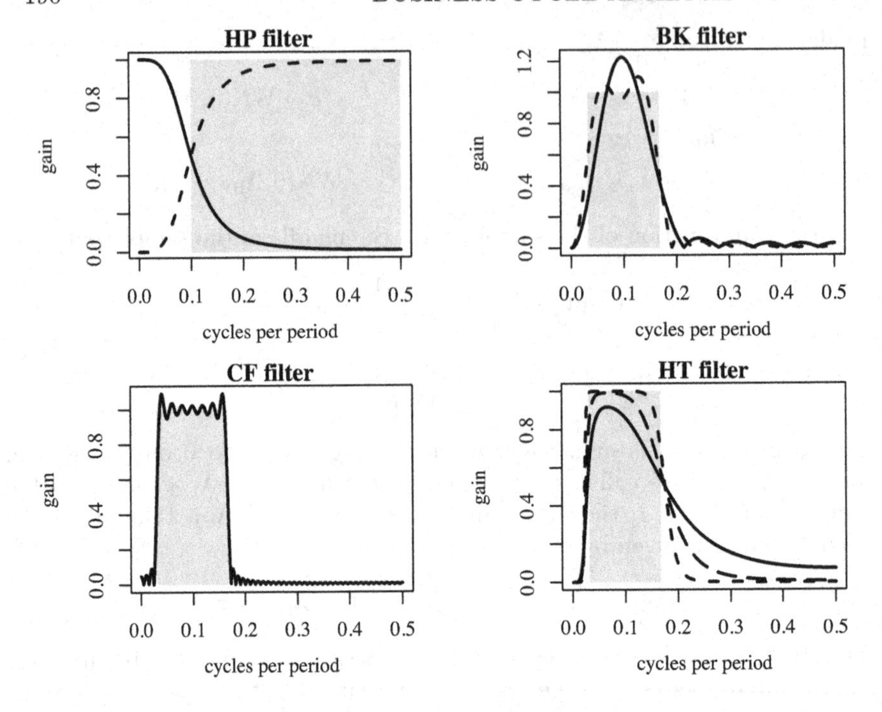

Figure 8.4 *Gain functions of: i) HP filter with* $\lambda_{\text{HP}} = 0.146$*, ii) BK filter with* $K = 8$ *(line),* $K = 12$*, iii) CF filter for random walk processes for the central observation of a time series with* $n = 101$*, iv) HT filter with* $\rho = 0.9$*,* $\lambda_c = 0.02 \cdot 2\pi$*,* $q_\zeta = 0.001$*,* $q_\kappa = 1$ *for with orders* (p, q) *equal to* $(2, 2)$ *(line),* $(2, 4)$ *(long dashes),* $(2, 4)$ *(short dashes). The grey areas highlight the ideal filter gain.*

The gain of the BK band-pass filter is given by

$$G_{\text{BK}}(\lambda) = \sum_{j=-K}^{K} h_j \cos(\lambda j),$$

and its approximation to the ideal filter improves as K increases. However, a large K can be a serious problem as the filter can be run only on the time span $t = K + 1, K + 2, \ldots, n - K$ and so the filtered series will be $2K$ observation shorter than the original. Figure 8.4 (top right) depicts the gain of the BK filter to extract the signal in the frequency range $[2\pi/32, 2\pi/6]$ (i.e., the business cycle frequency range for quarterly data) for filters with $K = 8$ and $K = 12$.

Christiano and Fitzgerald (2003) improve the BK filter in two directions: they show how to build a full-length asymmetric filter that always yields the best approximation to the ideal filter; in computing the filter weights, the (quadratic) loss function of the approximation is weighted by the spectral density of a process that well represents the dynamics of the time series. In this way they avoid the loss of data at the beginning and at the end of the sample path and the filter is always based on all observations. Furthermore, the weighting of the loss function lets the filter be more accurate where there is actually some variance to extract and less accurate where there is no significant movement to extract. If the pseudo-spectrum of a random walk is a decent approximation of the shape of the time series periodogram, then the Christiano–Fitzgerald (CF) filter for the t-th period is defined by the weights $h_j = b_j$, for $j = -t+2, -t+3, \ldots, n-t-1$, where the b_j have already been defined for the BK filter, and for $j = -t+1$ and $j = n-t$ are given by

$$h_{-t+1} = -\frac{1}{2}b_0 - \sum_{j=1}^{t-1} b_{j-1},$$

$$h_{n-t} = -\frac{1}{2}b_0 - \sum_{j=1}^{n-t} b_{j-1},$$

where the sums are to be considered only when the final index is larger than the initial index. So, for each observation a different filter is computed using the weights b_j of the BK filter, with the exception of the first and the last observation of the time series which receive a weight which is different from the corresponding b_j. Since the filter is time-varying, Figure 8.4 (bottom-left) depicts the gain for the central observation of a time series of length $n = 101$.

Now, consider the following trend plus cycle plus noise UCM proposed by Harvey and Trimbur (2003)

$$\begin{aligned}
Y_t &= \mu_{p,t} + \psi_{q,t} + \epsilon_t \\
\mu_{p,t+1} &= \mu_{p,t} + \mu_{p-1,t} \\
\mu_{p-1,t+1} &= \mu_{p-1,t} + \mu_{p-2,t} \\
&\ldots \\
\mu_{1,t+1} &= \mu_{1,t} + \zeta_t \\
\begin{bmatrix} \psi_{q,t+1} \\ \psi^*_{q,t+1} \end{bmatrix} &= \rho\mathbf{R}(\lambda_c) \begin{bmatrix} \psi_{q,t} \\ \psi^*_{q,t} \end{bmatrix} + \begin{bmatrix} \psi_{q-1,t} \\ 0 \end{bmatrix} \\
\begin{bmatrix} \psi_{q-1,t+1} \\ \psi^*_{q-1,t+1} \end{bmatrix} &= \rho\mathbf{R}(\lambda_c) \begin{bmatrix} \psi_{q-1,t} \\ \psi^*_{q-1,t} \end{bmatrix} + \begin{bmatrix} \psi_{q-2,t} \\ 0 \end{bmatrix}, \\
&\ldots \\
\begin{bmatrix} \psi_{1,t+1} \\ \psi^*_{1,t+1} \end{bmatrix} &= \rho\mathbf{R}(\lambda_c) \begin{bmatrix} \psi_{1,t} \\ \psi^*_{1,t} \end{bmatrix} + \begin{bmatrix} \kappa_{1,t} \\ 0 \end{bmatrix},
\end{aligned}$$

where $\{\epsilon_t\}$, $\{\zeta_t\}$ and $\{\kappa_t\}$ are white noise sequences and $\mathbf{R}(\lambda_c)$ is a rotation matrix with angle λ_c. The component $\mu_{p,t}$ is called p-order trend and $\psi_{q,t}$ q-order cycle. For $p = 2$ and $q = 1$ we have the integrated random walk plus cycle plus noise already used in the previous chapters, even though the cycle is slightly different because only one source of disturbance is present (i.e., there is no κ^*_t). If the cycle is extracted by the smoother in the central part of a long time series, the gain of the Harvey–Trimbur (HT) filter is given by

$$G_{p,q}(\lambda) = \frac{q_\kappa g_\psi(\lambda)^q}{q_\zeta[2 - 2\cos(\lambda)]^{-p} + q_\kappa g_\psi(\lambda)^q + 1},$$

where $q_\kappa = \sigma_\kappa^2/\sigma_\epsilon^2$, $q_\zeta = \sigma_\zeta^2/\sigma_\epsilon^2$ and

$$g_\psi(\lambda) = \frac{1 + \rho^2\cos(\lambda_c)^2 - 2\rho\cos(\lambda_c)\cos(\lambda)}{1 + \rho^4 + 4\rho^2\cos(\lambda_c) - 4(\rho + \rho^3)\cos(\lambda_c)\cos(\lambda) + 2\rho^2\cos(2\lambda)}.$$

This gain function is depicted in Figure 8.4 (bottom-right): the cut is rather steep on the low frequencies and smoother on the high frequencies. Since economic time series have movements mostly concentrated on low and mid-low frequencies – because of the trend and of the business cycle – the frequency cut must be sharp in this low range, while the damping can be softer on higher frequencies where the spectral density is low. The gain depends on many parameters, but the advantage of this filter is that it comes with a credible model and, thus, all these quantities can be estimated on the data. You can also decide to fix some parameters (e.g., the peak frequency λ_c) and estimate the remaining ones. As for p and q, in order for the model to be credible, p should be either 1 or 2 (i.e., the data are generated by an I(1) or an I(2) process), while q can be chosen freely: the larger q, the smoother the cycle. However, Harvey and Trimbur (2003) show that on real economic time series values of q larger than 2 do not produce significant improvements in the smoothness of the extracted component. Harvey and Trimbur (2003) also consider the form of the stochastic cycle with both disturbances κ_t and κ_t^* and write:

> Our experience is that the balanced form seems to give a slightly better fit insofar as maximum likelihood estimation tends to yield more plausible values for the period.

Example 8.2 (U.S. business cycle).
We applied the four filters to the quarterly time series of U.S. real GDP dating from 1947:Q1 to 2014:Q3 (Figure 8.5). The time series is seasonally adjusted and the filters were applied to the logarithm of the data.

The HT filter is of order $(2, 2)$ based on the following parameters estimated by maximum likelihood on the data: $\rho = 0.75$, $\lambda_c = 0.00063$, $q_\zeta = 0.011$, $q_\kappa = 6.637$. The peak frequency is extremely low and the not so high value of ρ makes the gain of the filter spread over a wide range of frequencies. So it is not surprising that the signal extracted by the HP filter with $\lambda_{\text{HP}} = 1600$ is close to the HT filtered series. Nonetheless, the signal extracted by the HT filter is also rather close to BK filtered series with $K = 12$, and also similar to the CF filtered data. The correlation for the pairs HT-HP and HT-BK is 0.94 while the HT-CF correlation is 0.77.

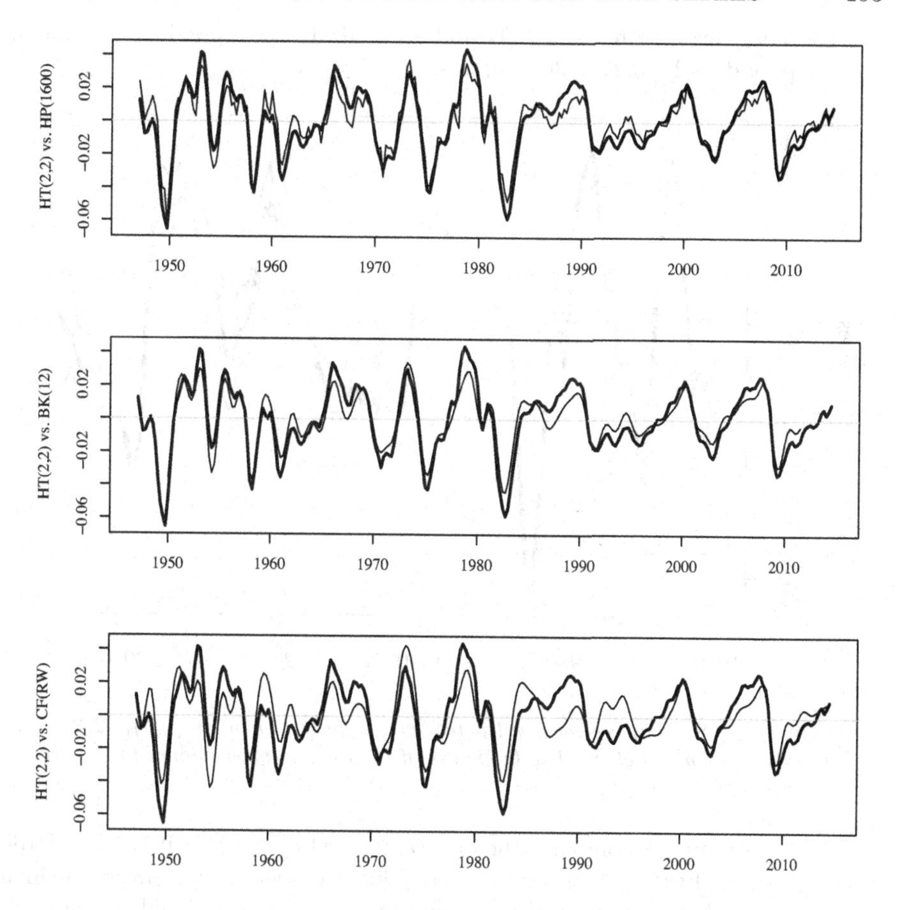

Figure 8.5 *Cycle extracted with the HT filter of order* $(2, 2)$ *with estimated parameters (thin line) plotted versus cycles extracted by i) the high-pass HP filter with* $\lambda = 1600$, *ii) the BK filter with* $K = 12$, *iii) the random-walk optimal CF filter.*

By playing with the order of the HT filter and with the other parameters, we could adjust it to make it looking more like the ideal filter for the desired frequency band, but what makes the HT filter more interesting is its being model-based. For example, we can use the underlying UCM to forecast future values of the business cycle.

By observing the behaviour of the cycles in the graph, the cycles starting from the seventies look somewhat different from the previous ones. So, we estimated the model underlying the HT filter again on this sub-sample by setting the orders equal to to $(2, 1)$ (i.e., the IRW + cycle + WN encountered previously in this book), $(2, 2)$ and $(2, 4)$. The estimates of the cycle parameters for the $(2, 1)$-order model was much more meaningful: $\rho = 0.94$ and $\lambda_c = 0.21$, corresponding to a modal period of $2\pi/0.21 \approx 30$ quarters, or 7.5 years.

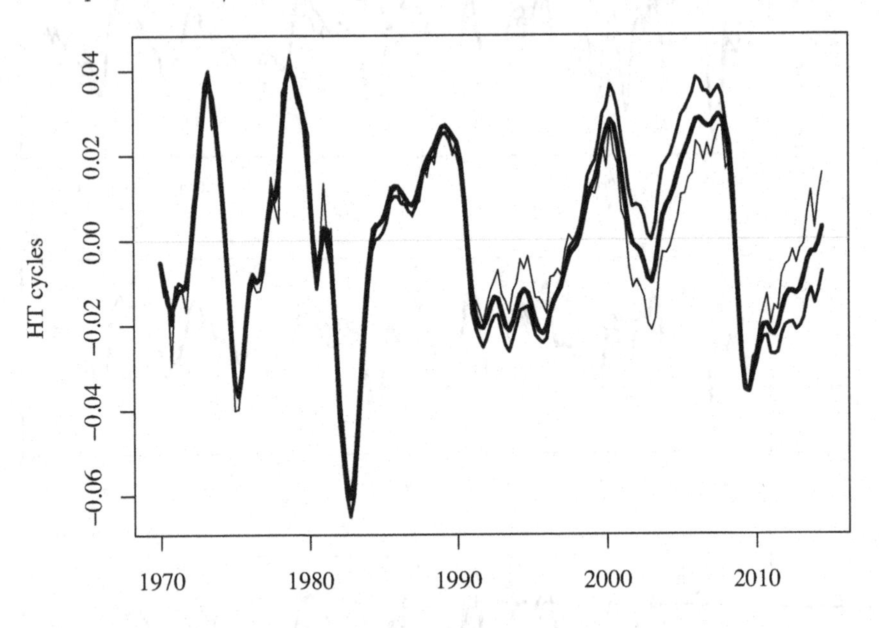

Figure 8.6 *U.S. business cycle extracted by HT filters of order* $(2,1)$, $(2,2)$, $(2,4)$ *with estimated parameters. The thickness of the lines is proportional to the order of the cycle.*

Figure 8.6 compares the cycle extracted by the HT filter for different values of the cycle order. The three lines are very close, but it is evident that when the cycle order is higher the extracted signal is smoother. However, even the cycle of order one – the one usually implemented in most software packages – does a very good job.

8.3 Extracting the business cycle from a pool of time series

A very nice feature of the UCM is that the cycle signal can be extracted in a very natural way from a pool of time series. Indeed, from the definition of business cycle of Burns and Mitchell (1946) reported at page 170 of this book, the natural way to measure these fluctuations is by using more than one time series.

The similar cycles shown in Section 7.2 can be adapted to one common cycle extraction in different ways. The simplest approach is through the common cycle model, where one cycle, possibly of order $q > 1$, is scaled to fit all the time series in the model. If we include also the order 2 trend, we can write the UCM as

$$\begin{aligned}
\underbrace{\mathbf{Y}_t}_{N\times 1} &= \underbrace{\boldsymbol{\mu}_t}_{N\times 1} + \underbrace{\boldsymbol{\Theta}}_{N\times 2} \underbrace{\boldsymbol{\psi}_{q,t}}_{2\times 1} + \underbrace{\boldsymbol{\epsilon}_t}_{N\times 1} \\
\boldsymbol{\mu}_{t+1} &= \boldsymbol{\mu}_t + \boldsymbol{\beta}_t \\
\boldsymbol{\beta}_{t+1} &= \boldsymbol{\beta}_t + \boldsymbol{\zeta}_t \\
\boldsymbol{\psi}_{q,t+1} &= \rho\mathbf{R}(\lambda_c)\boldsymbol{\psi}_{q,t} + \boldsymbol{\psi}_{q-1,t} \\
\boldsymbol{\psi}_{q-1,t+1} &= \rho\mathbf{R}(\lambda_c)\boldsymbol{\psi}_{q-1,t} + \boldsymbol{\psi}_{q-2,t} \\
&\dots \\
\boldsymbol{\psi}_{1,t+1} &= \rho\mathbf{R}(\lambda_c)\boldsymbol{\psi}_{1,t} + \boldsymbol{\kappa}_t,
\end{aligned} \tag{8.6}$$

with $\boldsymbol{\zeta}_t \sim \mathrm{WN}(\mathbf{0}, \boldsymbol{\Sigma}_\zeta)$ with N elements, $\boldsymbol{\kappa}_t \sim \mathrm{WN}(\mathbf{0}, \mathbf{I}_2)$, $\mathbf{R}(\lambda_c)$ rotation matrix of angle λ_c, and $\boldsymbol{\Theta}$ matrix with N loadings in the first column and N zeros in the second column. If the common cycle is not perfectly in phase over all the N time series, then the cycle shift proposed by Rünstler (2004) can be easily implemented by redefining the $\boldsymbol{\Theta}$ matrix as

$$\boldsymbol{\Theta} = \begin{bmatrix} \delta_1 & 0 \\ \delta_2 \cos(\xi_2\lambda_c) & \delta_2 \sin(\xi_2\lambda_c) \\ \vdots & \vdots \\ \delta_N \cos(\xi_N\lambda_c) & \delta_N \cos(\xi_N\lambda_c) \end{bmatrix},$$

so that the common cycle is synchronised with the first time series, while all the other time series are allowed to shift the common cycle by ξ_i periods, with $i = 2, \dots, N$.

Alternatively, you can use the similar cycles of equation (7.7), using one of the following configurations.

1. Estimate the UCM with a full rank covariance matrix $\boldsymbol{\Sigma}_\kappa$ and define the common cycle as the first principal component of the covariance matrix or of the corresponding correlation matrix $\mathbf{C}_\kappa$ (i.e., compute the first eigenvector of the matrix and use it as a vector of weights to build the smoothed common cycle from the individual smoothed cycles). In this way, the common cycle is not directly imposed into the model, but derived from the correlation characteristics of the estimated $\boldsymbol{\Sigma}_\kappa$.
2. Estimate the UCM with cycle covariance matrix

$$\boldsymbol{\Sigma}_\kappa = \begin{bmatrix} \delta_1 \\ \delta_2 \\ \vdots \\ \delta_N \end{bmatrix} \begin{bmatrix} \delta_1 & \delta_2 & \dots & \delta_N \end{bmatrix} + \begin{bmatrix} \sigma_1^2 & 0 & \dots & 0 \\ 0 & \sigma_2^2 & \dots & 0 \\ \vdots & \vdots & \ddots & 0 \\ 0 & 0 & \dots & \sigma_N^2 \end{bmatrix}, \tag{8.7}$$

which can be interpreted as deriving from a common cycle plus idiosyncratic cycles model, with all the cycles reciprocally. So, the only source of correlation among the similar cycles is the common cycle, and the vector of loadings $\boldsymbol{\delta}$ relates the common cycle to each time series. The common cycle can be recovered using the projection

$$\psi_t^{(c)} = \boldsymbol{\delta}^\top \boldsymbol{\Sigma}_\kappa^{-1} \boldsymbol{\psi}_t = \boldsymbol{w}^\top \boldsymbol{\psi}_t.$$

Since the common cycle has an arbitrary variance, you can rescale the vector $\boldsymbol{w}$ as you wish by multiplying it by an arbitrary constant. The generic element of $\boldsymbol{w}$ can be proved to be proportional to δ_j/σ_j^2 and so, given the arbitrary of the scale, you can just set

$$\boldsymbol{w}^\top = \begin{bmatrix} \frac{\delta_1}{\sigma_1^2} & \frac{\delta_2}{\sigma_2^2} & \cdots & \frac{\delta_N}{\sigma_N^2} \end{bmatrix}.$$

3. Estimate an UCM with N uncorrelated similar cycles and one common cycle (with unitary variance) related to the observed time series through a vector of loadings $\boldsymbol{\delta}$. This approach is equivalent to the previous one but it presents advantages and disadvantages. The main disadvantage is that the state vector has to be augmented with one more stochastic cycle. The advantages are that (i) the common cycle is automatically derived by the Kalman filter/smoother; (ii) the common cycle can be shifted using the matrix $\boldsymbol{\Theta}$ defined above.

Example 8.3 (Multivariate methods applied to the four time series used by the Conference Board to compute the coincident index).
We estimated the common cycle of the four time series described in Example 7.2 using three of the methods discussed above. In particular, we estimated a pure order-1 common cycle model with no shift as in equation (8.6).

Secondly, we estimated a model with similar cycles and computed the first eigenvector that, then, was used as weight vector to build the common cycle as a linear combination of the standardised similar cycles. The four weight values were almost the same.

Thirdly, we estimated a model as in Point 2 of the above list, were the four variances of the idiosyncratic cycles (multiplied by 10,000) were estimated to $[0.00 \quad 1.04 \quad 2.63 \quad 1.10]$. Since the first variance is zero, it means that the cycle of the time series Employment does not have an idiosyncratic part and so the common cycle is coincident with the cycle of that time series.

Figure 8.7 reports, from top to bottom, the common cycles estimated with the three methods. The first two methods produced almost identical cycles (their correlation is 0.97), while the application of the third method resulted in a smoother common cycle highly correlated with the other two extracted cycles (the correlations with the first and second cycles are, respectively, 0.94 and 0.84).

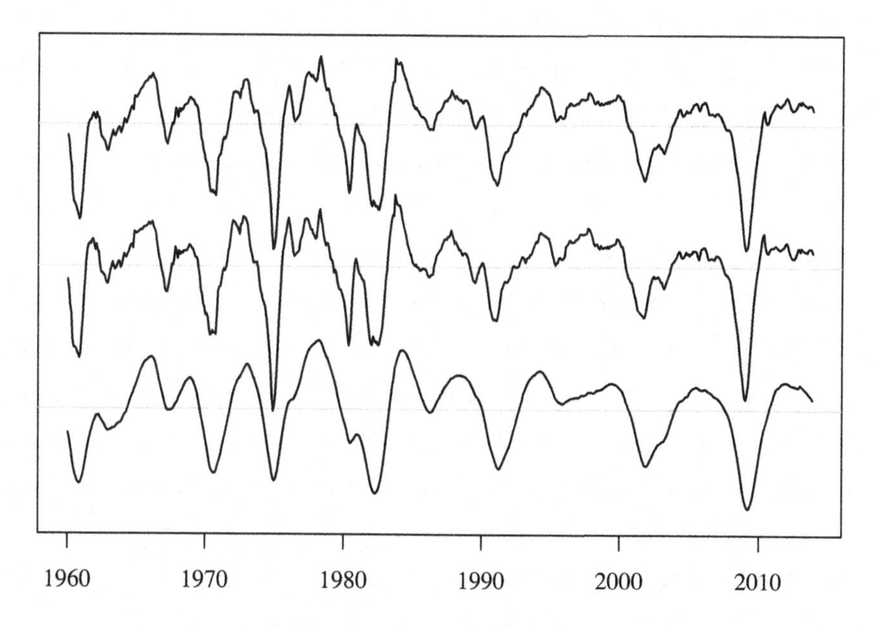

Figure 8.7 *Common cycles extracted from the four Conference Board time series using three different methods. From top to bottom: i) pure common cycle, ii) similar cycles weighted with the first eigenvector values, iii) similar cycle with covariance matrix as in equation (8.7).*

Interesting applications of the multivariate UCM approach to extract business cycles can be found in Pelagatti (2005), Azevedo et al. (2006), Koopman and Azevedo (2008) and Pelagatti and Negri (2010). This last article shows how the amplitude of the cycle can be monitored in any phase of the business cycle.

Chapter 9

Case Studies

In this chapter we show how to put together the notions given in the rest of the book by working out in detail three case studies.

9.1 Impact of the point system on motor vehicle injuries in Italy

In July 2003, Italy introduced the point system in its traffic laws. Every Italian driver receives 20 points and every notified moving traffic violation reduces the number of points by an amount proportional to the severity of the violated rule. The number of points can also slowly increase if the driver is not notified of any violation for a sufficiently long time span. The aim of this reform was to make Italian drivers more law-abiding with the consequent reduction in the number and severity of road accidents.

Was this a successful reform? The general opinion in Italy is that the point system had a positive impact at the very beginning, but when it was clear that the law enforcement was not very effective, the initial effect faded out.

Here, we want to assess the impact of the point system on the time series of the number of injuries due to motor vehicle accidents (Figure 9.1). In particular, we want to measure both the transitory and permanent effect of the reform on that time series. The UCM we are going to use is the basic structural model (BSM) enriched with an AR(1)-like transfer function and a level shift.

First, we just estimate a BSM to see if the level shift is evident from the auxiliary residuals plot. The standardised auxiliary residuals of the level component plotted in Figure 9.2 can be read as t-tests for the null of no level shift. There is a clear down-shift on July 2003.

Then, we estimate a model with a level shift starting on July 2003 and an AR(1)-like transfer function model, so that the total impact on the time series is given by

$$z_t = x_t + \lambda TS_t$$

with

$$x_t = \delta_1 x_{t-1} + \omega_0 D_t,$$

where TS_t is a variable that takes the value 0 for $t <$ July 2003 and 1 for $t \geq$ July 2003, while D_t is one on July 2003 and zero otherwise.

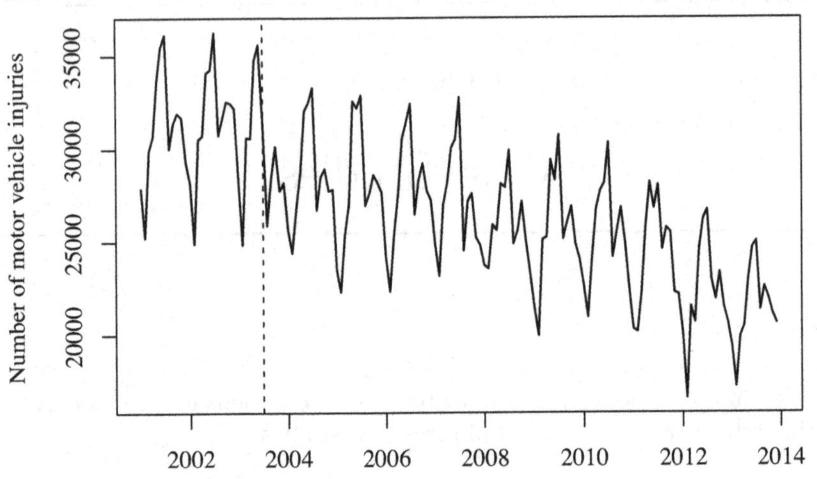

Figure 9.1 *Monthly number of motor vehicle injuries in Italy. The point system started in July 2003 (vertical dashed line).*

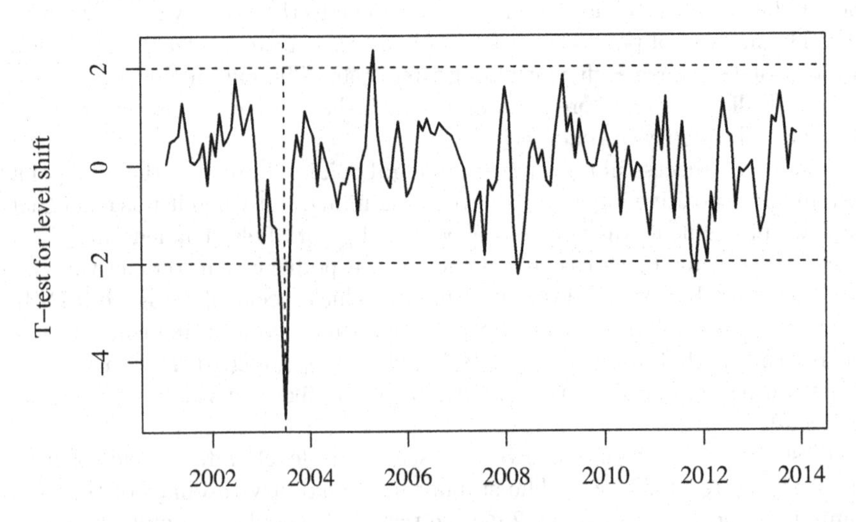

Figure 9.2 *t-test for a level shift computed using the standardised auxiliary residuals of the level component. The point system started in July 2003 (vertical dashed line).*

The estimate of the permanent level shift is $\hat{\lambda} = -2399$ with a standard error of 46. Thus, it seems that the new regulation reduced the monthly number of motor vehicle injuries by some 2400 units. The estimates of the transfer function coefficients are $\hat{\omega}_0 = -2826$ and $\hat{\delta} = 0.61$. This means that on July 2003, the introduction of the point system reduced the number of injuries by an additional 2826 units, but this transitory effect faded out quickly and was virtually zero after one year. The combined effects of the estimated transitory and permanent influence of the point system is depicted in Figure 9.3: the top-left graph depicts the smoothed trend summed to the intervention effects, and the top-right graph represents the total monthly impact of the intervention.

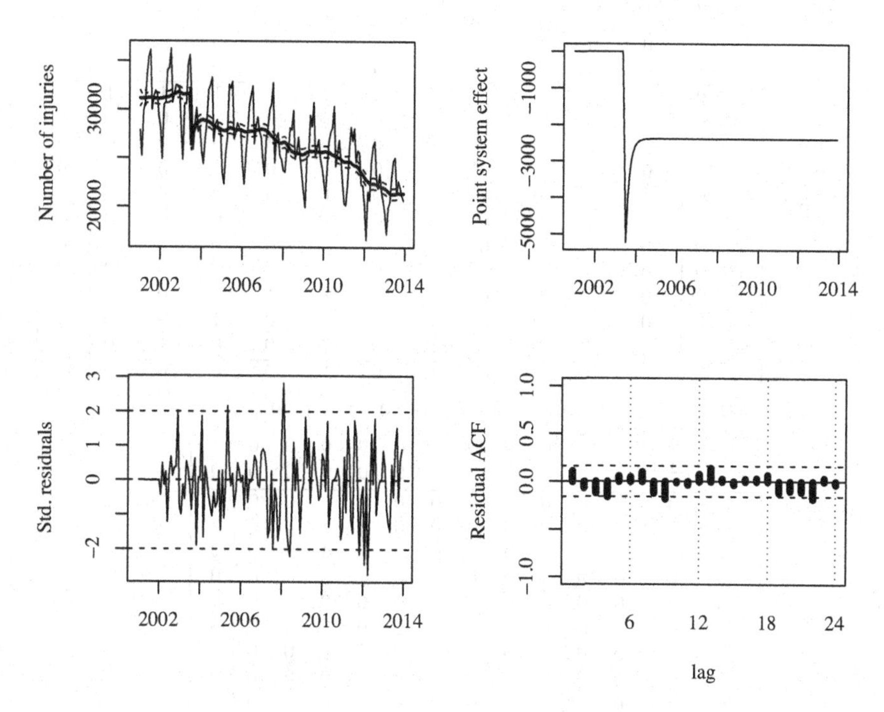

Figure 9.3 *Top-left: level of incident including intervention. Top-right: permanent and transitory effect of the point system on the number of monthly injuries. Bottom-left: standardised residuals. Bottom-right: sample autocorrelation function of standardised residuals.*

The two bottom graphs of Figure 9.3 represent the standardised residuals (i.e., estimated innovations) of the model and their sample autocorrelations with a 95% confidence interval for the zero. The first plot reveals a possible outlier on February 2008 (which we are going to ignore since it is of limited influence and not very interesting for our goal), however, the residuals seems to behave as white noise and so no other dynamic component seems necessary.

When the intervention (transfer function + level shift) is added to the model, the log-likelihood increases from -1213.04 to -1207.28 and, thus, the log-likelihood ratio statistic is $LR = 11.52$ with a p-value smaller than 0.01, which lets us reject the BSM in favour of the model with intervention[1].

[1]The log-likelihood ratio statistic is computed as $LR = 2(l_1 - l_0)$, with l_0 maximum log-likelihood of the constrained model and l_1 maximum log-likelihood of the unconstrained

The system matrices of the final model are provided below: notice that because of the structure (i.e., the future form) of the transition equation, the dummy variable for the event D_t takes the value 1 on June 2003 (not on July 2003) and 0 otherwise.

State equation matrices

$$
\mathbf{T} = \left[\begin{array}{cc|ccccccccccc|c}
1 & 1 & 0 & 0 & 0 & 0 & 0 & 0 & 0 & 0 & 0 & 0 & 0 & 0 \\
0 & 1 & 0 & 0 & 0 & 0 & 0 & 0 & 0 & 0 & 0 & 0 & 0 & 0 \\
\hline
0 & 0 & -1 & -1 & -1 & -1 & -1 & -1 & -1 & -1 & -1 & -1 & -1 & 0 \\
0 & 0 & 1 & 0 & 0 & 0 & 0 & 0 & 0 & 0 & 0 & 0 & 0 & 0 \\
0 & 0 & 0 & 1 & 0 & 0 & 0 & 0 & 0 & 0 & 0 & 0 & 0 & 0 \\
0 & 0 & 0 & 0 & 1 & 0 & 0 & 0 & 0 & 0 & 0 & 0 & 0 & 0 \\
0 & 0 & 0 & 0 & 0 & 1 & 0 & 0 & 0 & 0 & 0 & 0 & 0 & 0 \\
0 & 0 & 0 & 0 & 0 & 0 & 1 & 0 & 0 & 0 & 0 & 0 & 0 & 0 \\
0 & 0 & 0 & 0 & 0 & 0 & 0 & 1 & 0 & 0 & 0 & 0 & 0 & 0 \\
0 & 0 & 0 & 0 & 0 & 0 & 0 & 0 & 1 & 0 & 0 & 0 & 0 & 0 \\
0 & 0 & 0 & 0 & 0 & 0 & 0 & 0 & 0 & 1 & 0 & 0 & 0 & 0 \\
0 & 0 & 0 & 0 & 0 & 0 & 0 & 0 & 0 & 0 & 1 & 0 & 0 & 0 \\
0 & 0 & 0 & 0 & 0 & 0 & 0 & 0 & 0 & 0 & 0 & 1 & 0 & 0 \\
\hline
0 & 0 & 0 & 0 & 0 & 0 & 0 & 0 & 0 & 0 & 0 & 0 & 0 & \delta_1
\end{array}\right],
$$

$$
\mathbf{d}_t = \left[\begin{array}{cc|ccccccccccc|c} \lambda D_t & 0 & 0 & 0 & 0 & 0 & 0 & 0 & 0 & 0 & 0 & 0 & 0 & \omega_0 D_t \end{array}\right]^\top,
$$

$$
\mathbf{Q} = \left[\begin{array}{cc|ccccccccccc|c}
\sigma^2_\eta & 0 & 0 & 0 & 0 & 0 & 0 & 0 & 0 & 0 & 0 & 0 & 0 & 0 \\
0 & \sigma^2_\zeta & 0 & 0 & 0 & 0 & 0 & 0 & 0 & 0 & 0 & 0 & 0 & 0 \\
\hline
0 & 0 & \sigma^2_\omega & 0 & 0 & 0 & 0 & 0 & 0 & 0 & 0 & 0 & 0 & 0 \\
0 & 0 & 0 & 0 & 0 & 0 & 0 & 0 & 0 & 0 & 0 & 0 & 0 & 0 \\
0 & 0 & 0 & 0 & 0 & 0 & 0 & 0 & 0 & 0 & 0 & 0 & 0 & 0 \\
0 & 0 & 0 & 0 & 0 & 0 & 0 & 0 & 0 & 0 & 0 & 0 & 0 & 0 \\
0 & 0 & 0 & 0 & 0 & 0 & 0 & 0 & 0 & 0 & 0 & 0 & 0 & 0 \\
0 & 0 & 0 & 0 & 0 & 0 & 0 & 0 & 0 & 0 & 0 & 0 & 0 & 0 \\
0 & 0 & 0 & 0 & 0 & 0 & 0 & 0 & 0 & 0 & 0 & 0 & 0 & 0 \\
0 & 0 & 0 & 0 & 0 & 0 & 0 & 0 & 0 & 0 & 0 & 0 & 0 & 0 \\
0 & 0 & 0 & 0 & 0 & 0 & 0 & 0 & 0 & 0 & 0 & 0 & 0 & 0 \\
0 & 0 & 0 & 0 & 0 & 0 & 0 & 0 & 0 & 0 & 0 & 0 & 0 & 0 \\
0 & 0 & 0 & 0 & 0 & 0 & 0 & 0 & 0 & 0 & 0 & 0 & 0 & 0 \\
\hline
0 & 0 & 0 & 0 & 0 & 0 & 0 & 0 & 0 & 0 & 0 & 0 & 0 & 0
\end{array}\right],
$$

model. Under regularity conditions, the asymptotic distribution of this statistic is Chi-square with degrees of freedom equal to the number of constrained parameters (in our case $df = 3$ because we obtain the BSM from the model with intervention by setting $\lambda = \delta_1 = \omega_0 = 0$).

Initial state vector matrices

$$\boldsymbol{a}_{1|0} = \left[\begin{array}{cc|ccccccccccc|c} 0 & 0 & 0 & 0 & 0 & 0 & 0 & 0 & 0 & 0 & 0 & 0 & 0 & 0 \end{array}\right]^{\top},$$

$$\mathbf{P}_{1|0} = \left[\begin{array}{cc|ccccccccccc|c}
\infty & 0 & 0 & 0 & 0 & 0 & 0 & 0 & 0 & 0 & 0 & 0 & 0 & 0 \\
0 & \infty & 0 & 0 & 0 & 0 & 0 & 0 & 0 & 0 & 0 & 0 & 0 & 0 \\
\hline
0 & 0 & \infty & 0 & 0 & 0 & 0 & 0 & 0 & 0 & 0 & 0 & 0 & 0 \\
0 & 0 & 0 & \infty & 0 & 0 & 0 & 0 & 0 & 0 & 0 & 0 & 0 & 0 \\
0 & 0 & 0 & 0 & \infty & 0 & 0 & 0 & 0 & 0 & 0 & 0 & 0 & 0 \\
0 & 0 & 0 & 0 & 0 & \infty & 0 & 0 & 0 & 0 & 0 & 0 & 0 & 0 \\
0 & 0 & 0 & 0 & 0 & 0 & \infty & 0 & 0 & 0 & 0 & 0 & 0 & 0 \\
0 & 0 & 0 & 0 & 0 & 0 & 0 & \infty & 0 & 0 & 0 & 0 & 0 & 0 \\
0 & 0 & 0 & 0 & 0 & 0 & 0 & 0 & \infty & 0 & 0 & 0 & 0 & 0 \\
0 & 0 & 0 & 0 & 0 & 0 & 0 & 0 & 0 & \infty & 0 & 0 & 0 & 0 \\
0 & 0 & 0 & 0 & 0 & 0 & 0 & 0 & 0 & 0 & \infty & 0 & 0 & 0 \\
0 & 0 & 0 & 0 & 0 & 0 & 0 & 0 & 0 & 0 & 0 & \infty & 0 & 0 \\
0 & 0 & 0 & 0 & 0 & 0 & 0 & 0 & 0 & 0 & 0 & 0 & \infty & 0 \\
\hline
0 & 0 & 0 & 0 & 0 & 0 & 0 & 0 & 0 & 0 & 0 & 0 & 0 & 0
\end{array}\right].$$

Observation equation matrices

$$\mathbf{Z} = \left[\begin{array}{cc|ccccccccccc|c} 1 & 0 & 1 & 0 & 0 & 0 & 0 & 0 & 0 & 0 & 0 & 0 & 0 & 1 \end{array}\right],$$

$$\mathbf{c} = 0,$$

$$\mathbf{H} = 0.$$

ML estimates of the unknown parameters

$$\sigma_\eta^2 = 55\,317.93,\ \sigma_\zeta^2 = 18.30,\ \sigma_\omega^2 = 18\,069.42,\ \sigma_\epsilon^2 = 502\,564.31,$$

$$\lambda = -2\,398.64,\ \omega_0 = -2\,825.97,\ \delta_1 = 0.61.$$

9.2 An example of benchmarking: building monthly GDP data

The first few lines of Durbin and Quenneville (1997) contain the following description of the *benchmarking* problem in time series analysis.

> A common problem faced by official statistical agencies is the adjustment of monthly or quarterly time series which have been obtained from sample surveys to make them consistent with more accurate values obtained from other sources. These values can be aggregates or individual values at arbitrary points along the series. The sources can be censuses or more accurate sample surveys or administrative data or some combination of these. This adjustment process is called benchmarking and the more accurate values are called benchmarks. Typically, the benchmarks are either yearly totals or values observed at a particular time-point each year.

Here, we want to build a monthly GDP time series for U.S., using the official quarterly GDP as benchmark and the industrial production index as a proxy for the dynamics of the unobservable monthly GDP measurements. Indeed, the GDP is rarely used as a basis for coincident business cycle indicators because its quarterly frequency does not allow a timely and precise dating of the business cycle phases. For example, the well-known Conference Board Coincident Economic Index is an index whose growth rate is computed as the linear combination of the growth rates of four monthly time series that include the (seasonally adjusted) industrial production index[2].

We model the (seasonally adjusted) monthly U.S. GDP time series and the (seasonally adjusted) industrial production index (IPI) as integrated random walks plus similar cycles; but while the IPI is observable, the GDP is observable only through its sums over three consecutive months, measured every three months.

$$\begin{aligned} IPI_t &= \mu_{1,t} + \psi_{1,t} + \epsilon_{1,t}, \\ gdp_t &= \mu_{2,t} + \psi_{2,t} + \epsilon_{2,t}, \\ GDP_t &= gdp_t + gdp_{t-1} + gdp_{t-2}, \end{aligned}$$

where gdp_t is the unobservable monthly GDP time series, GDP_t is the observable quarterly GDP and $\{\mu_{1,t}, \mu_{2,t}\}$ is a bivariate random walk (cf. Section 7.1), $\{\psi_{1,t}, \psi_{2,t}\}$ are two similar cycles (cf. Section 7.2) and $\{\epsilon_{1,t}, \epsilon_{2,t}\}$ is a bivariate white noise process. Notice that GDP_t is observable only every three months, but this is not a problem as the Kalman filter deals with missing observations in a very natural way, and so does the smoother.

The graphs in Figure 9.4 depict the (smoothed) trend-cycle component of the monthly GDP together with the NBER business cycle dating. The turning points one could infer from the monthly GDP are very close, when not identical, to the official ones.

Figure 9.5 compares the trend-cycle component of the monthly GDP rescaled as index (Jan 1959 = 100) with the Conference Board Coincident Economic Index (CEI). The business cycle movements are very similar, and indeed the correlation of the monthly rate of growth of the two series is 0.8, however, the average rate of growth is higher in the GDP: 3.0% per annum for the GDP and 2.3 for the CEI. In particular, the GDP shows a geometric growth as opposed to the linear growth of the CEI. Furthermore, the CEI rate of growth is really hard to interpret, being obtained as average of the rates of growth of four very different time series.

An even higher coherence of the monthly GDP with the CEI could be achieved, if all four time series used in the construction of the CEI were set as benchmarks in the model above.

---

[2]The other three time series are: manufacturing and trade sales, personal income less transfer payments, employees on non-agricultural payrolls.

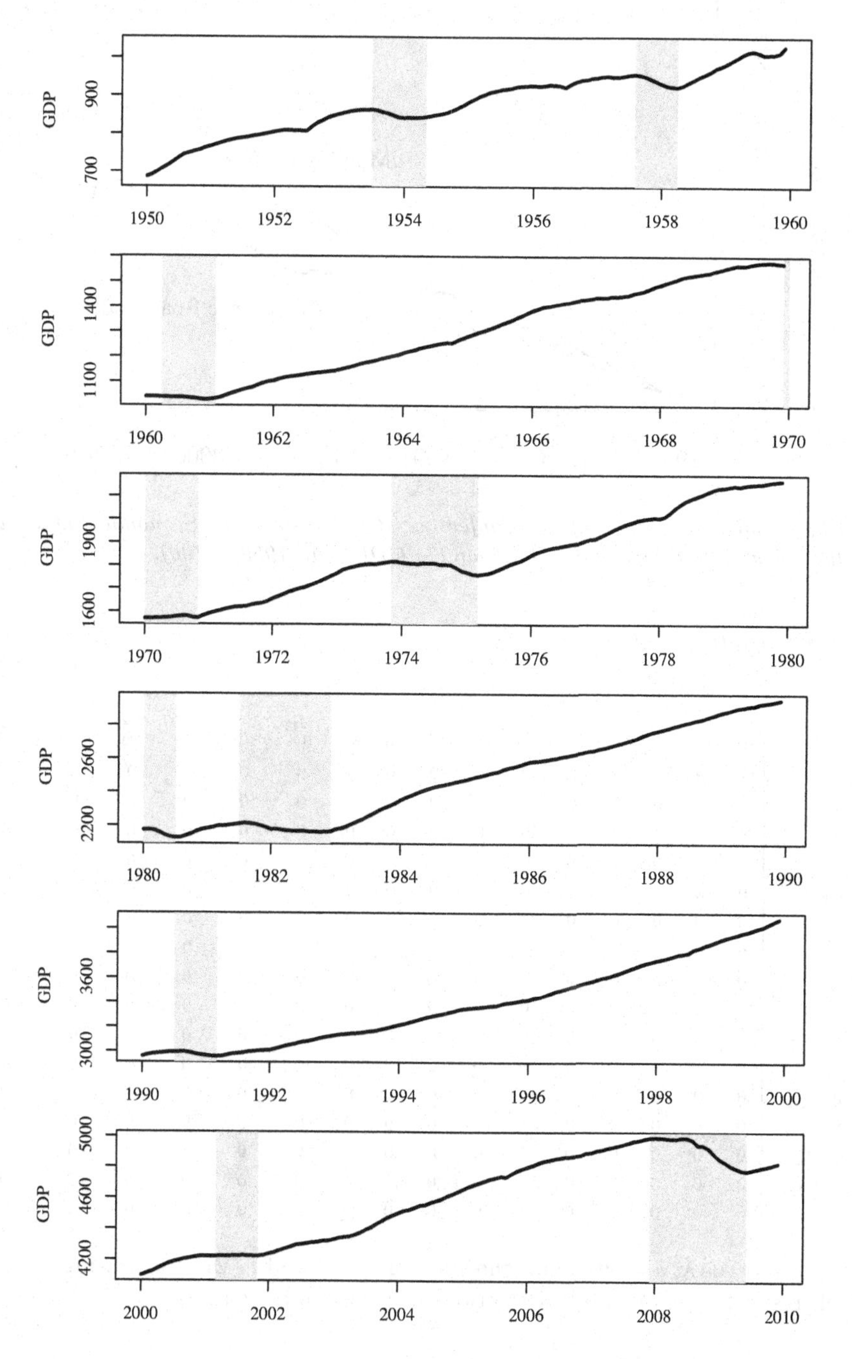

Figure 9.4 *Smoothed trend-cycle of the monthly GDP. The shaded areas represent economic contractions according to the NBER business cycle dating.*

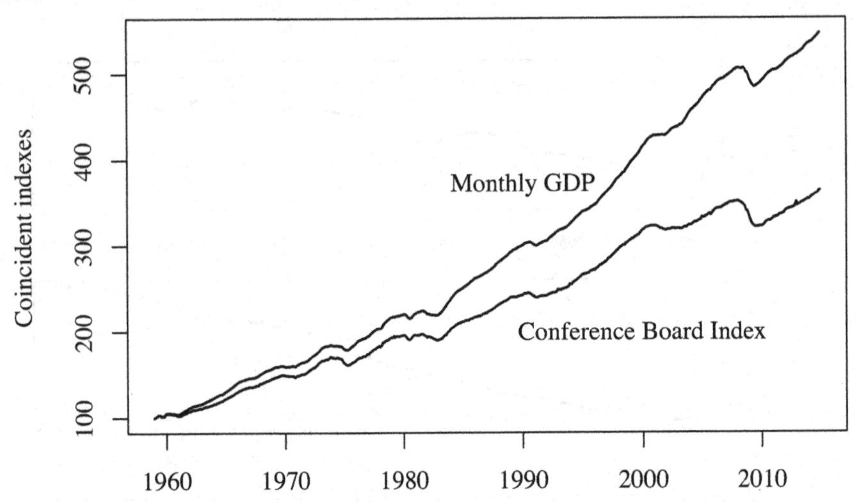

Figure 9.5 *Comparison of the Conference Board Coincident Economic Index with the trend-cycle component of the monthly GDP (Jan 1959 = 100).*

State equation matrices

$\mathbf{T} =$

	μ_1	μ_2	β_1	β_2	ψ_1	ψ_2	ψ_1^*	ψ_2^*	ε_1	ε_2	$\mu_2^{(-1)}$	$\mu_2^{(-2)}$	$\psi_2^{(-1)}$	$\psi_2^{(-2)}$	$\epsilon_2^{(-1)}$	$\epsilon_2^{(-2)}$
μ_1	1	0	1	0	0	0	0	0	0	0	0	0	0	0	0	0
μ_2	0	1	0	1	0	0	0	0	0	0	0	0	0	0	0	0
β_1	0	0	1	0	0	0	0	0	0	0	0	0	0	0	0	0
β_2	0	0	0	1	0	0	0	0	0	0	0	0	0	0	0	0
ψ_1	0	0	0	0	c	0	s	0	0	0	0	0	0	0	0	0
ψ_2	0	0	0	0	0	c	0	s	0	0	0	0	0	0	0	0
ψ_1^*	0	0	0	0	$-s$	0	c	0	0	0	0	0	0	0	0	0
ψ_2^*	0	0	0	0	0	$-s$	0	c	0	0	0	0	0	0	0	0
ε_1	0	0	0	0	0	0	0	0	0	0	0	0	0	0	0	0
ε_2	0	0	0	0	0	0	0	0	0	0	0	0	0	0	0	0
$\mu_{2(-1)}$	0	1	0	0	0	0	0	0	0	0	0	0	0	0	0	0
$\mu_{2(-2)}$	0	0	0	0	0	0	0	0	0	0	1	0	0	0	0	0
$\psi_{2(-1)}$	0	0	0	0	0	1	0	0	0	0	0	0	0	0	0	0
$\psi_{2(-2)}$	0	0	0	0	0	0	0	0	0	0	0	0	1	0	0	0
$\epsilon_{2(-1)}$	0	0	0	0	0	0	0	0	0	1	0	0	0	0	0	0
$\epsilon_{2(-2)}$	0	0	0	0	0	0	0	0	0	0	0	0	0	0	1	0

with $c = \cos\lambda$, $s = \sin\lambda$ and the symbols (-1) and (-2) representing the lag of, respectively, one and two periods of that component.

$$\mathbf{Q} =$$

$$
\begin{array}{c|cccccccccccccccc}
 & \mu_1 & \mu_2 & \beta_1 & \beta_2 & \psi_1 & \psi_2 & \psi_1^* & \psi_2^* & \epsilon_1 & \epsilon_2 & \mu_2^{(-1)} & \mu_2^{(-2)} & \psi_2^{(-1)} & \psi_2^{(-2)} & \epsilon_2^{(-1)} & \epsilon_2^{(-2)} \\
\hline
\mu_1 & 0 & 0 & 0 & 0 & 0 & 0 & 0 & 0 & 0 & 0 & 0 & 0 & 0 & 0 & 0 & 0 \\
\mu_2 & 0 & 0 & 0 & 0 & 0 & 0 & 0 & 0 & 0 & 0 & 0 & 0 & 0 & 0 & 0 & 0 \\
\beta_1 & 0 & 0 & \sigma^\zeta_{11} & \sigma^\zeta_{12} & 0 & 0 & 0 & 0 & 0 & 0 & 0 & 0 & 0 & 0 & 0 & 0 \\
\beta_2 & 0 & 0 & \sigma^\zeta_{21} & \sigma^\zeta_{22} & 0 & 0 & 0 & 0 & 0 & 0 & 0 & 0 & 0 & 0 & 0 & 0 \\
\psi_1 & 0 & 0 & 0 & 0 & \sigma^\kappa_{11} & \sigma^\kappa_{12} & 0 & 0 & 0 & 0 & 0 & 0 & 0 & 0 & 0 & 0 \\
\psi_2 & 0 & 0 & 0 & 0 & \sigma^\kappa_{21} & \sigma^\kappa_{22} & 0 & 0 & 0 & 0 & 0 & 0 & 0 & 0 & 0 & 0 \\
\psi_1^* & 0 & 0 & 0 & 0 & 0 & 0 & \sigma^\kappa_{11} & \sigma^\kappa_{12} & 0 & 0 & 0 & 0 & 0 & 0 & 0 & 0 \\
\psi_2^* & 0 & 0 & 0 & 0 & 0 & 0 & \sigma^\kappa_{21} & \sigma^\kappa_{22} & 0 & 0 & 0 & 0 & 0 & 0 & 0 & 0 \\
\epsilon_1 & 0 & 0 & 0 & 0 & 0 & 0 & 0 & 0 & \sigma^\epsilon_{11} & \sigma^\epsilon_{12} & 0 & 0 & 0 & 0 & 0 & 0 \\
\epsilon_2 & 0 & 0 & 0 & 0 & 0 & 0 & 0 & 0 & \sigma^\epsilon_{21} & \sigma^\epsilon_{22} & 0 & 0 & 0 & 0 & 0 & 0 \\
\mu_{2(-1)} & 0 & 0 & 0 & 0 & 0 & 0 & 0 & 0 & 0 & 0 & 0 & 0 & 0 & 0 & 0 & 0 \\
\mu_{2(-2)} & 0 & 0 & 0 & 0 & 0 & 0 & 0 & 0 & 0 & 0 & 0 & 0 & 0 & 0 & 0 & 0 \\
\psi_{2(-1)} & 0 & 0 & 0 & 0 & 0 & 0 & 0 & 0 & 0 & 0 & 0 & 0 & 0 & 0 & 0 & 0 \\
\psi_{2(-2)} & 0 & 0 & 0 & 0 & 0 & 0 & 0 & 0 & 0 & 0 & 0 & 0 & 0 & 0 & 0 & 0 \\
\epsilon_{2(-1)} & 0 & 0 & 0 & 0 & 0 & 0 & 0 & 0 & 0 & 0 & 0 & 0 & 0 & 0 & 0 & 0 \\
\epsilon_{2(-2)} & 0 & 0 & 0 & 0 & 0 & 0 & 0 & 0 & 0 & 0 & 0 & 0 & 0 & 0 & 0 & 0
\end{array}
$$

Initial state vector matrices

$$
\mathbf{a}_{1|0}^\top = \begin{bmatrix} \mu_1 & \mu_2 & \beta_1 & \beta_2 & \psi_1 & \psi_2 & \psi_1^* & \psi_2^* & \epsilon_1 & \epsilon_2 & \mu_2^{(-1)} & \mu_2^{(-2)} & \psi_2^{(-1)} & \psi_2^{(-2)} & \epsilon_2^{(-1)} & \epsilon_2^{(-2)} \\ 0 & 0 & 0 & 0 & 0 & 0 & 0 & 0 & 0 & 0 & 0 & 0 & 0 & 0 & 0 & 0 \end{bmatrix}
$$

$$\mathbf{P}_{1|0} =$$

$$
\begin{array}{c|cccccccccccccccc}
 & \mu_1 & \mu_2 & \beta_1 & \beta_2 & \psi_1 & \psi_2 & \psi_1^* & \psi_2^* & \epsilon_1 & \epsilon_2 & \mu_2^{(-1)} & \mu_2^{(-2)} & \psi_2^{(-1)} & \psi_2^{(-2)} & \epsilon_2^{(-1)} & \epsilon_2^{(-2)} \\
\hline
\mu_1 & \infty & 0 & 0 & 0 & 0 & 0 & 0 & 0 & 0 & 0 & 0 & 0 & 0 & 0 & 0 & 0 \\
\mu_2 & 0 & \infty & 0 & 0 & 0 & 0 & 0 & 0 & 0 & 0 & 0 & 0 & 0 & 0 & 0 & 0 \\
\beta_1 & 0 & 0 & \infty & 0 & 0 & 0 & 0 & 0 & 0 & 0 & 0 & 0 & 0 & 0 & 0 & 0 \\
\beta_2 & 0 & 0 & 0 & \infty & 0 & 0 & 0 & 0 & 0 & 0 & 0 & 0 & 0 & 0 & 0 & 0 \\
\psi_1 & 0 & 0 & 0 & 0 & \sigma^\psi_{11} & \sigma^\psi_{12} & 0 & 0 & 0 & 0 & 0 & 0 & 0 & 0 & 0 & 0 \\
\psi_2 & 0 & 0 & 0 & 0 & \sigma^\psi_{21} & \sigma^\psi_{22} & 0 & 0 & 0 & 0 & 0 & 0 & \gamma_1 & \gamma_2 & 0 & 0 \\
\psi_1^* & 0 & 0 & 0 & 0 & 0 & 0 & \sigma^\psi_{11} & \sigma^\psi_{12} & 0 & 0 & 0 & 0 & 0 & 0 & 0 & 0 \\
\psi_2^* & 0 & 0 & 0 & 0 & 0 & 0 & \sigma^\psi_{21} & \sigma^\psi_{22} & 0 & 0 & 0 & 0 & 0 & 0 & 0 & 0 \\
\epsilon_1 & 0 & 0 & 0 & 0 & 0 & 0 & 0 & 0 & \sigma^\epsilon_{11} & \sigma^\epsilon_{12} & 0 & 0 & 0 & 0 & 0 & 0 \\
\epsilon_2 & 0 & 0 & 0 & 0 & 0 & 0 & 0 & 0 & \sigma^\epsilon_{21} & \sigma^\epsilon_{22} & 0 & 0 & 0 & 0 & 0 & 0 \\
\mu_{2(-1)} & 0 & 0 & 0 & 0 & 0 & 0 & 0 & 0 & 0 & 0 & \infty & 0 & 0 & 0 & 0 & 0 \\
\mu_{2(-2)} & 0 & 0 & 0 & 0 & 0 & 0 & 0 & 0 & 0 & 0 & 0 & \infty & 0 & 0 & 0 & 0 \\
\psi_{2(-1)} & 0 & 0 & 0 & 0 & 0 & \gamma_1 & 0 & 0 & 0 & 0 & 0 & 0 & \sigma^\psi_{22} & \gamma_1 & 0 & 0 \\
\psi_{2(-2)} & 0 & 0 & 0 & 0 & 0 & \gamma_2 & 0 & 0 & 0 & 0 & 0 & 0 & \gamma_2 & \sigma^\psi_{22} & 0 & 0 \\
\epsilon_{2(-1)} & 0 & 0 & 0 & 0 & 0 & 0 & 0 & 0 & 0 & 0 & 0 & 0 & 0 & 0 & \sigma^\epsilon_{22} & 0 \\
\epsilon_{2(-2)} & 0 & 0 & 0 & 0 & 0 & 0 & 0 & 0 & 0 & 0 & 0 & 0 & 0 & 0 & 0 & \sigma^\epsilon_{22}
\end{array}
$$

where we name $\boldsymbol{\Sigma}_\psi$ the matrix with generic element $\{\sigma^\psi_{ij}\}$, and similarly the other disturbance covariance matrices. Notice that from Section 7.2 we know that the marginal covariance matrix of the similar cycle is given by $\boldsymbol{\Sigma}_\psi =$

$\mathbf{\Sigma}_\kappa(1-\rho)^{-1}$. Furthermore, γ_h represents the autocovariance function of $\{\psi_t\}$, which is given by $\gamma_h = \rho^h \cos(h\lambda)$, as seen in Section 3.3.

For the numerical maximisation of the log-likelihood we parametrised the cycles' variability using $\mathbf{\Sigma}_\psi$ and obtained $\mathbf{\Sigma}_\kappa$ as $\mathbf{\Sigma}_\psi(1-\rho)$. This way, the stochastic cycle is either stationary $(0 \le \rho < 1)$ or deterministic $(\rho = 1)$.

In order to guarantee positive semi-definite covariance matrices, we maximised the log-likelihood with respect to lower triangular matrices, say $\mathbf{C}$, such that

$$\mathbf{C}\mathbf{C}^\top = \mathbf{\Sigma},$$

thus, each matrix is fully determined by three parameters that are free to vary over $\mathbb{R}^3$.

Observation equation matrices

$$\mathbf{Z} =$$

$$\begin{bmatrix} \mu_1 & \mu_2 & \beta_1 & \beta_2 & \psi_1 & \psi_2 & \psi_1^* & \psi_2^* & \varepsilon_1 & \varepsilon_2 & \mu_2^{(-1)} & \mu_2^{(-2)} & \psi_2^{(-1)} & \psi_2^{(-2)} & \epsilon_2^{(-1)} & \epsilon_2^{(-2)} \\ 1 & 0 & 0 & 0 & 1 & 0 & 0 & 0 & 1 & 0 & 0 & 0 & 0 & 0 & 0 & 0 \\ 0 & 1 & 0 & 0 & 0 & 1 & 0 & 0 & 0 & 1 & 1 & 1 & 1 & 1 & 1 & 1 \end{bmatrix},$$

$$\mathbf{H} = 0.$$

ML estimates of the unknown parameters The cycle parameters are $\hat{\rho} = 0.989$ and $\lambda = 0.021$, which make the cycle behave almost as a persistent AR(1). The covariance (the upper triangular part contains correlations) matrices are

$$\mathbf{\Sigma}_\zeta = \begin{bmatrix} 0.012 & 0.998 \\ 0.233 & 4.561 \end{bmatrix}, \quad \mathbf{\Sigma}_\kappa = \begin{bmatrix} 0.102 & 0.526 \\ 1.167 & 48.070 \end{bmatrix}, \quad \mathbf{\Sigma}_\epsilon = \begin{bmatrix} 0.009 & -0.652 \\ -0.823 & 176.780 \end{bmatrix}.$$

Given the geometric growth of the two time series and the volatility that increases with the level of the two series, modelling the data in logarithm would have resulted in more interpretable parameter estimates for the cycle, but it would not be possible to impose in the state space form that the sum of three consecutive monthly GDP values equal the actual quarterly GDP. Indeed, the model we are building is not meant for forecasting, but for benchmarking, and the results seem very reasonable when compared with the Conference Board index and the NBER dating.

As the correlograms in Figure 9.6 show, there seems to be no significant residual autocorrelation and, thus, we can consider the model satisfactory. Notice that the residuals are monthly for the IPI series and quarterly for the GD, as we observe only one value every three months. There seems to be some autocorrelation of the IPI residuals on lags that are multiples of 12. This signals a not-so-perfect seasonal adjustment of the original IPI time series.

9.3 Hourly electricity demand

Almost all players in the power markets have to deal with predictions of the electricity demand. In most markets the data on electricity demand are

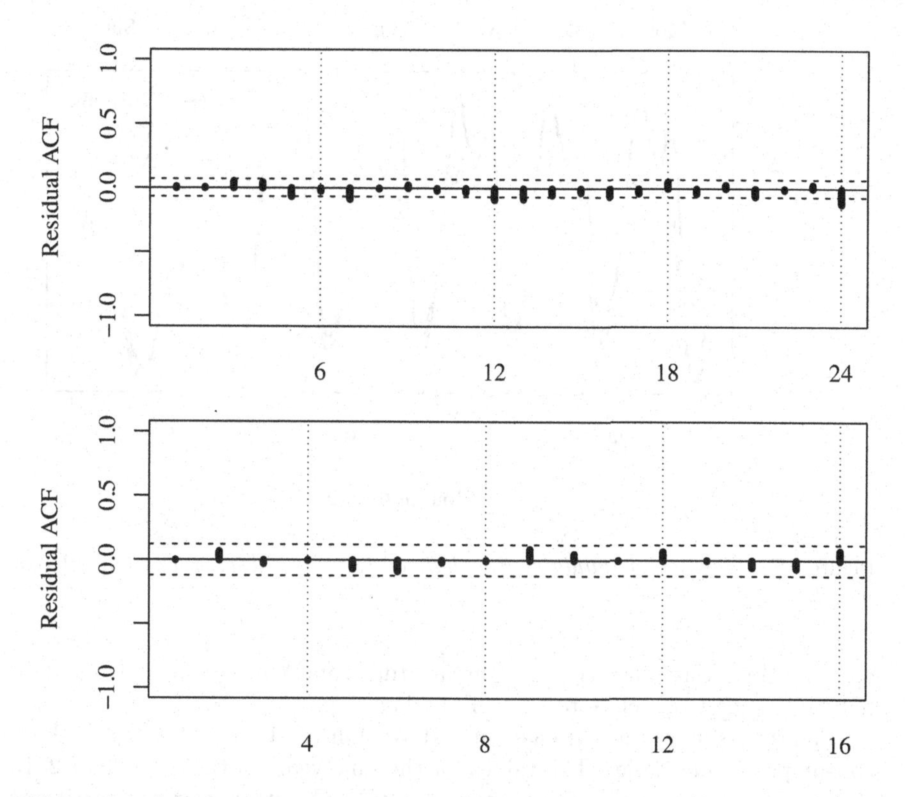

Figure 9.6 *Sample autocorrelations of the model's standardised residuals for the Industrial Production Index (top) and GDP (bottom).*

available on an hourly or semi-hourly basis. What makes these time series difficult to model is the presence of multiple seasonal periodicities that interact with each other. Indeed, there is a within-day seasonal pattern associated with a higher demand during the day and a lower consumption in the night, which is, in its turn, influenced by the day of the week, as on Saturday and Sunday the demand is lower and the daily pattern shape is different. Both these seasonal patterns (i.e., within-day and within-week) evolve periodically according to the period of the year in which they are measured. In fact, the quantity and duration of the solar radiation, as well as the temperature change over the year, and this implies a different use of electric lights, heaters and air conditioners, which heavily influence the demand for electricity.

Figure 9.7 depicts weekly patterns of the hourly demand of electricity for Italy taken in three different times of the year. The November pattern tends to be above the other two and clearly shows peaks in the morning and, particularly, in the evening when the sunlight is absent. The August pattern is below the other two since in Italy most industrial activities and many offices close during this month. For the same reason (lower economic activity), the

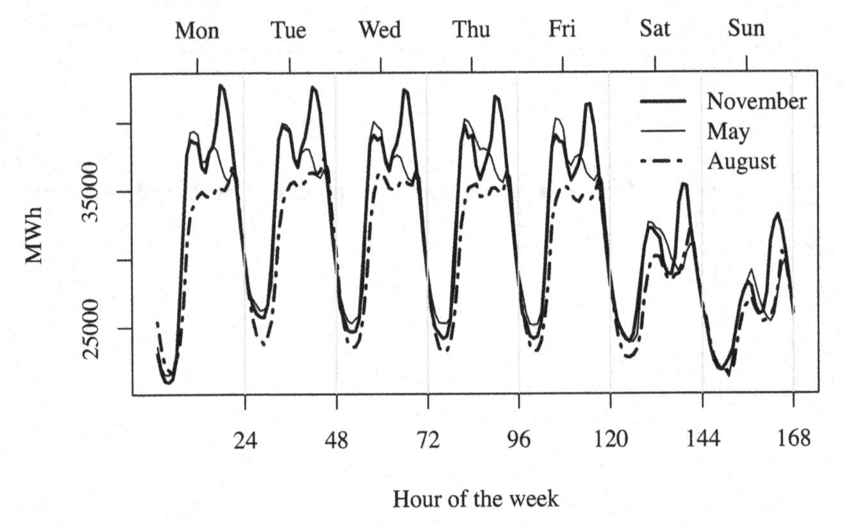

Figure 9.7 *One week of Italian hourly electricity demand observations for different months.*

consumption is significantly lower on Saturdays and Sundays. The demand on holidays tends to be close to that of Sundays.

Figure 9.8 shows a low-pass filtered version of the trend-adjusted daily consumption time series. The values for the nine years in the span 2005–2013 are superimposed to visually capture the typical within-year periodicity of the time series. The thick line represents the mean of all the curves in the plot. There is a clear decrease in the demand during vacation periods such as Christmas/New Year/Three Kings holidays, Easter and summer break (which in Italy is mostly August). The peak of consumption is generally reached in the second half of July when air conditioning is working at its maximum power.

Figure 9.9 compares the mean yearly seasonal pattern of Figure 9.8 with a least-squares approximation based on 16 sinusoids at frequencies $\{2\pi/365.25, 4\pi/365.25, \ldots, 32\pi/365.25\}$ (recall that the average duration of one year is 365 days and 6 hours). The approximation is virtually indistinguishable from the real pattern. This suggests that the use of 16 time-varying sinusoids at those frequencies should do excellent work in modelling the within-year seasonal component.

There is a vast literature on electricity demand (also load or consumption) forecasting since this problem is present in every country both on the micro and macro level. Models that follow the UCM approach can be found in Harvey and Koopman (1993), Dordonnat et al. (2008), Dordonnat et al. (2012). Some authors fit a single model to the hourly time series, while others fit a different model to each hour of the day. Here, we will follow this second approach, which avoids the complications of building a component for

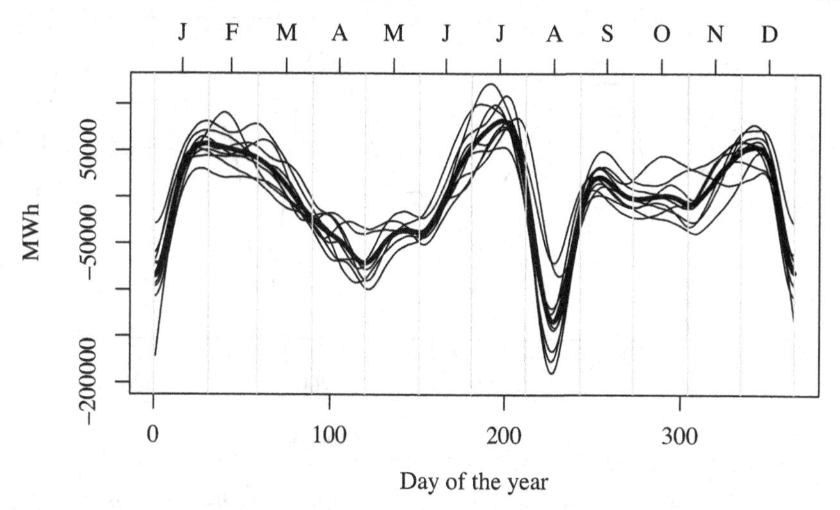

Figure 9.8 *Superposition of nine years of low-pass filtered daily demand of electricity in Italy after subtraction of a trend component. The dark curve is the mean of all the yearly curves.*

the time-varying within-day seasonal pattern. Thus, for each hour of the day we build a daily UCM with a random walk level component, a within-year seasonal component composed by 16 (nonstationary) stochastic cycles at frequencies $\lambda_{y,j} = 2\pi j/365.25$, $j = 1, 2, \ldots, 16$, and a within-weak seasonal component composed of three stochastic sinusoids at frequencies $\lambda_{w,j} = 2\pi j/7$, $j = 1, 2, 3$. Some dummy variables are then used to adjust the effect of holidays, long weekends, ...

So, for each hour $h = 1, 2, \ldots, 24$ we fit the daily UCM

$$Y_{h,t} = \mu_t + \gamma_t^{(w)} + \gamma_t^{(y)} + \sum_i \delta_i X_{i,t} + \epsilon_t,$$

where μ_t is a random walk level component, $\gamma_t^{(w)}$ is the within-week seasonal component $\gamma_t^{(w)} = \gamma_{1,t}^{(w)} + \gamma_{2,t}^{(w)} + \gamma_{3,t}^{(w)}$ with the γs on the rhs of the equal sign taken from the first element of the two-dimensional vectors

$$\boldsymbol{\gamma}_{j,t+1}^{(w)} = \mathbf{R}(2\pi j/7)\,\boldsymbol{\gamma}_{j,t}^{(w)} + \boldsymbol{\kappa}_{j,t}^{(w)}, \quad j = 1, 2, 3,$$

where $\mathbf{R}(\cdot)$ is the rotation matrix (3.6). For the within-year seasonal component we use a slightly different approach: we build the 16 sinusoids (16 sine and 16 cosine functions),

$$c_{j,t} = \cos\left(\frac{2\pi jt}{365.25}\right), \quad s_{j,t} = \sin\left(\frac{2\pi jt}{365.25}\right), \quad j = 1, \ldots, 16; t = 1, \ldots, n$$

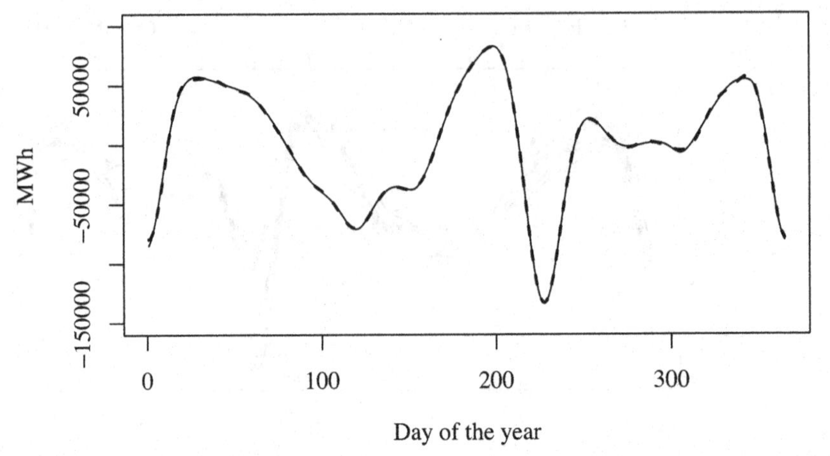

Figure 9.9 *Mean yearly seasonal pattern (solid line) with its approximation (dashed line) based on 16 sinusoids (the base frequency and its first 15 harmonics).*

and adopt them as 32 additional regressors with random walk coefficients:

$$\gamma_t^{(y)} = \sum_{j=1}^{16} (\beta_{j,t} c_{j,t} + \beta_{j+1,t} s_{j,t})$$

with

$$\beta_{j,t+1} = \beta_{j,t} + \zeta_{j,t}, \quad j = 1, 2, \ldots, 32,$$

and $\{\zeta_{j,t}\}$ uncorrelated white noise sequences with common variance σ_ζ^2. This alternative representation of a trigonometric seasonal component is equivalent to the one based on stochastic sinusoids (Proietti, 2000)[3].

As for the regressors in the model, we used the dummies listed in Table 9.1. The two observations on March 17th and 18th were set to missing, because they are related to the celebrations of the 150th anniversary of the Italian unification.

We estimated the 24 daily UCM on the actual electricity load observed in Italy from 1st January 2009 to 31st December 2014 (2191 observations per model). The largest absolute values of the standardised residuals (i.e., the hardest values to predict) are found during vacation times: especially in the central days of August, and around Christmas and Easter. Probably a more thorough analysis of the patterns during these periods could improve the model performance. The in-sample one-step-ahead prediction MAPE is around 2.2%, with a peak of 2.6% at the 18th hour and a trough of 1.8% at the 23th hour. In all 24 models' innovations there is some residual autocorrelation, which is probably due to the non-constancy of the model variances during

[3]The reason we are using this alternative representation in this example is to show the reader how to force software packages that do not allow multiple seasonal components.

Table 9.1 *Dummy variables for the electricity load model.*

Variable	Description
Dec24	December 24th
Dec25	December 25th
Dec26	December 26th
Jan1	January 1st
Jan6	January 6th
Aug15	August 15th
EasterSat	Last Saturday before Easter
EasterSun	Easter Sunday
EasterMon	Easter Monday
EasterTue	First Tuesday after Easter
Holidays	Other Italian holidays (Apr 25, May 1, Jun 2, Nov 1, Dec 8)
HolySat	1 if one of the above holidays is on Saturday
HolySun	1 if one of the above holidays is on Sunday
BridgeDay	Bridge day: Monday before a holiday or Friday after a holiday
EndYear	School Christmas vacations

different times of the year rather than to the lack of short-range memory components such as ARMA. We reduced, although we did not eliminate, the amount of residual autocorrelation by letting the three sinusoidal components of the weekly seasonal pattern have different variances. Indeed, the variance gets lower as the frequency of the sinusoid increases.

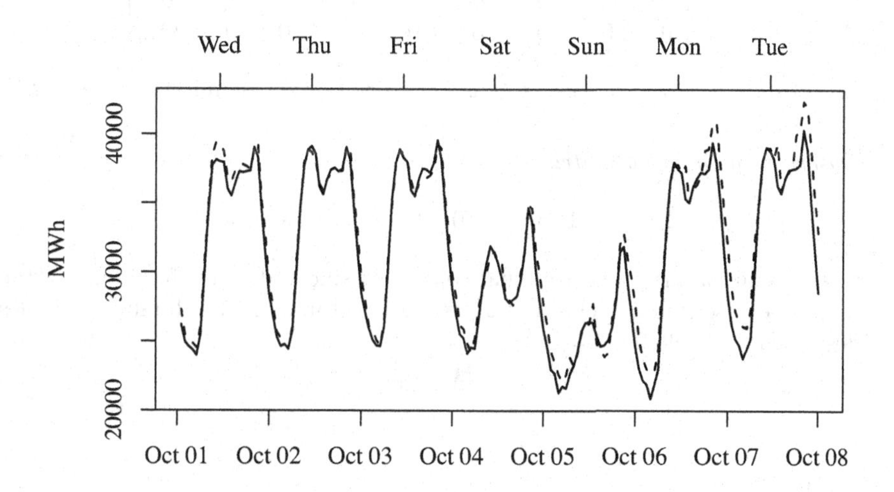

Figure 9.10 *Actual values and out-of-sample forecast of the load in the week from the 1st October 2014 to the 7th October 2014.*

Figure 9.10 depicts one week of hourly out-of-sample forecasts together with the actual values for the first seven days of October 2014. The forecasts seem accurate especially up to four days ahead. Notice that most firms in this field need to produce forecasts in the range 1 to 7 days ahead either at macro level (country and regions) or for their major buyers like large industrial plants.

State equation matrices

$$\mathbf{T} = \begin{bmatrix} 1 & 0 & 0 & 0 & 0 & 0 & 0 & \mathbf{0}^\top \\ 0 & c_1 & s_1 & 0 & 0 & 0 & 0 & \mathbf{0}^\top \\ 0 & -s_1 & c_1 & 0 & 0 & 0 & 0 & \mathbf{0}^\top \\ 0 & 0 & 0 & c_2 & s_2 & 0 & 0 & \mathbf{0}^\top \\ 0 & 0 & 0 & -s_2 & c_2 & 0 & 0 & \mathbf{0}^\top \\ 0 & 0 & 0 & 0 & 0 & c_3 & s_3 & \mathbf{0}^\top \\ 0 & 0 & 0 & 0 & 0 & -s_3 & c_3 & \mathbf{0}^\top \\ \mathbf{0} & \mathbf{0} & \mathbf{0} & \mathbf{0} & \mathbf{0} & \mathbf{0} & \mathbf{0} & \mathbf{I}_{47} \end{bmatrix},$$

with $c_j = \cos(2\pi j/7)$ and $s_j = \sin(2\pi j/7)$,

$$\mathbf{Q} = \begin{bmatrix} \sigma_\eta^2 & 0 & 0 & 0 & 0 & 0 & 0 & \mathbf{0}^\top & \mathbf{0}^\top \\ 0 & \sigma_{\omega_1}^2 & 0 & 0 & 0 & 0 & 0 & \mathbf{0}^\top & \mathbf{0}^\top \\ 0 & 0 & \sigma_{\omega_1}^2 & 0 & 0 & 0 & 0 & \mathbf{0}^\top & \mathbf{0}^\top \\ 0 & 0 & 0 & \sigma_{\omega_2}^2 & 0 & 0 & 0 & \mathbf{0}^\top & \mathbf{0}^\top \\ 0 & 0 & 0 & 0 & \sigma_{\omega_2}^2 & 0 & 0 & \mathbf{0}^\top & \mathbf{0}^\top \\ 0 & 0 & 0 & 0 & 0 & \sigma_{\omega_3}^2 & 0 & \mathbf{0}^\top & \mathbf{0}^\top \\ 0 & 0 & 0 & 0 & 0 & 0 & \sigma_{\omega_3}^2 & \mathbf{0}^\top & \mathbf{0}^\top \\ \mathbf{0} & \mathbf{0} & \mathbf{0} & \mathbf{0} & \mathbf{0} & \mathbf{0} & \mathbf{0} & \mathbf{I}_{32} & \mathbf{0}^\top \\ \mathbf{0} & \mathbf{0} & \mathbf{0} & \mathbf{0} & \mathbf{0} & \mathbf{0} & \mathbf{0} & \mathbf{0} & \mathbf{O}_{15} \end{bmatrix}.$$

Initial state vector matrices $\boldsymbol{a}_{1|0}$ is a vector of 52 zeros and $\mathbf{P}_{1|0} = \tau\mathbf{I}_{52}$ with $\tau \to \infty$.

Observation equation matrices

$$\mathbf{Z} = \begin{bmatrix} 1 & 1 & 0 & 1 & 0 & 1 & 0 & \boldsymbol{c}_t^\top & \boldsymbol{s}_t^\top & \boldsymbol{x}_t^\top \end{bmatrix},$$

where $\boldsymbol{c}_t$ and $\boldsymbol{s}_t$ are 16-vectors that collect the sine and cosine functions for the within-year seasonal cycles and $\boldsymbol{x}_t$ is a 15-vector with the dummy variables described in Table 9.1.

$$\mathbf{H} = \sigma_\epsilon^2.$$

Chapter 10

Software for UCM

One of the obstacles to the spread of UCM among practitioners was the unavailability of these models in mainstream software systems. Now things have changed, and for estimating and forecasting many UCM there is no more need for writing complicated code.

Of course, even nowadays, if the time series analyst needs complete control and flexibility over the model, coding is still the way to go. However, numerical programming languages now offer a range of functions to make the implementation of UCM quicker and simpler.

In this chapter, we briefly review some software systems that offer ready-to-use functions for unobserved component modelling, and some numerical programming languages that have packages that facilitate the implementation and estimation of models in state space form.

This chapter considers many packages but not all the software that can deal with state space models, although I am reasonably sure that all packages having ready-to-use procedures for UCM are covered. For example, software packages such as MATLAB® (by MathWorks), Gauss (by Aptech), RATS (by Estima) and Python offer functions to deal with models in state space form.

For more information on the software packages treated here or on packages not covered in this chapter, the reader should refer to the special issue of the *Journal of Statistical Software* (vol. 41), titled "Statistical Software for State Space Methods", which is freely available from the journal's site[1].

10.1 Software with ready-to-use UCM procedures

For a long time there has been a lack of software packages for the quick and simple estimation of UCM. The first package to appear was STAMP, developed by Prof. Harvey, the author of the 1989 path-breaking book on UCM (or *Structural Time Series Models*). The first version of STAMP was released in 1989, and to know more about its capabilities you can read the

[1]The journal web address is `www.jstatsoft.org`, the special issue web address is `www.jstatsoft.org/v41` .

review by Diebold (1989). However, this package did not spread much outside the academic community, despite its quality.

Since version 9.0, the SAS Institute introduced in the ETS module of its widespread software system the UCM procedure[2] which allows the simple estimation of a wide range of univariate UCM. Stata introduced a similar procedure starting from version 12. The popularity of these two software systems is certainly helping the spread of UCM among practitioners.

At the time of writing this book, the only package with ready-to-use procedures for multivariate UCM is STAMP, whose flexibility and speed have dramatically increased over its 25+ years.

Table 10.1 summarises some of the features of the UCM software packages.

Table 10.1 *UCM software features.*

Feature	SAS/ETS 9.4	STAMP 8.2	Stata 13
Diffuse Kalman filter	yes	yes	yes
Auxiliary residuals	no	yes	no
Autom. outlier treatment	yes	yes	no
Time-varying regression*	RW	RW IRW AR1	no
Non-linear regression	spline	no	no
Multivariate UCM	no	yes	no
GUI	no	yes	yes
Runs on†	W L	W L M	W L M

*RW = random walk, IRW = integrated random walk, AR1 = AR(1).
†W = MS Windows, L = Linux, M = Mac iOS.

In addition to the packages treated in this section, there is the function `StructTS()` in the R language, which allows the estimation and forecasting of the basic structural model (BSM) only (or subsets of it), without the possibility of including regressors.

10.1.1 SAS/ETS

At this writing, the newest version of SAS is 9.4, and since version 8.9 the ETS module offers the PROC UCM tailored for univariate time series modelling by UCM.

The components allowed are the following: local linear trend, stochastic cycles (more than one allowed), stochastic dummy seasonal component, stochastic trigonometric seasonal component, AR(1) process. Moreover, for the modelling of multiple seasonal patterns, the PROC UCM offers the so-called *block seasonal* component which is a stochastic dummy component whose behaviour is determined by the parameters `nblocks` and `blocksize`. This component changes value every `blocksize` time points and completes its cycle in a period of `nblocks`×`blocksize` time points. For example, it can be used to model an

[2]Actually, the PROC UCM was already available in version 8.9 but is was labelled *experimental*.

hourly time series whose average daily level changes according to the weekday: in this case a seasonal component of period 24 and a block seasonal component of `blocksize`= 24 and `nblocks`= 7 are to be used. Notice that the shape of the within-day seasonal component is not influenced by the block seasonal component, but only its mean level, and thus this component is not fit for applications where there is an interaction between the two seasonal patterns as in the application illustrated in Section 9.3.

As for the regressors, there is a wide range of possibilities: fixed coefficient and random-walk coefficient linear regressors, fixed coefficient regression on the lagged dependent variable, and time-invariant nonlinear regression using splines, which includes the possible use of a non-stochastic periodic (seasonal) component approximated by a spline function.

The procedure allows the estimation, filtering/smoothing, forecasting and plotting of the model's components. An automatic treatment of additive outliers and level shifts is also available.

The syntax of the procedure is rather simple, but it does not allow the transformation of the time series, which has to be prepared using the DATA STEP or other procedures. For example, the following code estimates a basic structural model (with trigonometric seasonal component) on the logarithm of the airline time series and forecasts it up to 12 periods ahead.

SAS/ETS code for BSM (with trigonometric seasonal)

```
/* data step that takes the airline time series from
   the sashelp library and put in the the work library
   with the additional 'logair' variable */

data airline;
   set sashelp.air;                  * select dataset;
   logair = log(air);                * create log of 'air';
run;

ods graphics on;                     * for good looking plots;

proc ucm data=airline plots=all; * dataset to work on;
   id date interval = month;         * 'date' as index;
   model logair;                     * time series to work on;
   level;                            * level component;
   slope;                            * slope component;
   season length=12 type=trig;       * trig. seasonal, period 12;
   irregular;                        * observation error;
   estimate plot=panel;              * estimate model;
   forecast lead=12;                 * forecast up to 12 periods;
run;
```

```
ods graphics off;
```

The output of the PROC UCM is customisable and can be extremely rich. In the example plot we asked to produce all plots (`plots=all`), and this option produces some 30 graphs. The outlier option is by default on, and in this particular series the following output is obtained.

Outlier Summary

Obs	DATE	Break Type	Estimate	Std Error	ChiSq	DF	Pr>ChiSq
135	MAR1960	AO	−0.11680	0.0319588	13.36	1	0.0003
29	MAY1951	AO	0.08497	0.0283466	8.98	1	0.0027

The output includes information on the starting values (which can be set by the user) of the numerical estimation, the maximum likelihood estimates, the unobservable component inference at the last date of the series.

The estimates are based on the diffuse Kalman filter and detailed information on the likelihood function at its maximum is provided, as the following table shows.

Likelihood-Based Fit Statistics

Statistic	Value
Full Log Likelihood	229.37
Diffuse Part of Log Likelihood	−4.97
Non−Missing Observations Used	144
Estimated Parameters	4
Initialized Diffuse State Elements	13
Normalized Residual Sum of Squares	131
AIC (smaller is better)	−450.7
BIC (smaller is better)	−439.2
AICC (smaller is better)	−450.4
HQIC (smaller is better)	−446.1
CAIC (smaller is better)	−435.2

The parameter estimates are always accompanied by the t-statistic for the null hypothesis that the parameter equals zero. This can be misleading because, as we saw in Section 5.4, the t-statistic distribution is non-standard when the parameter is a variance.

Final Estimates of the Free Parameters

Component	Parameter	Estimate	Std Error	t Value	Pr> $\|t\|$
Irregular	Error Variance	0.00012951	0.0001294	1.00	0.3167
Level	Error Variance	0.00069945	0.0001903	3.67	0.0002
Slope	Error Variance	1.9727E−12	1.07124E−9	0.00	0.9985
Season	Error Variance	0.00006413	0.00004383	1.46	0.1435

The textual output is completed with a battery of residual based fit statistics. As for the plot, as already mentioned, you can choose from a large menu of graphs, but one of the most useful for the diagnostic analysis of the model, is the one obtained by writing `plot=panel` in the `estimate` statement, which is depicted in Figure 10.1.

For details you can refer to the SAS/ETS documentation which is freely available on the SAS Institute support site (`support.sas.com`) and to the article by Selukar (2011) on the *Journal of Statistical Software*.

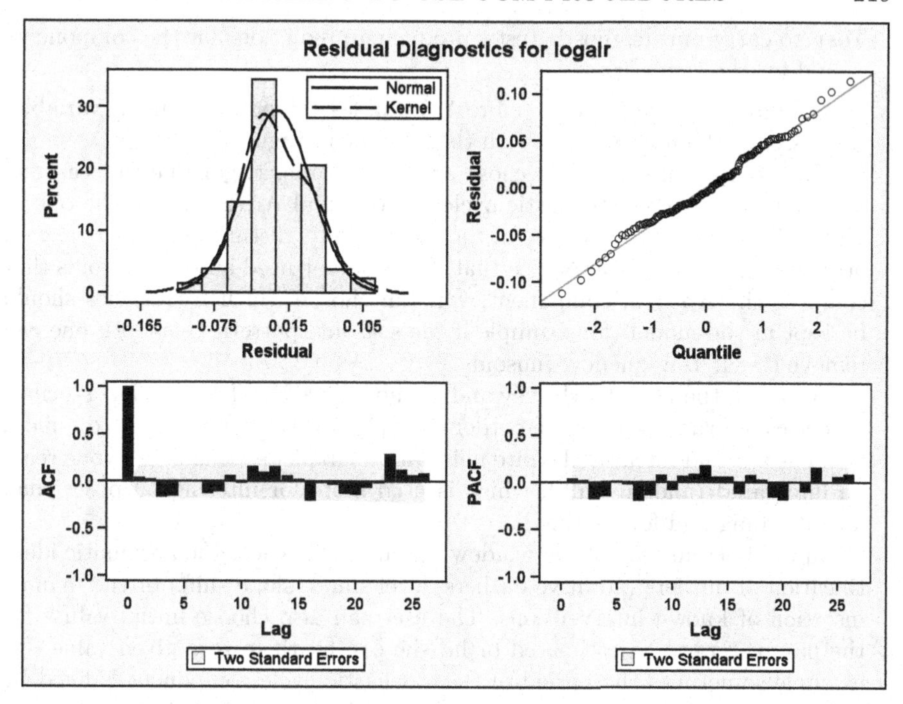

Figure 10.1 *Residual diagnostics plot of the SAS/ETS PROC UCM.*

10.1.2 STAMP

STAMP (Structural Time series Analyser, Modeller and Predictor) is the UCM software with the longest history. At the time of writing this book, the newest version is 8.3, but version 9.0 should be released during 2015.

STAMP is a module of the OxMetrics system, which includes the powerful and efficient Ox language, and can be run either using a GUI or in batch mode though a simple scripting language, or also using the Ox language. STAMP is the only ready-to-use software for UCM that estimates also multivariate models.

STAMP is written by professors A.C. Harvey and S.J. Koopman, and this means that the algorithms are reliable and efficient: indeed, most software package manuals cite the works of Koopman as base for their algorithms (in particular for the diffuse Kalman filter and the smoother).

As for all OxMetrics packages, the actions are organised in four steps:

Formulate for choosing the response variable(s) and the regressors from a dataset already loaded in OxMetrics,

Model for selecting the components and various options that define the UCM,

Estimate to determine the sample to work on and the estimation method,

Test to carry out diagnostic tests, producing predictions for the components and for the time series.

So, after choosing one or more `Y` variables (responses) and `X` variables (regressors), the user is faced with the window in Figure 10.2.

The components you can chose are level, slope, trigonometric seasonal, irregular, up to three stochastic cycles, AR(1) and AR(2). When the option 'Fixed' is selected the variance of the component's disturbance is set to zero, otherwise it is estimated. Notice that if you do not need all the sinusoids that compose the seasonal component, you can choose which frequencies should be kept in the model (for example if the seasonal pattern is smooth one can remove the high-frequency sinusoids).

Based on the work by Harvey and Trimbur (2003), STAMP offers stochastic cycles of order up to 4. The order d ($\in \{1, 2, 3, 4\}$) slope is used to build a trend as a d-fold integrated white noise (for example, setting $d = 2$ you create an integrated random walk), which is used more for filtering purposes than for modelling and forecasting.

In the bottom part of the window, the user can select the automatic identification of outliers (additive outliers, level shifts, slope shift) or the manual insertion of known interventions. The user can also choose initial values for the parameters to be estimated or he/she can fix them to a given value (for example sometimes the period of the stochastic cycle component is fixed by the user to match some economic definition of business cycle).

If in the 'Formulate' window one or more regressors are chosen, then the user can choose the behaviour of the regression coefficients, which can be: fixed, random walk, smooth spline (= integrated random walk), autoregression (= AR(1) with non-zero mean).

Finally, if more `Y` variables are chosen in the 'Formulate' window, then the user can select 'Multivariate settings...' and obtain a window as the one in Figure 10.3. Under the column 'Var matrix' you can choose the structure of the covariance matrix of that multivariate component:

full all variances and covariances of the disturbances of that component are freely estimated,

scalar there is only one such component that drives all the time series,

diagonal all the disturbances of that component are mutually uncorrelated,

ones all disturbances are perfectly correlated,

common+diag there is one common component plus as many uncorrelated components as the number of time series (i.e., the covariances of the disturbances all depend only on one common component).

All the columns starting from the second refer to a different `Y` variable. For each of these variables you can choose:

in that variable has that component,

out that variable does not have that component,

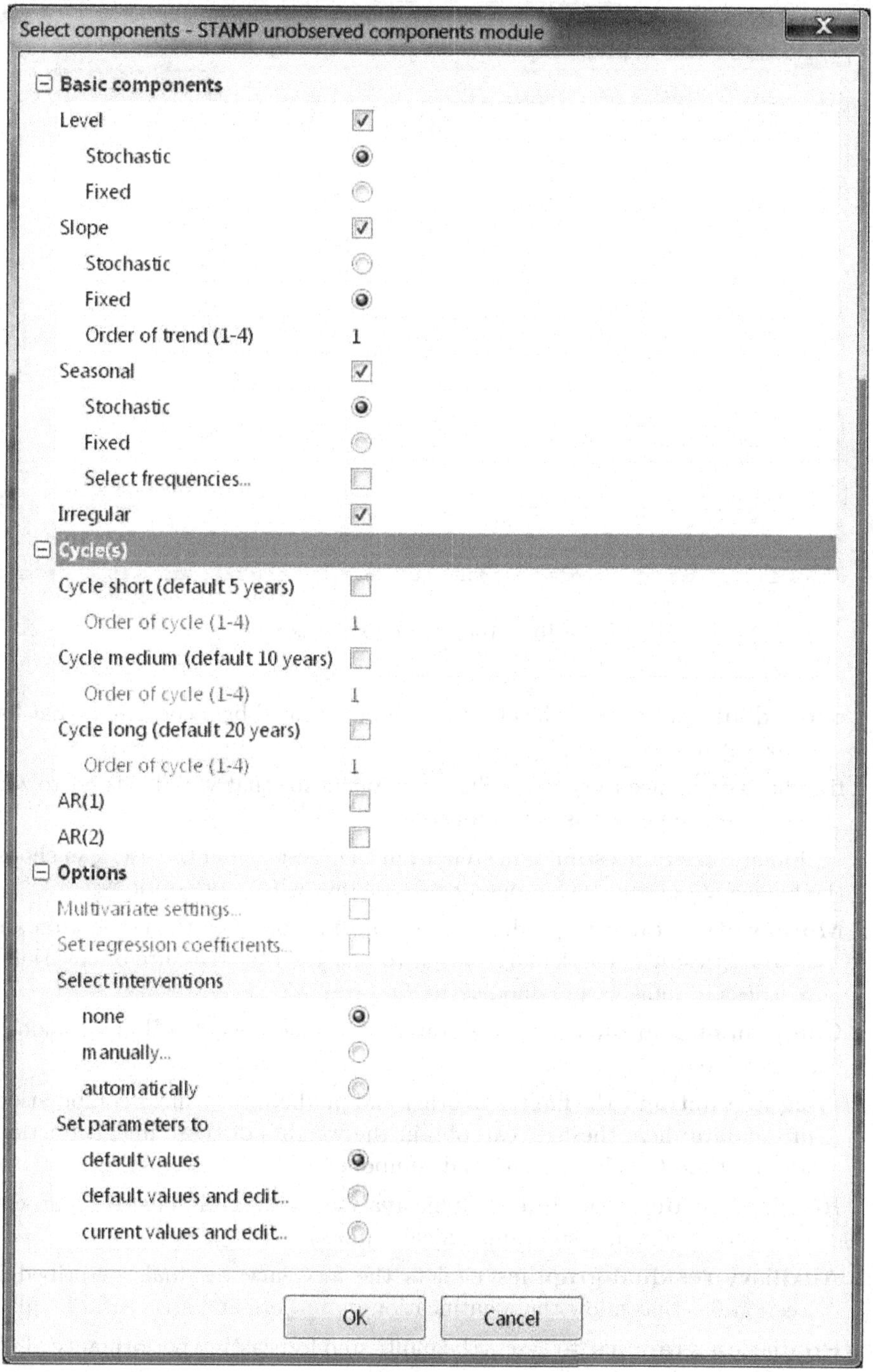

Select components - STAMP unobserved components module

Basic components
Level
Stochastic
Fixed
Slope
Stochastic
Fixed
Order of trend (1-4) 1
Seasonal
Stochastic
Fixed
Select frequencies...
Irregular
Cycle(s)
Cycle short (default 5 years)
Order of cycle (1-4) 1
Cycle medium (default 10 years)
Order of cycle (1-4) 1
Cycle long (default 20 years)
Order of cycle (1-4) 1
AR(1)
AR(2)
Options
Multivariate settings...
Set regression coefficients...
Select interventions
none
manually...
automatically
Set parameters to
default values
default values and edit...
current values and edit...

OK Cancel

Figure 10.2 *Model window in which the components are chosen.*

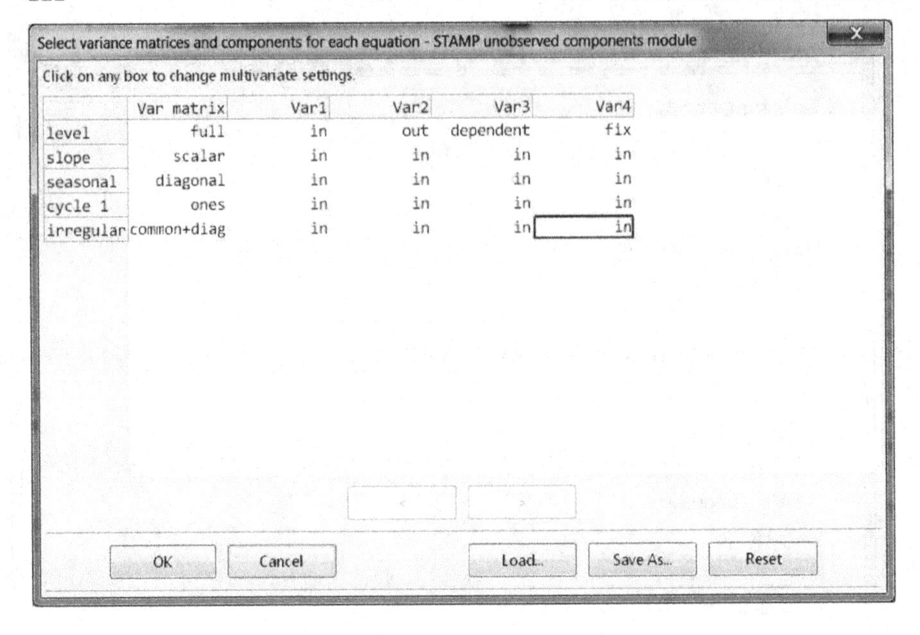

Figure 10.3 *Multivariate settings window.*

dependent that variable has that component scaled by a coefficient that has to be estimated,

fix the disturbance that drives that component for that variable is set to zero (i.e., the component is deterministic).

Finally, after the estimation step, from the 'Test' menu the user can choose the following options, which open new windows with more options:

More written output produces more information from the estimation and fit statistics; it is particularly useful to analyse the structure of covariance matrices in multivariate models;

Component graphics produces graphs of the components (filtered, smoothed or predicted);

Weight functions the filter, smoother and predictor are linear combinations of the data; here the user can obtain the weights of these linear functions for any time $t \in \{1, 2, \ldots, n\}$ and component;

Residual graphics produces graphic analyses of the residuals (i.e., autocorrelation spectrum, histogram, cumsum plots),

Auxiliary residual graphics depicts the auxiliary residuals described in Section 6.4 that allow the identification of outliers and structural breaks,

Prediction graphics to assess the multi-step forecasting performance of the UCM,

Forecasting produces out-of-sample forecasts;

Store in database to store components and residuals in the current OxMetrics database.

As described above, STAMP is capable of producing any textual or graphic output the user may desire. The graph produced by default is shown in Figure 10.4 and depicts the smoothed components. The first information provided in the textual output is listed below,

```
Summary statistics
                   Lairline
 T                   144.00
 p                   3.0000
 std.error         0.027298
 Normality           1.0329
 H(41)              0.63239
 DW                  1.9015
 r(1)              0.045263
 q                   24.000
 r(q)             -0.024242
 Q(q,q-p)            25.186
 Rs^2               0.52485
```

and is to be read as:

`T` length of the time series,

`p` number of estimated parameters, `std.error` one-step-ahead prediction error standard deviation,

`Normality` Bowman–Shenton (also Jarque–Bera) statistic, it is to compare with a chi-square distribution with 2 df (the 5% critical value is 5.99),

`H(h)` heteroskedasticity test statistic, built as the ratio of the mean of the squares of the first `h`≈`T/3` residuals to the same quantity compute on the last `h` residuals in the sample, it is approximately distributed as an $F(h, h)$,

`DW` Durbin–Watson statistic,

`r(q)` order-`q` residual autocorrelation,

`Q(q,q-p)` Box–Ljung statistic based on the first q autocorrelations, it should be compared with a chi square distribution with $q - p$ df,

`Rs2` or `Rd2` or `R2` appropriate coefficient of determination: if the model is stationary, the classical `R2` is computed, if the model has a stochastic trend component, but no seasonal component, the `Rd2` is the ratio of the variance explained by the UCM to the variance of the differenced series, finally, if the model has a seasonal component the `Rs2` is computed as the ratio of the variance explained by the model to the variance of the seasonally differenced time series.

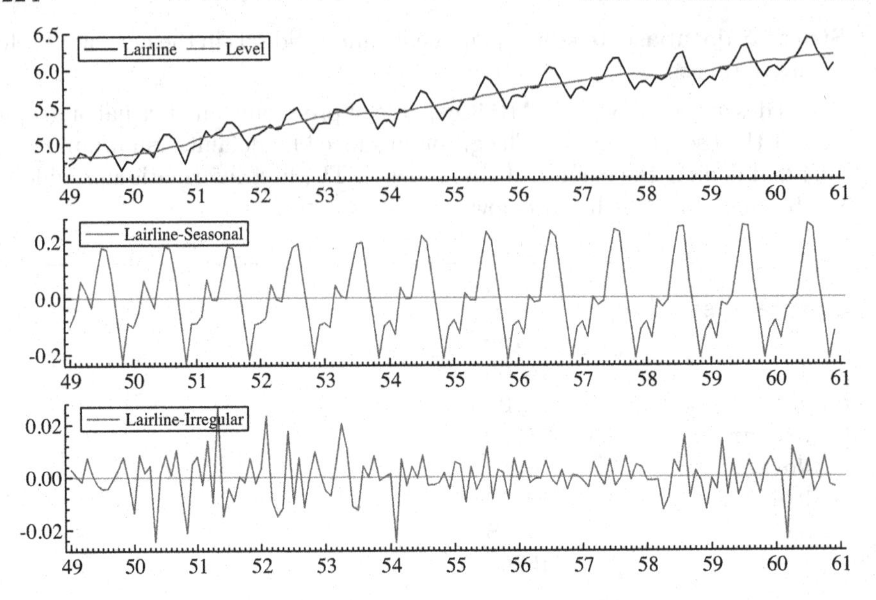

Figure 10.4 *Component graphs produced by STAMP by default.*

I personally find very eloquent the output produced by the 'Prediction graphics', which is exemplified in Figure 10.5. In this particular plot prediction of the time series up to 24-step ahead are compared with the real data. The CUSUM plot reveals instability in the model, when the cumulated standardised residuals exceed one of the two bounds.

The batch code for the estimation of the BSM (with trigonometric seasonal component) can be generated by STAMP after a model has been estimated and it is rather easy to interpret.

STAMP Batch code for estimating a BSM (with trigonometric seasonal) on the log-airline data

```
module("STAMP");
package("UCstamp");
usedata("AIRLINE.in7");
system
{
    Y = Lairline;
}
setcmp("level", 1, 0.000298277, 0, 0);
setcmp("slope", 1, 0, 0, 0);
setcmp("seasonal", 12, 3.55769e-006, 0, 0);
setcmp("irregular", 0, 0.000234355, 0, 0);
```

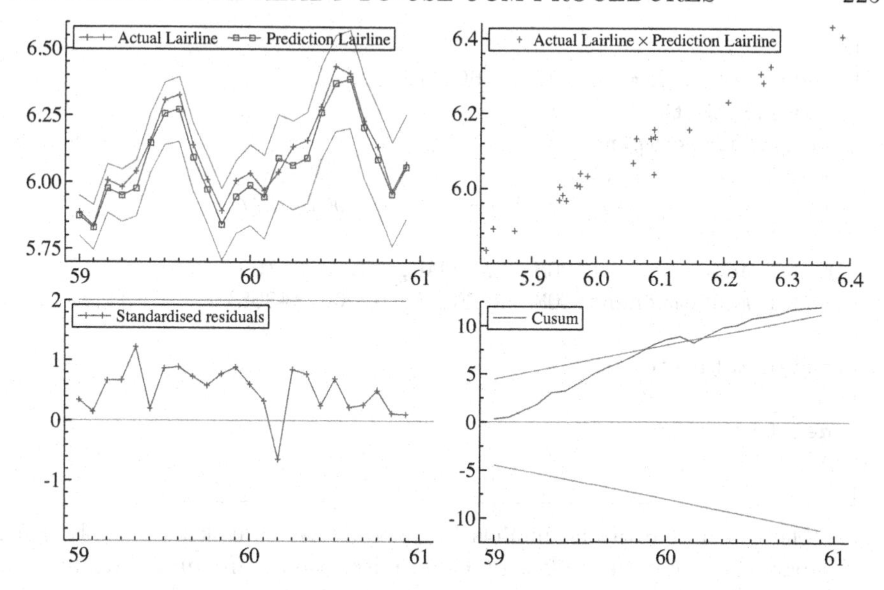

Figure 10.5 *Prediction graphs produced by STAMP.*

```
setmodel();
estimate("ML", 49, 1, 60, 12);
```

Alternatively, if you are acquainted with the Ox language, you can let STAMP create Ox code for carrying out the same estimation as in the GUI environment.

STAMP-generated Ox code for estimating a BSM (with trigonometric seasonal) on the log-airline data

```
#include <oxstd.oxh>
#import <packages/stamp/stamp_ox_uc>

main()
{
decl model = new UCstamp();

model.Load("C:\\Users\\pelagatti\\Documents\\
            OxMetrics7\\data\\AIRLINE.in7");
model.Deterministic(-1);

model.Select(Y_VAR, {"Lairline", 0, 0});
```

```
model.SetSelSample(49, 1, 60, 12);
model.SetMethod(M_ML);
// Specify components
model.StartStamp();
model.AddComponent(COM_LEVEL, 1, 0.000298277);
model.AddComponent(COM_SLOPE, 1, 0);
model.AddComponent(COM_SEASONAL, 12, 3.55769e-006);
model.AddComponent(COM_IRREG, 0, 0.000234355);

model.Estimate();

delete model;
}
```

The estimation of the models is generally very quick and stable, when compared to competing software. Good initial values are automatically computed by STAMP using few steps of the EM algorithm and, for the numerical maximisation of the log-likelihood function, analytical scores for the variances are used.

For more information about STAMP, refer to the *Journal of Statistical Software* article by Mendelssohn (2011) or to the STAMP manual (Koopman et al., 2009). The official web site of the software package is `www.stamp-software.com`.

10.1.3 Stata

At the time of writing, the latest version of Stata is 13.0 and since version 12 the `ucm` procedure has been in the system.

In Stata is it possible to use a GUI to define the UCM or to write the corresponding syntax. In particular, the main window, depicted in Figure 10.6, allows the selection of the pair trend-irregular component (referred to as *idiosyncratic component* in Stata), of a stochastic dummy seasonal component, and of up to three stochastic cycles of order $m \in \{1, 2, 3\}$. The user can also use regressors (independent variables), but their coefficients are not allowed to vary over time.

The combinations of trends and irregular components available in the cascading combobox are the following (Stata names are mapped to the names used in this book).

Random walk model trend is random walk and no irregular component is present,

No trend or idiosyncratic component no trend component, no irregular component,

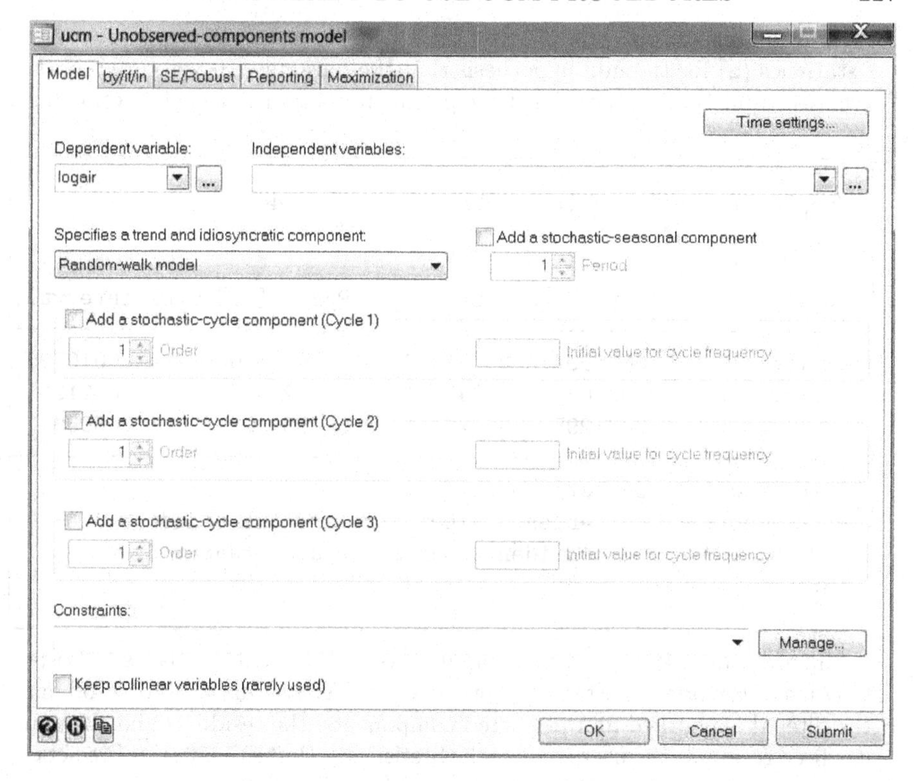

Figure 10.6 *Stata window to define an UCM.*

No trend model but includes idiosyncratic component only the irregular component is present,

Deterministic constant with idiosyncratic component constant plus irregular component,

Local-level model trend is random walk and the irregular component is present,

Deterministic-trend model with idiosyncratic component linear trend plus irregular component,

Local-level model with deterministic trend trend is random walk with drift and the irregular component is present,

Random-walk-with-drift model trend is random walk with drift and no irregular component is present,

Local-linear-trend model local linear trend plus irregular component,

Smooth-trend model integrated random walk plus irregular component,

Random-trend model trend is a local linear trend and no irregular component is present.

The default output produced by the UCM procedure is very concise, and t-statistics (`z`) for the null hypothesis that the parameter is zero are reported even when misleading (i.e., when the parameter is a variance, cf. Section 5.4).

```
Sample: 1 - 144 Number of obs    =         144
Log likelihood =  217.42043

logair          Coef.    Std.Err. z     P>z    [95% Conf.Interval]
--------------------------------------------------------------------
var(level)     .0006994 .0001903 3.67 0.000  .0003264 .0010725
var(seasonal) .0000641 .0000438 1.46 0.072  0         .00015
var(logair)    .0001295 .0001294 1.00 0.158  0         .000383
--------------------------------------------------------------------
Note: Model is not stationary.
Note: Tests of variances against zero are one sided,
and the two-sided confidence intervals are truncated at
zero.
```

Information criteria can be computed using the menu Statistics > Postestimation > Reports and Statistics (or the `estat ic` statement). For producing the filtered, smoothed and predicted components, the residuals and the standardised residuals the user can use the Statistics > Postestimation > Prediction, residuals, etc. (or the `predict` statement with its options).

The code is automatically generated by Stata even when the GUI is used. For example, the code for fitting a random walk with drift plus stochastic dummy seasonal component plus irregular, generating the information criteria, producing smoothed components, plotting the time series with the trend is reported below.

Stata code to fit an UCM to the log-airline data

```
// Use remote airline data
   use http://www.stata-press.com/data/r13/air2.dta
// Compute new varable with log of airline data
   generate logair = ln(air)
// Fit an UCM with drifted random walk, wn, monthly seas.
   ucm logair, model(lldtrend) seasonal(12)
// creates 3 new var containing smoothed components
   predict smo_trend, trend smethod(smooth)
   predict smo_seas, seasonal smethod(smooth)
// creates variable with std residuals
   predict stdres, rstandard
// compute AIC and BIC
```

```
   estat ic
// plot log-airline with smoothed trend
   twoway (tsline smo_trend) (tsline logair)
```

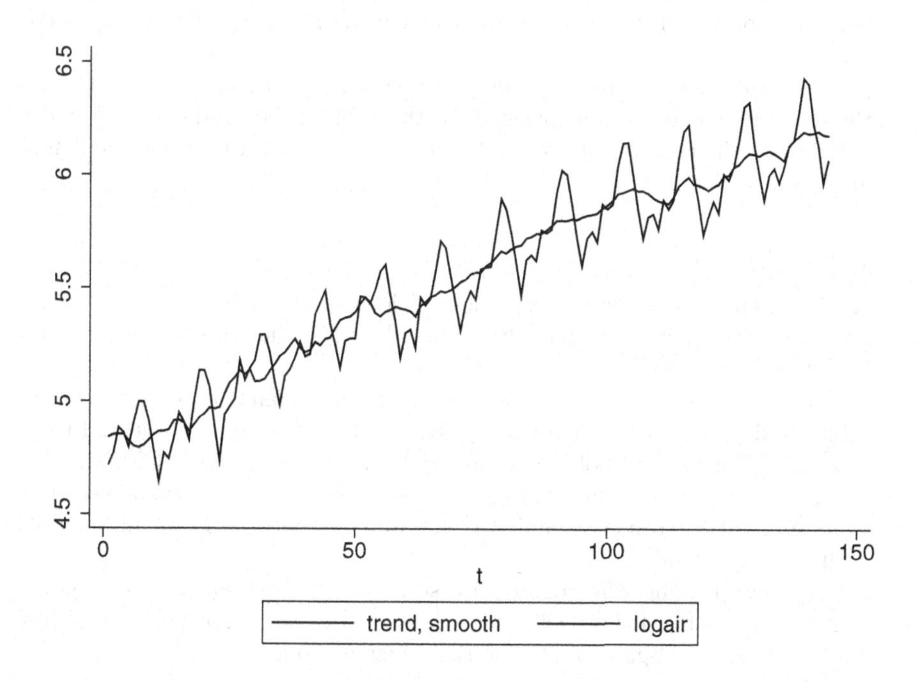

Figure 10.7 *Stata window to define an UCM.*

The graphical output produced by the above code is shown in Figure 10.7. Stata uses a diffuse Kalman filter for the nonstationary components, and for the numerical maximisation of the log-likelihood it computes analytical derivatives with respect to the unknown variances. Despite these good theoretical properties, when the variance of a component is estimated zero or very close to zero by other software packages, sometimes the Stata optimiser fails to converge unless that variance is previously set to zero. For instance, as a standard example, we tried to fit a BSM to the log-airline time series, but we were not able to obtain estimates until we imposed a random walk with drift as trend component.

For more information refer to the official documentation and manuals available at the Stata web site `www.stata.com` .

10.2 Software for generic models in state space form

There are many languages and software packages that offer algorithms for Kalman filtering, and there is no hope (and probably no sense) that one book can cover them all. Moreover, it goes beyond the aims of this chapter to get into details in describing the Kalman-filter related functions of the software packages covered in this section. On the contrary, for each of the considered software packages the aim of this section is providing enough information to put the reader who is approaching state space modelling in condition to chose the software which is more consistent with his/her needs and tastes. Thus, we will discuss the features of each software and show how a BSM can be defined and estimated.

10.2.1 EViews

At the moment of writing these lines, EViews is at version 8.1, even though no significant innovations have been introduced in the state space modelling object since version 4.

The people at EViews have found an excellent compromise between flexibility and ease of use when they projected the state space object. In fact, although EViews does not feature ready-to-use functions for UCM, stating a model in state space form, obtaining Gaussian maximum likelihood estimates of its unknown parameters and carrying out inference on the components is really a piece of cake.

For example, once the 'logair' time series is in the EView workfile, in order to implement a BSM the user just needs to create a state space object (Object > New Object > SSpace) and type the following code.

Code for the EViews' sspace object that defines a BSM for the log-airline time series

```
' Initial values for the disturbances' std. deviations
@param c(1) 0.5 c(2) 0.1 c(3) 0.1 c(4) 0.1

' Observation equation
logair = mu + gamma + [var=c(1)^2]

' State equations
' -- Local linear trend
@state mu = mu(-1) + beta(-1) + [var=c(2)^2]
@state beta = beta(-1) + [var=c(3)^2]

' -- Stochastic dummy seasonal component
@state gamma = -gamma(-1)-gamma(-1)-gamma2(-1)-gamma3(-1)-
               gamma4(-1)-gamma5(-1)-gamma6(-1)-gamma7(-1)-
```

```
                    gamma8(-1)-gamma9(-1)-gamma10(-1)+
                    [var=c(4)^2]
@state gamma1 = gamma(-1)
@state gamma2 = gamma1(-1)
@state gamma3 = gamma2(-1)
@state gamma4 = gamma3(-1)
@state gamma5 = gamma4(-1)
@state gamma6 = gamma5(-1)
@state gamma7 = gamma6(-1)
@state gamma8 = gamma7(-1)
@state gamma9 = gamma8(-1)
@state gamma10 = gamma9(-1)
```

In the EViews workspace, an object of class *coefficients* named 'C' is always present, and one can refer to its elements by `C(1)`, `C(2)`, ... The user can also define other *coefficients* objects in case he/she wants more mnemonic names for the unknown parameters.

The first line of code begins with the keyword `@param` followed by the parameters (of class *coefficients*) that are going to be used in the model and initial values for them. This part is not necessary if the user has already assigned initial values directly in the vector of parameters `C` in the workspace.

The next line defines the observation equation where one or more time series in the workspace (`logair` in this example) are defined as linear combinations of unobservable components plus a white noise disturbance, whose variance is defined in the brackets. In this case we set `[var=c(1)^2]`, and the variance of the irregular component is equal to the square of the unknown parameter `C(1)`, which, if the eventual minus sign is ignored, represents the standard deviation of that white noise sequence. The variables `mu` and `gamma` are not defined yet, but if the interpreter finds anywhere in the object a state variable with that name, it does not produce any error.

The next lines, all starting with the keyword `@state` define the state variables of the model and their transition equations. The rule is simple: on the rhs of the equal sign the state variables can only appear linearly and lagged by one period. Again, if a disturbance is needed, its variance is defined using the syntax `[var=...]`.

You can insert regressors both in the state equation, in the observation equations and even in the definition of the variances. They just need to be present somewhere in the workspace. Furthermore, disturbances are, by default, mutually uncorrelated, but the user can name them using the `@ename ...` statement, use them in the state and measurement equations and define their variances and covariances with constructs such as `@evar var(e1)=...` and `@evar cov(e1,e2)=...`

When the user presses the *Estimate* button the EViews parser builds the system matrices, maximises the log-likelihood with respect to the unknown parameters and produces an output like the following.

```
        Coefficient Std.Error  z-Statistic Prob.

C(1)    -0.000658   0.537497   -0.001224   0.9990
C(2)     0.015998   0.003968    4.031957   0.0001
C(3)     0.000576   0.000662    0.869122   0.3848
C(4)     0.106442   0.010390   10.24428    0.0000

        Final State   Root MSE   z-Statistic  Prob.

MU         6.160639   0.032578   189.1048     0.0000
BETA       0.007089   0.003253     2.178964   0.0293
GAMMA     -0.309250   0.132060    -2.341732   0.0192
GAMMA1    -0.085122   0.027287    -3.119571   0.0018
GAMMA2    -0.181490   0.022460    -8.080423   0.0000
GAMMA3    -0.010801   0.019730    -0.547445   0.5841
GAMMA4     0.087230   0.017699     4.928523   0.0000
GAMMA5     0.258519   0.016421    15.74330    0.0000
GAMMA6     0.280695   0.015741    17.83246    0.0000
GAMMA7     0.131473   0.015447     8.511457   0.0000
GAMMA8     0.008837   0.015357     0.575453   0.5650
GAMMA9    -0.012268   0.015348    -0.799330   0.4241
GAMMA10   -0.100852   0.015341    -6.573851   0.0000

Log likelihood -51.16069   Akaike info criterion 0.766121
Parameters 4               Schwarz criterion     0.848616
Diffuse priors 13          Hannan-Quinn criter.  0.799642
```

Notice that t-ratios are automatically computed; it is the user who must be aware if they are meaningful or not, because he/she is the only one to know if zero is on the frontier of the parameter space.

The filtered, predicted or smoothed components and their disturbances can be plotted or saved into the workspace using the buttons View and Proc. The button Forecast produces forecasts of the response variable(s).

All these operations can be run in batch mode: in fact the scripting language of EViews can be used to create and manage state space objects.

As for the Kalman filter related algorithms, EViews tries to determine which components are stationary and which are not. In the first case it uses the marginal moments to initialise the filter, while in the second case it uses a diffuse Kalman filter. However, the estimates obtained by the state space object in EViews tend to be different from those obtained with other software packages (cf. Table 5.2).

For more details, the reader should refer to the article by Van den Bossche (2011) and the dedicated chapter in EViews user's guide. The official web site is `www.eviews.com` .

10.2.2 Gretl

Gretl's web site (`gretl.sourceforge.net`) introduces the software package with these lines:

> [Gretl] is a cross-platform software package for econometric analysis, written in the C programming language. It is free, open-source software. You may redistribute it and/or modify it under the terms of the GNU General Public License (GPL) as published by the Free Software Foundation.

At the moment of writing, the latest version is 1.9.92, but as in many open-source projects the rightmost digits of the version number tend to increase quickly. Gretl can be used both through its GUI or by using a scripting language. The main developers of Gretl are Professors Allin Cottrell (Wake Forest University) and Riccardo "Jack" Lucchetti (Università Politecnica delle Marche), who are also very active at providing assistance to Gretl's users through the mailing lists (`gretl.sourceforge.net/lists.html`).

Most of Gretl's numerous features are available through menus, but state space models have to be defined and estimated using the syntax. For setting up a model in state space form and drawing the related inference, one needs to define the state space form through the `kalman` block of instructions, perform the estimation using the `mle` block and smoothing the components though the `ksmooth()` function. At the moment, there are no ready-to-use functions to build the auxiliary residual, but it is not hard to write code for that following Section 6.4 (see also Lucchetti, 2011). The following code illustrates the definition and estimation of a BSM model on the log-airline data.

Gretl code to define and estimate a BSM on the log-airline data

```
open bjg.gdt        # open airline data
genr lair = log(g) # create log-airline

# initial values for std. deviations
scalar sdeta = 0.1
scalar sdzeta = 0.1
scalar sdomega = 0.1
scalar sdeps = 0.1

# State equation matrices
```

```
matrix T = zeros(13,13)
# -- Local linear trend
T[1,1] = 1
T[1,2] = 1
T[2,2] = 1
# -- Stochastic dummy seasonal component
T[3,3:13] = -1
T[4:13,3:12] = I(10)
matrix Q = zeros(13,13)

# Observation equation matrices
matrix Z = zeros(13,1)
Z[1,1] = 1
Z[3,1] = 1
matrix H = {sdeps^2}

# Initial conditions
matrix a1 = zeros(13,1)
a1[1,1] = lair[1]

# Definition of the state space form
kalman
    obsy lair
    obsymat Z
    statemat T
    obsvar H
    statevar Q
    inistate a1
end kalman --diffuse

# Maximum likelihood estimation
mle ll = ERR ? NA : $kalman_llt
    # assign variances to right matrix positions
    Q[1,1] = sdeta^2
    Q[2,2] = sdzeta^2
    Q[3,3] = sdomega^2
    H[1,1] = sdeps^2
    ERR = kfilter()
    params sdeta sdzeta sdomega sdeps
end mle

# Smoothing
matrix ret = ksmooth()
```

```
# Plot series and trend
series trend = ret[,1]              # series from smooth trend
setinfo lair -n "log-airline"       # label to plot for lair
setinfo trend -n "smoothed trend" # label to plot for trend
gnuplot lair trend time --with-lines --output=display
```

In the first part of the script, the data are loaded and transformed and the system matrices are created. The `kalman` block defines the state space form by assigning the system matrices using the keywords listed in Table 10.2. Some of these matrices can be made time-varying if the model requires so.

Table 10.2 *Kalman block keywords.*

Keyword	Matrix	Description
obsy	$\mathbf{Y}$	Matrix of response variables
obsymat	$\mathbf{Z}$	Observation matrix
obsvar	$\mathbf{H}$	Cov. matrix of observation disturbances
obsx	$\mathbf{X}$	Matrix of regressors
obsxmat	$\mathbf{B}$	Regression coefficient matrix
statemat	$\mathbf{T}$	Transition matrix
statevar	$\mathbf{Q}$	Cov. matrix of state disturbances
stconst	$\boldsymbol{d}$	Constant vector in transition eq.
initstate	$\boldsymbol{a}_{1\|0}$	Mean vector of initial states
initvar	$\mathbf{P}_{1\|0}$	Cov. matrix of initial states

The `mle` block defines the optimisation problem, which is then solved at runtime. The statement following the `mle` keyword is to be interpreted as: define the variable `ll` which will contain the vector of log-likelihood contributions for each time point t, but if the value in the variable `ERR` returned by the Kalman filter function `kfilter()` is not zero, then assign missing values to the log-likelihood. Otherwise fill the vector with the log-likelihood contributions generated by `kfilter()` and accessed through `$kalman_llt`. In the body of the `mle` block, the user assigns the parameters to be estimated defined by the `params` keyword to the right elements in system matrices.

Finally, the matrix `ret` is assigned the smoothed estimates of the state variables though the function `ksmooth()`. In case the variances and covariances of the smoothed state variables are needed, they can be recovered from the same function passing a pointer to a matrix as argument: `ksmooth(&V)`, where `V` has been previously defined as matrix.

The last four lines plot the log-airline time series with the smoothed trend component. Omitting the graph, the output generated by the above code is the following.

```
Using numerical derivatives
Tolerance = 1.81899e-012

Function evaluations: 464
Evaluations of gradient: 54

Model 1: ML, using observations 1949:01-1960:12 (T = 144)
ll = ERR ? NA : $kalman_llt
Standard errors based on Outer Products matrix

             estimate     std. error        z         p-value
  --------------------------------------------------------------
  sdeta       0.0254459     0.00320654     7.936      2.09e-015 ***
  sdzeta      4.39216e-08  10.3360         4.249e-09  1.0000
  sdomega    -0.00809239    0.00210525    -3.844      0.0001    ***
  sdeps      -0.0125477     0.00458295    -2.738      0.0062    ***

Log-likelihood          112.6082   Akaike criterion    217.2164
Schwarz criterion      -205.3371   Hannan-Quinn       -212.3893
```

Notice that in order to interpret the estimates as standard deviations one has to omit the minus signs, as these parameters enter the objective function only though their squares (variances). Furthermore, t-ratios are automatically generated by the `mle` block, but the user should be aware that, as in this case, their distribution can be non-standard (cf. Section 5.4).

As for the Kalman filter related algorithms, Gretl's implementation is quick and stable, even though there is no exact treatment of initial diffuse conditions. When the `--diffuse` option is used (as in the above script) the variances of the initial states are set to 10^7 and the ordinal Kalman filter is run.

For more details, the reader should refer to the article by Lucchetti (2011), the dedicated chapter in Gretl user's guide (Cottrell and Lucchetti, 2015) and the official web site `gretl.sourceforge.net` .

10.2.3 Ox/SsfPack

Ox (Doornik, 2007) is a very efficient language for matrix algebra and numerical programming developed by Jurgen Doornik (University of Oxford). SsfPack (Koopman et al., 1999, 2008) is an Ox package for state space modelling mainly developed by Siem Jan Koopman (VU University Amsterdam), who is probably the world's leading researcher in state space related algorithms, as his numerous publications on the topic prove.

Ox and SsfPack are commercial software packages, but there are versions of them (Ox Console and SsfPack Basic) that are free to download and use under the following conditions (only for Ox Console):

[J.A. Doornik] grants a non-exclusive licence without charge to any in-

> dividual engaged in academic research, study or teaching, subject to the following restrictions: Ox Console should be clearly cited in any electronic or non-electronic publication for which it was used; Ox Console may not be used for externally funded, sponsored or commercial research; Ox Console may not be used at non-educational institutions (including governmental and non-profit institutions), except for evaluation purposes.

The commercial versions are named, respectively, Ox Professional and SsfPack Extended. The latest versions at the time of writing this book are 7.0 for Ox and 3.0 for SsfPack. In this review we will use only functions from SsfPack Basic. The impovement of SsfPack Extended with respect to SsfPack Basic are listed by Koopman et al. (2008):

> (i) an exact treatment for the diffuse initial conditions [. . .], (ii) allowing for sparse structures in the state space matrices in order to avoid multiplications by zero and unity values, and (iii) the univariate treatment of filtering and smoothing computations for multivariate state space models. These modifications [. . .] lead to computationally faster and numerically more stable computations. [. . .] New ready-to-use functions are added [. . .] for more advanced and professional analyses of time series.

The syntax of Ox is very similar to that of the C language, and so if the reader has some experience with C or its derivatives, he/she will not have problems reading the following code, for estimating a BSM on the log-airline time series. For readability reasons we split the code in different boxes.

Ox/SsfPark code for a BSM for the log-airline data: preamble

```
#include <oxstd.oxh>  // standard library
#include <oxdraw.oxh> // graphic library
#import  <maximize>  // optimisation library
#include <packages/ssfpack/ssfpack.h> // ssfpack

// For efficiency gain we use some global variables
// for the state-space matrices
static decl mPhi, mOmega, mSigma, yt, dVar;
```

This preamble loads the needed libraries and defines five global variables that will contain the state space system matrices, the time series and a scale parameter. In general good coding practice discourages the use of global variables, but efficiency reasons justify this choice here.

Ox/SsfPark code for a BSM for the log-airline data: objective function

```
// Objective function definition to feed the optimiser
loglik(const vP, const adFunc, const avScore, const amHess)
{
  mOmega[1][1] = vP[0]^2;    // var(zeta)/var(eta)
  mOmega[2][2] = vP[1]^2;    // var(omega)/var(eta)
  mOmega[13][13] = vP[2]^2; // var(epsilon)/var(eta)
  // we use the likelihood with a scale parameter,
  // var(eta), concentrated out
  // dVar will contain var(eta)
  return SsfLikConc(adFunc,&dVar,yt,mPhi,mOmega,mSigma);
}
```

The structure of this function is imposed by the optimisation function `MaxBFGS()`, but the last two formal parameters are not used here. This code returns, through the pointer `adFunc`, the value of the log-likelihood at parameter values passed though `vP`. The SsfPack function `SsfLikConc()` returns the value of the concentrated likelihood: in fact it can be shown (Harvey, 1989; Koopman et al., 2008) that a scale factor in the likelihood can be computed in closed form from the standardised residuals. This allows to perform the optimisation with respect to one parameter less. One of the variances in the state space matrices has to be fixed to one, and, then, all the estimated covariance matrices have to be multiplied by the scale factor that `SsfLikConc()` returns though the pointer `dVar`. This is why in this code we assign the variances only to the slope, seasonal and irregular components' (rescaled) disturbances.

Notice that both the objective function and `SsfLikConc()` pass the log-likelihood value through the pointer `adFunc`, while both functions return one if successful and zero otherwise.

As in the C language, the Ox interpreter runs the code contained in the `main()` function. The following blocks of code make up the body of the `main()` function in our little program.

Ox/SsfPark code for a BSM for the log-airline data: first block in the function `main()`

```
decl mData = loadmat("AIRLINE.in7");
yt = log(mData)'; // take log of the data and transpose

// Create a matrix to define the components in GetSsfStsm
decl mStsm = <CMP_LEVEL,       1,  0, 0, 0;
              CMP_SLOPE,       2,  0, 0, 0;
              CMP_SEAS_DUMMY, 3, 12, 0, 0;
              CMP_IRREG,       4,  0, 0, 0>;
// Create the state space system
```

```
GetSsfStsm(mStsm, &mPhi, &mOmega, &mSigma);
```

Here the data are loaded, the log-airline series is computed and the ready-to-use function `GetSsfStsm` builds the state space system matrices for an UCM. SsfPack uses the following notation for the state space form:

$$\texttt{mPhi} = \begin{bmatrix} \mathbf{T} \\ \mathbf{Z} \end{bmatrix}, \quad \texttt{mOmega} = \begin{bmatrix} \mathbf{Q} & \mathbf{G}^\top \\ \mathbf{G} & \mathbf{H} \end{bmatrix}, \quad \texttt{mSigma} = \begin{bmatrix} \mathbf{P}_{1|0} \\ \boldsymbol{a}_{1|0}^\top \end{bmatrix}, \quad \texttt{mDelta} = \begin{bmatrix} \mathbf{d} \\ \mathbf{c} \end{bmatrix}.$$

If `mDelta` is not defined SsfPack assumes it is a zero vector. For models with time-varying matrices four more matrices are to be assigned (see Koopman et al., 1999). The function `GetSsfStsm()` reads the matrix `mStsm` where the components are defined and returns the system matrices `mPhi`, `mOmega` and `mSigma` through their addresses. Beside the four components used in our code, `GetSsfStsm()` admits also the trigonometric seasonal component and stochastic cycles up to order 4. Of course, the user is free to assign the system matrices directly without using ready-to-use functions such as `GetSsfStsm()`.

Ox/SsfPark code for a BSM for the log-airline data: estimation

```
// Starting values
decl vPar = ones(3,1);

// Maximisation of the log-likelihood
decl dFunc;
MaxBFGS(loglik, &vPar, &dFunc, 0, 1); // optimisation
// Run loglik to assign the parameters to system matrices
loglik(vPar, &dFunc, 0, 1);

// Multiply cov matrix by var(eta) that was concentrated out
mOmega = mOmega * dVar;
```

The above snippet sets some initial values (ones), assuming that σ_η^2 is concentrated out and, thus, all other variances are initially set equal to that variance. The function `MaxBFGS`, maximises the objective function `loglik` with respect to the parameters `vPar`, whose current value is used as starting value for the BFGS optimiser. The values of the final estimation and log-likelihood are passed through the addresses of `vP` and `dFunc`. In the next line, our `loglik` function is called to make sure the global variables `mOmega`, `dVar` are assigned according to the ML estimates[3].

[3]I am not certain the last call of the objective function carried out by `MaxBFGS` is done at the final estimates.

Then, the covariance matrix `mOmega` is rescaled with respect to the scale factor `dVar` (here equal to σ_η^2) that was concentrated out of the likelihood[4].

Ox/SsfPark code for a BSM for the log-airline data: inference for the unobserved components

```
  // State inference
  decl mStSmo, mStPre, mDsSmo;
  // -- state smoothing
  SsfMomentEst(ST_SMO,  &mStSmo, yt, mPhi, mOmega, mSigma);
  // -- state predictions
  SsfMomentEst(ST_PRED, &mStPre, yt, mPhi, mOmega, mSigma);
  // -- disturbance smoothing
  SsfMomentEst(DS_SMO,  &mDsSmo, yt, mPhi, mOmega, mSigma);

  // Residuals
  decl vStdRes, mAuxRes;
  // -- standardised residuals
  vStdRes = (yt - mStPre[13][])./sqrt(mStPre[27][]);
  // -- auxiliary residuals
  mAuxRes = mDsSmo[0:13][]./sqrt(mDsSmo[14:27][]);
}
```

In these few last lines, the function `SsfMomentEst` is used to run the state smoother (`ST_SMO`), the state one-step predictor (`ST_PRED`) and the disturbance smoother (`DS_SMO`). The function returns the requested projections and their MSE through the matrix pointer in the second formal parameter. The time series are written in the rows of that pointed matrix and follow the same ordering as in the rows of the matrix `mPhi`: for example, for the one-step predictions,

$$\texttt{mStPre}[][\texttt{t}] = \begin{bmatrix} \boldsymbol{a}_{t|t-1} \\ \hat{\boldsymbol{y}}_{t|t-1} \\ \text{diag}(\mathbf{P}_{t|t-1}) \\ \text{diag}(\mathbf{F}_t) \end{bmatrix}.$$

From these matrices, standardised residuals and auxiliary residuals can be easily computed.

The algorithms in SsfPack (and even more in SsfPack Extended) are extremely efficient, and the optimisation can be made even faster and more stable by using the function `SsfLikSco()` that computes also the scores of the

[4]In this case we do not have to rescale the covariance matrix in `mSigma` because all states have diffuse initial conditions. A diffuse initial condition for the i-th state variable is coded with a value -1 in the corresponding diagonal element of the covariance matrix in texttttmSigma.

Gaussian log-likelihood function with respect to the elements in the system covariance matrix `mOmega` as explained by Koopman and Shephard (1992).

For more details, the reader should refer to the article by Pelagatti (2011), the dedicated SsfPack user's guide (Koopman et al., 2008) and the official web site `http://www.ssfpack.com/`. The books by Durbin and Koopman (2001) and Commandeur and Koopman (2007) on state space modes also provide many code snippets in Ox/SsfPack.

10.2.4 R

R, at version 3.1.2 at this writing, is one of the most popular softwares for data analysis and its popularity is still growing fast[5]. It is released under the terms of the GNU General Public License and so it is an open source software.

Algorithms for state space models are available in the `stat` package, contained in the basic distribution, and in many other packages available from the CRAN, the online Comprehensive R Archive Network: `dse` (Gilbert, 2009), `sspir` (no longer maintained, but you can find old versions on the CRAN), `dlm` (Petris et al., 2009; Petris, 2009, 2010), `FKF` (Luethi et al., 2015), `KFAS` (Helsk, 2015), `astsa` (Shumway and Stoffer, 2006; Stoffer, 2015).

Many of these packages are described in Tusell (2011). Here we limit our attention to the functions in the `stat` package and to the packages `FKF` and `KFAS`. This choice is due to the efficiency of the algorithms and to the similarity of approach with this book. The `dlm` package is also well written, but more oriented towards the Bayesian analysis of time series in the spirit of West and Harrison (1989). The Kalman filtering algorithms in `sspir` and `astsa` are written in R (and not in a low-level language such as C or Fortran) and, thus, they are much slower (cf. Tusell, 2011, Table 2); moreover, they do not allow missing values in the response variable $\{\boldsymbol{Y}_t\}$.

The functions for state space modelling you find in the default installation of R are (the names are self-explanatory)

```
KalmanLike(y, mod, nit = 0L, update = FALSE)
KalmanRun(y, mod, nit = 0L, update = FALSE)
KalmanSmooth(y, mod, nit = 0L)
KalmanForecast(n.ahead = 10L, mod, update = FALSE)
```

where `y` is a vector of observations and `mod` a list of matrices defining the state space form. The limitations of these functions are that (i) they work only on univariate time series, (ii) allow only for time-homogeneous systems, (iii) there is no exact treatment of the initial conditions, (iv) the vectors $\boldsymbol{c}_t$ and $\boldsymbol{d}_t$ of our (actually Harvey's) version of the state space form cannot be used.

The ready-to-use function

[5]See this post for an interesting analysis of the popularity of software for data analysis `r4stats.com/articles/popularity/`.

```
StructTS(x, type = c("level", "trend", "BSM"), init = NULL,
        fixed = NULL, optim.control = NULL)
```

is based on the functions above, but its scope is limited to the models: random walk plus noise (`type="level"`), local linear trend plus noise (`type="trend"`) and BSM (`type="BSM"`).

The library `FKF` (Fast Kalman Filter, version 0.1.3), implements only two functions: one for Kalman filtering and one for plotting the results. There are no functions for smoothing, and this is really its main limitation. The Kalman filtering function is

```
fkf(a0, P0, dt, ct, Tt, Zt, HHt, GGt, yt, check.input = TRUE)
```

where the mapping of the symbols is almost one-to-one with those used in this book, with the exception of `GGt` $= \mathbf{Q}_t$. As its name promises, this function (coded in C) is extremely fast. We estimated the BSM using the functions `fkf()` and `KalmanLike()`: the former took one half of the time of the latter. Furthermore, according to the speed comparison to Tusell (2011, Table 2), the KFK package results faster than all the other R packages in the great majority of his experiments.

The package `KFAS` (version 1.0.4 at the moment of writing), is the fastest evolving package. Indeed, the description of the package you can find in the article by Petris and Petrone (2011) does not apply to the current version anymore. The state space form you can use in the model is

$$
\begin{aligned}
Y_t &= \mathbf{Z}_t \boldsymbol{\alpha}_t + \boldsymbol{\epsilon}_t \\
\boldsymbol{\alpha}_{t+1} &= \mathbf{T}_t \boldsymbol{\alpha}_t + \mathbf{R}_t \boldsymbol{\nu}_t,
\end{aligned}
$$

with $\boldsymbol{\epsilon}_t \sim \mathrm{WN}(\mathbf{0}, \mathbf{H}_t)$, $\boldsymbol{\nu}_t \sim \mathrm{WN}(\mathbf{0}, \mathbf{Q}_t)$ so that the state disturbance covariance matrix is now $\mathbf{R}_t^\top \mathbf{Q}_t \mathbf{R}_t$ and not $\mathbf{Q}_t$ as in the rest of this book. All the system matrices are to be contained in matrix objects when time-homogeneous, and in array of matrices when time-varying.

The basic function to assemble a state space form is

```
SSMcustom(Z, T, R, Q, a1, P1, P1inf, index, n)
```

where the first six formal parameters are the four system matrices and the vector and matrix of initial conditions. The matrix `P1inf` is for communicating which state variable should be given a diffuse initial condition: the diagonal element of such a variable should be one, while all other elements are set to zero. The formal parameter `index` indicates to which time series these matrices refer in multivariate models. This information has a meaning when we want to build a multivariate model by assembling different univariate models. The formal parameter `n` should be assigned the number of time points in the sample (missing values are allowed), and it is used only for internal dimension checks.

In building the system matrices for an UCM, the user can take advantage of the following ready-to-use functions:

```
SSMarima(ar = NULL, ma = NULL, d = 0, Q, stationary = TRUE,
         index, n, ynames)
SSMcycle(period, type, Q, index, a1, P1, P1inf, n, ynames)
SSMseasonal(period, sea.type = c("dummy", "trigonometric"),
            type, Q, index, a1, P1, P1inf, n, ynames)
SSMtrend(degree = 1, type, Q, index, a1, P1, P1inf, n, ynames)
```

which are used to build, respectively, an ARMA, a stochastic cycle, a seasonal, and a trend component. The different components built using these functions can be assembled using the function

```
SSModel(formula, data, H, u, distribution)
```

where `formula` is a formula object used to compose the model's components (see below), `data` is an optional parameter to pass a `data.frame` object containing the relevant time series, `H` is the covariance matrix (or array of covariance matrices) of the observation errors, and `u` and `distribution` are used when a non-Gaussian state space form is to be specified (approximated). The formula object represents the complete UCM as

```
formula <- cbind(y1,y2,...) ~ -1 + x1 + x2 + ... +
           SSMfunc(...) + SSMfunc(...) + ... + SSMfunc(...)
```

where `cbind(y1,y2,...)` is the matrix of dependent variables, `-1` is to be used to omit the constant, `+ x1 + x2 + ...` represent an arbitrary number of regressors, and `SSMfunc(...)` is a placeholder for any of the five component functions seen above (`SSMcustom` can be one).

The estimation is carried out by the function

```
fitSSM(model, inits, updatefn, checkfn, ...)
```

where `model` is an object of class `SSModel`, as returned by `SSModel()`, `inits` is a vector of starting values for the unknown parameters and `updatefn` is a user-defined function of form `updatefn(pars, model,...)` that places the parameters in vector `pars` in the model `model` and returns the updated model. If `updatefn` is not supplied, a default function is used, which estimates the values marked as `NA` in time invariant covariance matrices `Q` and `H`.

Finally, filtering and smoothing is carried out by the function

```
KFS(model, filtering, smoothing, ...)
```

where `model` is an object of class `SSModel` and the options for `filtering` are "state", "signal", "mean", "none", while the options for `smoothing` are the same ones plus "disturbance" for computing the auxiliary residuals.

Using the ready-to-use function the implementation of a BSM model is very simple.

```
library(KFAS) # loads the KFAS library
logair <- log(AirPassengers) # compute log-airline time series
nobs <- length(logair) # length of time series
```

```
// formula that defines the UCM
ff <- logair~-1+
      SSMtrend(degree=2,Q=list(NA,NA),n=nobs)+
      SSMseasonal(period=12,sea.type="dummy", Q=NA,n=nobs)

mod    <- SSModel(formula=ff, H=NA) // make SSModel object
fit <- fitSSM(mod, inits=rep(0.1,4)) // fit model

// smoothing the components
smooth <- KFS(fit$model, filtering="none", smoothing="state")

// plotting log-airline and trend
plot(logair)
lines(ts(smooth2$alphahat[,"level"],start=c(1949,1),freq=12))
```

As Tusell (2011, Table 2) shows, the basic Kalman filtering algorithm of the KFAS package is very efficient, but using `fitSSM` introduces a significant overhead that makes the estimation slower than it could be using global variables for the system matrices and, so, avoiding the continuous generation and deletion of possibly large system matrices.

The features of the KFAS package go beyond the inference for Gaussian state space models. In fact, there are functions to carry out approximate inference for exponential family state space forms using simulation (importance sampling).

10.2.5 Stata

In addition to the UCM procedure presented above, Stata (version 13 at the moment of writing) can also estimate models in general state space form.

The state space form can be defined both using the GUI or writing a script. The state equations are defined line by line using statements like

```
(state_variable [lagged state_variables] [indep_variables],
 state [noerror noconstant])
```

where, if the name of the state variable is `foo`, its lag is identified by `L.foo`, and the keywords `noerror` and `noconstant` indicate that the state equation should contain no disturbance and/or no constant. Similarly, the observation equations are defined line by line using statements like

```
(dep_variable [state_variables] [indep_variables]
 [, noerror noconstant])
```

where the interpretation of the keywords is now straightforward.

Alternatively, the state space can be defined using the following *error form*: for the state equations,

```
(state_variable [lagged state_variables] [indep_variables]
 [state_errors], state [noconstant])
```

and for the observation equations,

```
(dep_variable [state_variables] [indep_variables] [obs_error]
 [, noconstant])
```

If the coefficients linking the variables on the rhs to those on the lhs of the equations are not unknown (as form many components), then this information is to be given through the definition of constraints. A constraint is to be defined with the command

```
constraint define # [lhs_var]rhs_var=value
```

where # is to be substituted with a unique number that identifies that constraint in the Stata session, `lhs_var` refers to the variable on the lhs of the equal sign and `rhs_var` to the variable on the rhs; finally `value` is to be substituted with the value of the coefficient (often 1 in UCM).

```
use http://www.stata-press.com/data/r13/air2.dta
generate logair = ln(air)

* Contstraints definition
* -- local linear trend
constraint define 1 [level]L.level=1
constraint define 2 [level]L.slope=1
* -- stochastic seasonal dummies
constraint define 3 [gamma]L.gamma=-1
constraint define 4 [gamma]L.gamma1=-1
constraint define 5 [gamma]L.gamma2=-1
constraint define 6 [gamma]L.gamma3=-1
constraint define 7 [gamma]L.gamma4=-1
constraint define 8 [gamma]L.gamma5=-1
constraint define 9 [gamma]L.gamma6=-1
constraint define 10 [gamma]L.gamma7=-1
constraint define 11 [gamma]L.gamma8=-1
constraint define 12 [gamma]L.gamma9=-1
constraint define 13 [gamma]L.gamma10=-1
constraint define 14 [gamma1]L.gamma=1
constraint define 15 [gamma2]L.gamma1=1
constraint define 16 [gamma3]L.gamma2=1
constraint define 17 [gamma4]L.gamma3=1
constraint define 18 [gamma5]L.gamma4=1
constraint define 19 [gamma6]L.gamma5=1
constraint define 20 [gamma7]L.gamma6=1
constraint define 21 [gamma8]L.gamma7=1
```

```
constraint define 22 [gamma9]L.gamma8=1
constraint define 23 [gamma10]L.gamma9=1

* State space form definition and estimation
sspace (level L.level L.slope, state noconstant)
       (slope L.slope, state noerror noconstant)
       (gamma L.gamma L.gamma1 L.gamma2 L.gamma3
       L.gamma4 L.gamma5 L.gamma6 L.gamma7 L.gamma8
       L.gamma9 L.gamma10, state noconstant)
       (gamma1 L.gamma, state noerror noconstant)
       (gamma2 L.gamma1, state noerror noconstant)
       (gamma3 L.gamma2, state noerror noconstant)
       (gamma4 L.gamma3, state noerror noconstant)
       (gamma5 L.gamma4, state noerror noconstant)
       (gamma6 L.gamma5, state noerror noconstant)
       (gamma7 L.gamma6, state noerror noconstant)
       (gamma8 L.gamma7, state noerror noconstant)
       (gamma9 L.gamma8, state noerror noconstant)
       (gamma10 L.gamma9, state noerror noconstant)
       (logair level gamma, noconstant),
       constraints(1/25)

* Compute the variables smo1, smo2, ..., with smoothed states
predict smo*, states smethod(smooth)

* plot log-airline data and smoothed trend
twoway (tsline logair) (tsline smo1)
```

In this case no particular form for the covariance matrix of the disturbances is specified and Stata assumes a diagonal matrix (identity matrix for the error form specification). If other covariance structures are needed, these can be passed though `covform()` statement, where in the parenthesis the following options are admitted: `identity`, `dscalar` (diagonal scalar), `diagonal`, `unstructured` (symmetric positive-definite matrix). The procedure automatically uses the algorithm proposed by de Jong (1991) for the exact treatment of diffuse initial conditions for nonstationary state variables. The estimation is not particularly quick and stable. In fact, we could not get estimates for the full BSM (the optimiser did not converge) and had to set the variance of the slope disturbance to zero (random walk with drift instead of local linear trend) to obtain a result.

Bibliography

Anderson, B. D. O. and J. B. Moore (1979). *Optimal Filtering.* Prentice-Hall.

Azevedo, J. V. E., S. J. Koopman, and A. Rua (2006). Tracking the business cycle of the euro area: A multivariate model-based bandpass filter. *Journal of Business & Economic Statistics 24*(3), 278–290.

Baxter, M. and R. G. King (1999, November). Measuring business cycles: approximate band-pass filters for economic time series. *The Review of Economics and Statistics 81*(4), 575–593.

Bloomfield, P. (2000). *Fourier Analysis of Time Series: An Introduction.* Wiley.

Box, G. and G. Jenkins (1976). *Time Series Analysis: Forecasting and Control.* San Francisco: Holden-Day.

Box, G. E. P. and D. A. Pierce (1970). Distribution of residual autocorrelations in autoregressive-integrated moving average time series models. *Journal of the American Statistical Association 65*(332), pp. 1509–1526.

Brockwell, P. J. and R. A. Davis (1991). *Time Series: Theory and Methods* (2nd edition). Springer.

Brockwell, P. J. and R. A. Davis (2002). *Introduction to Time Series and Forecasting* (2nd edition). Springer.

Burns, A. F. and W. C. Mitchell (1946). *Measuring Business Cycles.* National Bureau of Economic Research.

Caines, P. E. (1988). *Linear Stochastic Systems.* Wiley.

Christiano, L. J. and T. J. Fitzgerald (2003). The band pass filter. *International Economic Review 44*(2), 435–465.

Cobb, G. W. (1978). The problem of the Nile: Conditional solution to a changepoint problem. *Biometrika 65*(2), pp. 243–251.

Commandeur, J. J. F. and S. J. Koopman (2007). *An Introduction to State Space Time Series Analysis.* Oxford University Press.

Cottrell, A. and R. Lucchetti (2015, February). Gretl users guide. GNU Free Documentation License.

Davidson, J. (2000). *Econometric Theory.* Oxford: Blackwell Publishing Inc.

de Jong, P. (1988). A cross-validation filter for time series models. *Biometrika 75*(3), 594–600.

de Jong, P. (1988). The likelihood for a state space model. *Biometrika 75*, 165–169.

de Jong, P. D. (1991). The diffuse Kalman filter. *Annals of Statistics 19*, 1073–1083.

de Jong, R. M., C. Amsler, and P. Schmidt (2007). A robust version of the {KPSS} test based on indicators. *Journal of Econometrics 137*(2), 311–333.

Dickey, D. A. and W. A. Fuller (1979). Distribution of the estimators for autoregressive time series with a unit root. *Journal of the American Statistical Association 74*(366a), 427–431.

Diebold, F. X. (1989). Structural time series analysis and modelling package: A review. *Journal of Applied Econometrics 4*(2), 195–204.

Doornik, J. (2007). *Object-Oriented Matrix Programming Using Ox* (3rd ed.). London: Timberlake Consultants.

Dordonnat, V., S. J. Koopman, and M. Ooms (2012). Dynamic factors in periodic time-varying regressions with an application to hourly electricity load modelling. *Computational Statistics and Data Analysis 56*, 3134–3152.

Dordonnat, V., S. J. Koopman, M. Ooms, A. Dessertaine, and J. Collett (2008). An hourly periodic state space model for modelling French national electricity load. *International Journal of Forecasting 24*, 566–587.

Durbin, J. and S. J. Koopman (2001). *Time Series Analysis by State Space Methods.* Oxford University Press.

Durbin, J. and B. Quenneville (1997). Benchmarking by state space models. *International Statistical Review / Revue Internationale de Statistique 65*(1), pp. 23–48.

Elliott, G., T. J. Rothenberg, and J. H. Stock (1996, July). Efficient tests for an autoregressive unit root. *Econometrica 64*(4), 813–836.

Engle, R. F. and C. W. J. Granger (1987). Co-integration and error correction: Representation, estimation, and testing. *Econometrica 55*(2), 251–276.

Ghosh, D. (1989). Maximum likelihood estimation of the dynamic shock-error model. *Journal of Econometrics 41*, 121–143.

Gilbert, P. (2009). Brief users guide: Dynamic systems estimation (DSE). http://cran.r-project.org/web/packages/dse/vignettes/Guide.pdf.

Granger, C. and P. Newbold (1974). Spurious regressions in econometrics. *Journal of Econometrics 2*, 111–120.

Granger, C. W. J. and P. Newbold (1977). *Forecasting Economic Time Series.* New York: Academic Press.

Green, P. and B. W. Silverman (1993). *Nonparametric Regression and Generalized Linear Models: A Roughness Penalty Approach.* Chapman &

Hall.

Hamilton, J. D. (1994). *Time Series Analysis.* Princeton University Press.

Harrison, P. J. and C. F. Stevens (1976). Bayesian forecasting. *Journal of the Royal Statistical Society. Series B 38*, 205–247.

Harvey, A., S. J. Koopman, and N. Shephard (Eds.) (2004). *State Space and Unobserved Component Models.* Cambridge University Press.

Harvey, A. and T. Proietti (Eds.) (2005). *Readings in Unobserved Components Models.* Oxford University Press.

Harvey, A. C. (1989). *Forecasting, Structural Time Series Models and the Kalman Filter.* Cambridge University Press.

Harvey, A. C. (2006). Forecasting with unobserved components time series models. In G. Elliot, C. Granger, and A. Timmermann (Eds.), *Handbook of Economic Forecasting.* North Holland.

Harvey, A. C. and S. J. Koopman (1993). Forecasting hourly electricity demand using time-varying splines. *Journal of the American Statistical Association 88*, 1228–1237.

Harvey, A. C. and S. J. Koopman (1997). Multivariate structural time series models. In C. Heij, J. Schumacher, B. Hanzon, and C. Praagman (Eds.), *System Dynamics in Economics and Financial Models*, Chapter 9, pp. 269–285. Wiley.

Harvey, A. C. and S. J. Koopman (2009). Unobserved components models in economics and finance. *IEEE Control Systems Magazine* (December), 71–81.

Harvey, A. C. and T. M. Trimbur (2003, May). General model-based filters for extracting cycles and trends in economic time series. *The Review of Economics and Statistics 85*(2), 244–255.

Heesterman, A. R. G. (1990). *Matrices and Their Roots: A Textbook of Matrix Algebra.* World Scientific.

Helsk, J. (2015). Package KFAS. http://cran.r-project.org/web/packages/KFAS/KFAS.pdf.

Hodrick, R. J. and E. C. Prescott (1997, February). Postwar U.S. business cycles: An empirical investigation. *Journal of Money, Credit and Banking 29*(1), 1–16.

Holt, C. C. (1957). Forecasting trend and seasonality by exponentially weighted moving averages. Office of Naval Research INR 52, Carnegie Institute of Technology.

Holt, C. C. (2004). Forecasting seasonals and trends by exponentially weighted moving averages. *International Journal of Forecasting 20*(1), 5–10.

Hyndman, R., A. B. Koehler, J. K. Ord, and R. D. Snyder (2008). *Forecasting with Exponential Smoothing.* Springer.

Jenkins, G. M. and D. G. Watts (1968). *Spectral Analysis and Its Applications.* Holden Day.

Johansen, S. (1991). Estimation and hypothesis testing of cointegration vectors in Gaussian vector autoregressive models. *Econometrica 59*(6), 1551–1580.

Kalman, R. E. (1960). A new approach to linear filtering and prediction problems. *Journal of Basic Engineering, Transactions ASME, Series D 82*, 35–45.

Kalman, R. E. and R. S. Bucy (1961). New results in linear filtering and prediction theory. *Journal of Basic Engineering, Transactions ASME, Series D 83*, 95–108.

Kim, C.-J. and C. R. Nelson (1999). *State-Space Models with Regime Switching.* Cambridge MA: MIT Press.

Kitagawa, G. and W. Gersch (1996). *Smoothness Priors Analysis of Time Series.* Heidelberg, Berlin, New York: Springer.

Kohn, R. and C. F. Ansley (1987, January). A new algorithm for spline smoothing based on smoothing a stochastic process. *SIAM Journal of Scientific and Statistical Computing 8*(1), 33–48.

Kohn, R. and C. F. Ansley (1989). A fast algorithm for signal extraction, influence and cross-validation in state space models. *Biometrika 76*(1), 65–79.

Koopman, S., A. C. Harvey, J. A. Doornik, and N. Shephard (2009). *STAMP 8.2: Structural Time Series Analyser, Modeler, and Predictor.* London: Timberlake Consultants.

Koopman, S. J. (1993). Disturbance smoother for state space models. *Biometrika 80*(1), 117–126.

Koopman, S. J. (1997). Exact initial Kalman filter and smoother for non-stationary time series models. *Journal of the American Statistical Association 92*, 1630–1638.

Koopman, S. J. and J. V. E. Azevedo (2008). Measuring synchronization and convergence of business cycles for the euro area, UK and US. *Oxford Bulletin of Economics and Statistics 70*(1), 23–51.

Koopman, S. J. and N. Shephard (1992). Exact score for time series models in state space form. *Biometrika 79*, 823–826.

Koopman, S. J., N. Shephard, and J. A. Doornik (1999). Statistical algorithms for models in state space using SsfPack 2.2. *Econometrics Journal 2*, 113–166.

Koopman, S. J., N. Shephard, and J. A. Doornik (2008). *Statistical Algorithms for Models in State Space Form – SsfPack 3.0.* London: Timberlake Consultants.

Koopmans, L. H. (1995). *The Spectral Analysis of Time Series.* Elsevier.

Kwiatkowski, D., P. C. B. Phillips, P. Schmidt, and Y. Shin (1992). Testing the null hypothesis of stationarity against the alternative of a unit root: How sure are we that economic time series have a unit root? *Journal of Econometrics 54*(1-3), 159–178.

Ljung, G. M. and G. E. P. Box (1978). On a measure of lack of fit in time series models. *Biometrika 65*(2), 297–303.

Lucchetti, R. (2011). State space methods in gretl. *Journal of Statistical Software 41*(11), 1–22.

Luethi, D., P. Erb, and S. Otziger (2015). Package 'FKF'. http://cran.r-project.org/web/packages/FKF/FKF.pdf.

Lütkepohl, H. (2007). *New Introduction to Multiple Time Series Analysis.* Springer.

Makridakis, S. and M. Hibon (2000). The M3-competition: results, conclusions and implications. *International Journal of Forecasting 16*, 451–476.

McQuarrie, A. D. R. and C.-L. Tsai (1998). *Regression and Time Series Model Selection.* World Scientific.

Mendelssohn, R. (2011). The STAMP software for state space models. *Journal of Statistical Software 41*(2), 1–18.

Nyblom, J. and A. Harvey (2000, April). Tests of common stochastic trends. *Econometric Theory 16*(02), 176–199.

Okun, A. M. (1962). Potential GNP: Its measurement and significance. In *Proceedings of the Business and Economic Statistics Section of the American Statistical Association (reprinted as Cowles Foundation Paper 190).*

Pagan, A. (1980). Some identification and estimation results for regression models with stochastic varying coefficients. *Journal of Econometrics 13*, 341–363.

Pankratz, A. (1991). *Forecasting with Dynamic Regression Models.* Wiley.

Pelagatti, M. M. (2005). Business cycle and sector cycles. Available from the RePEc: http://econpapers.repec.org/RePEc:wpa:wuwpem:0503006.

Pelagatti, M. M. (2011). State space methods in Ox/SsfPack. *Journal of Statistical Software 41*(3), 1–25.

Pelagatti, M. M. and V. Negri (2010). The industrial cycle of Milan as an accurate leading indicator for the Italian business cycle. *Journal of Business Cycle Measurement and Analysis 2010*(2), 19–35.

Pelagatti, M. M. and P. K. Sen (2013). Rank tests for short memory stationarity. *Journal of Econometrics 172*(1), 90–105.

Petris, G. (2009). dlm: an R package for Bayesian analysis of dynamic linear models. http://cran.r-project.org/web/packages/dlm/vignettes/dlm.pdf.

Petris, G. (2010). An r package for dynamic linear models. *Journal of Statistical Software 36*(12), 1–16.

Petris, G. and S. Petrone (2011). State space models in r. *Journal of Statistical Software 41*(4), 1–25.

Petris, G., S. Petrone, and P. Campagnoli (2009). *Dynamic Linear Models with R.* Springer.

Phillips, P. C. B. (1986). Understanding spurious regressions in econometrics. *Journal of Econometrics 33*(3), 311–340.

Phillips, P. C. B. and S. Ouliaris (1990). Asymptotic properties of residual based tests for cointegration. *Econometrica 58*(1), 165–193.

Phillips, P. C. B. and P. Perron (1988). Testing for a unit root in time series regression. *Biometrika 75*(2), 335–346.

Pourahmadi, M. (2001). *Foundations of Time Series Analysis and Prediction Theory.* New York: Wiley.

Proietti, T. (2000). Comparing seasonal components for structural time series models. *International Journal of Forecasting 16*, 247–260.

Rünstler, G. (2004). Modelling phase shifts among stochastic cycles. *Econometrics Journal 7*(1), 232–248.

Said, S. E. and D. A. Dickey (1984). Testing for unit roots in autoregressive-moving average models of unknown order. *Biometrika 71*(3), 599–607.

Selukar, R. (2011). State space modeling using SAS. *Journal of Statistical Software 41*(12), 1–13.

Shorack, G. R. (2000). *Probability for Statisticians.* Springer.

Shumway, R. H. and D. S. Stoffer (2006). *Time Series Analysis and Its Applications with R Example* (2nd edition). Springer.

Stoffer, D. (2015). Package astsa. http://cran.r-project.org/web/packages/astsa/astsa.pdf.

Triacca, U. (2002). The partial autocorrelation function of a first order non-invertible moving average process. *Applied Economics Letters 9*, 13–15.

Trimbur, T. M. (2005). Properties of higher order stochastic cycles. *Journal of Timer Series Analysis 27*(1), 1–17.

Tusell, F. (2011). Kalman filtering in r. *Journal of Statistical Software 39*(2), 1–27.

Van den Bossche, F. A. M. (2011). Fitting state space models with EViews. *Journal of Statistical Software 41*(8), 1–16.

Wahba, G. (1978). Improper priors, spline smoothing and the problem of guarding against model errors in regression. *Journal of the Royal Statistical Society. Series B 40*(3), 364–372.

Wei, W. W. S. (2006). *Time Series Analysis: Univariate and Multivariate Methods* (2nd edition). Pearson.

West, M. and J. Harrison (1989). *Bayesian Forecasting and Dynamic Models.* Springer.

Winters, P. R. (1960). Forecasting sales by exponentially weighted moving averages. *Management Science 6*(3), 324–342.

Yaffee, R. and M. McGee (2000). *Introduction to Time Series Analysis and Forecasting with Applications of SAS and SPSS.* Academic Press.

Zavanella, B., M. Mezzanzanica, M. M. Pelagatti, S. C. Minotti, and M. Martini (2008). A two-step approach for regional medium-term skill needs forecasting. In *Regional Forecasting on Labour Markets.* Reiner Hampp Verlag.

Index

Printed in the United States
by Baker & Taylor Publisher Services